Sustainable Cities Reimagined

To assess urban sustainability performance, this book explores several clusters of cities, including megacities, cities of the Global South, European and North American cities, cities of the Middle East and North Africa, cities of Central and South East Asia, a city state of Singapore and a large group of global cities. It applies a multi-criteria approach using a panel of environmental, economic, social and smart indicators to assess progress and policies in global cities including London, New York, Hong Kong, San Francisco, Los Angeles, São Paolo, Rio de Janeiro, Buenos Aires, Paris, Berlin, Stockholm, Moscow, Beijing, Seoul, Singapore, Shanghai, Sydney, Tokyo and many others.

Additional attention is given to the issues of climate change, poverty and smart dimensions, with renewable energy and the drivers of urban CO_2 emissions playing the central role. This book is abundant with case studies considering strategies, policies and performance of the leading cities, including San Francisco, Stockholm and Seoul in greater depth, exploring how their successes can be used by other cities. The book identifies key linkages between different smart and sustainability dimensions as well as investment opportunities in cities with sustainability potential.

This book will be of great interest to policy makers, city and regional authorities as well as scholars and students of urban planning and sustainable development aiming to facilitate a sustainability transition in our cities around the world.

Stanislav E. Shmelev is Director of Environment Europe Ltd, Oxford, UK, and a Visiting Lecturer at the University of St Gallen, Switzerland.

Sustainable Cities Reimagined

Multidimensional Assessment and Smart Solutions

Edited by Stanislav E. Shmelev

First published 2020
by Routledge
2 Park Square, Milton Park, Abingdon, Oxon OX14 4RN

and by Routledge
52 Vanderbilt Avenue, New York, NY 10017

Routledge is an imprint of the Taylor & Francis Group, an informa business

British Library Cataloguing-in-Publication Data
A catalogue record for this book is available from the British Library

Library of Congress Cataloging-in-Publication Data
Names: Shmelev, Stanislav, editor.
Title: Sustainable cities reimagined : multidimensional assessment and smart solutions / Stanislav Shmelev, editor.
Description: First Edition. | New York : Routledge, 2019. | Includes bibliographical references and index. |
Identifiers: LCCN 2019029875 (print) | LCCN 2019029876 (ebook) | ISBN 9780367254216 (hardback) | ISBN 9780367254209 (paperback) | ISBN 9780429287725 (ebook)
Subjects: LCSH: Urban ecology (Sociology) | City planning—Environmental aspects.
Classification: LCC HT241 .S866 2019 (print) | LCC HT241 (ebook) | DDC 307.76—dc23
LC record available at https://lccn.loc.gov/2019029875
LC ebook record available at https://lccn.loc.gov/2019029876

ISBN: 978-0-367-25421-6 (hbk)
ISBN: 978-0-367-25420-9 (pbk)
ISBN: 978-0-429-28772-5 (ebk)

Typeset in Times New Roman
by Swales & Willis Ltd, Exeter, Devon, UK

Printed in the United Kingdom
by Henry Ling Limited

Contents

Figures

Tables

Boxes

Contributors

Stanislav E. Shmelev is Director of Environment Europe Ltd, Oxford, UK and a Visiting Lecturer at the University of St Gallen, Switzerland. Formerly a Senior Researcher at Oxford University and a Visiting Professor at the Universities of Geneva, Paris, Versailles, National University of Colombia and Kazakh National University, consultant to UNDP, UNEP and IUCN. Stanislav is an author and editor of bestsellers *Ecological Economics: Sustainability in Practice* (2012), *Green Economy Reader: Lectures in Ecological Economics and Sustainability* (2017) and editor and photographer of *Ecosystems: Complexity, Diversity and Nature's Contribution to Humanity* (2018).

Harrison Brook, MSc, University of Edinburgh, UK

Yelena Y. Chzhan, Sarsen Amanzholov East-Kazakhstan State University, Ust-Kamenogorsk, Republic of Kazakhstan

David Elliott, Professor Emeritus, The Open University

Zhanar M. Kadyrkhanova, Al-Farabi Kazakh National University, Almaty, Republic of Kazakhstan

Mohammad Kamruzzaman Palash is an urban planner who serves as the Urban Governance and Planning Coordinator for UNDP Bangladesh's National Urban Poverty Reduction Program.

Rimma K. Sagiyeva, Al-Farabi Kazakh National University, Almaty, Republic of Kazakhstan

Tobias Schnitzler, MA, University of Vienna, Austria

Irina A. Shmeleva, Head of Sustainable Cities Lab, Associate Professor, Institute of Design and Urban Studies, ITMO University, St Petersburg, Russia and Director, Institute of Sustainable Development Strategies NGO, St Petersburg, Russia

John W. Taylor, United Nations Development Programme

Ellie Tonks, MSc, University of Edinburgh, UK

Bei Zhang, MSc, University of Edinburgh, UK

1 Methods and indicators for urban sustainability assessment

Stanislav E. Shmelev and Irina A. Shmeleva

Introduction

Urban sustainability is defined as the multidimensional capacity of a city to operate successfully in economic, social and environmental domains simultaneously (Shmelev & Shmeleva, 2018). The subject of sustainable cities has been explored by Hall and Pfeiffer (2000), Hall et al. (2010), Hall (2014), Girardet (1993, 2004, 2014) and Shmelev and Shmeleva (2009). The multidimensional nature of an urban system defines a central analytical approach for sustainability assessment of cities used in this book, namely the methodology of multi-criteria decision aid (MCDA), following an approach outlined in our earlier work (Shmelev, 2017b). Urban sustainability is highlighted as one of the important dimensions in UNEP Green Economy Report (UNEP, 2011). UN Sustainable Development Goals include Goal 11 'Sustainable Cities and Communities', which aims to 'make cites and human settlements inclusive, safe, resilient and sustainable' (UN, 2015). Urban sustainability has been the focus of the recent Habitat III forum held in Quito, Ecuador in 2016 where the New Urban Agenda was firmly linked with UN Sustainable Development Goals (UN Habitat, 2016, 2015). The United for Smart and Sustainable Cities initiative (UNECE & ITU, 2016) pioneered a systemic thinking connecting urban smart and sustainable dimensions at the international scale.

In this chapter we will introduce several major methodologies for urban sustainability analysis, namely, material flow analysis, input–output analysis, optimization and multi-criteria decision aid, discuss sustainability indicators, review the content of Environment Europe™ Sustainable Cities database and apply multi-criteria decision aid to the set of the most prominent global megacities.

Urban sustainability analysis methods

A spectrum of tools and methods used in sustainability science includes material flows analysis, input–output analysis, optimization and multi-criteria decision aid (Shmelev, 2012). These methods, traditionally used in ecological economics, industrial ecology and operations research could be successfully applied to urban systems to improve their sustainability performance. Table 1.1 reviews applications of ecological-economic methods mentioned above to urban systems, illustrating how the chosen methods were applied to several sustainability issues in urban systems: water, resources, waste and CO_2 emission.

Material flows analysis (MFA) is a tool invented by Robert Ayres, accounting for the weight of resources extracted domestically, imported and accumulated, processed or recycled in the national economy, and then emitted into nature in the form of gaseous, liquid or solid residues, or exported (Eurostat, 2001), first applied to cities in a study focused on Hong Kong (Newcombe et al., 1978). Later on, studies focused on Hamburg, Vienna and Leipzig appeared (Hammer & Giljum, 2006), followed by Singapore (Schulz, 2007), Beijing (Zhang et al., 2009), Paris (Barles, 2009) and Lisbon (Rosado et al., 2014). This research enhanced our understanding of the material flows at the city scale and contributed towards filling the gap in data availability. The method requires statistical datasets that exist at the national scale and are still very rare at the urban and regional level. Water and carbon footprints could be considered as partial cases of MFA.

Water footprint[1] focuses on production-based and consumption-based water use, which illustrate the water requirements of an urban economy. Urban water footprint studies include applications for New York and Beijing (Jenerette et al., 2006), Oslo (Venkatesh & Brattebø, 2012), Berlin, Delhi and Lagos (Hoff et al., 2013) and Milan.

Input-output analysis is an economic tool designed by Wassily Leontief (Leontief 1970, 1936). It considers the economic system as a web of interconnected types of economic activity or sectors, namely agriculture, energy generation, oil and gas extraction, computer manufacturing, education, health care, etc. Existing environmental applications include studies focused on CO_2 emissions (Peters & Hertwich, 2006) and water use (Dietzenbacher & Velázquez, 2007) at the macro scale. Urban scale remains a promising new area of research, where input–output models were applied to assess the water use of such Chinese cities like Chongquing (Okadera et al., 2006) and Beijing (Zhang et al., 2012), urban metabolism with a focus on waste for Shenzhen (Ni et al., 2001), resources in Chongquing (Hui et al., 2006), Beijing (Zhou et al., 2010) and Suzhou (Liang & Zhang, 2012). One paper focused on the application of input–output analysis to E-waste recycling in Seattle (Leigh et al., 2012). Research on input–output analysis of CO_2 emissions for urban systems has been quite abundant: Sydney (Lenzen et al., 2004), Vienna (Ornetzeder et al., 2008), Copenhagen (Hallegatte et al., 2011), Suzhou (Liang et al., 2012), Beijing (Chen et al., 2013) and Helsinki (Ala-Mantila et al., 2013).

Optimization – a group of mathematical methods aimed at finding a minimum or a maximum of a certain goal function on a large constrained set of possible alternatives – is a tool widely applied in urban sustainability research. Optimization techniques include mixed integer programming, multi-objective optimization and linear programming and were applied in water management for Tabriz (Zarghami, 2010), Sydney (Mortazavi et al., 2012) and Shanghai (Lü et al., 2012); waste management in Genova (Minciardi et al., 2008), Palermo (Galante et al., 2010), Beijing (Xi et al., 2010; Dai et al., 2011), Taichung (Chang et al., 2012) and Mexico City (Santibañez-Aguilar et al., 2013).

Table 1.1 Studies focused on application of ecological-economic methods to water, resource waste and CO_2 emissions reduction issues in various cities[2]

Dimensions methods	*Water*	*Resources*	*Waste*	*Emissions CO_2*	*Smart city*
MFA	Milan (2014)[3] Berlin (2013)[4] Delhi (2013)[5] Lagos (2013)[6] Oslo (2011)[7] New York (2006)[8] Beijing (2009)[9]	Lisbon (2014)[10] Paris (2009)[11] Singapore (2008)[12] Hamburg (2006)[13] Leipzig (2006)[14] Vienna (2006)[15] Beijing (2009)[16] Hong Kong (1978)[17]	London (2011)	San Francisco (2004)[18] London (2009)[19] Paris (2011)[20] New York (2012)[21] Rio de Janeiro (2011)[22]	Stockholm (2015)[23] Atlanta and London (2013)[24]
Input-output	Beijing (2012)[25] Chongquing (2006)[26]	Beijing (2010)[27]	Suzhou (2012)[28] Seattle (2012)[29] Chongquing (2006)[30]	Helsinki (2013)[31] Beijing (2013)[32] Copenhagen (2011)[33] Sydney (2004)[34]	South Korea (2016)[35]
Optimization	Tabriz (2010)[36] Shanghai (2012)[37] Sydney (2012)[38]		Mexico (2013)[39] Taichung (2012)[40] Beijing (2011)[41]		Savona, Zaanstad, Sant Cugat (2017)[42]
MCDA	Granada (2012)[43] Athens (2012)[44] Berlin (2004)[45]		Beijing (2010)[46] Kampala (2013)[47] New York (2006)[48] Amsterdam (2006)[49] Moscow (2006)[50] Budapest (2006)[51] Barcelona (2006)[52]	Paris (1982)[53]	Oslo (2015)[54]

Multi-criteria decision aid (MCDA), originally developed by Bernard Roy (Roy, 1985, 1991, 1996), has been applied widely for urban sustainability assessments and decision support due to its ability to find compromise between conflicting goals and priorities (Munda, 1995, 2005a, 2005b). The MCDA tool ELECTRE was used to assess possible locations for new underground stations in Paris (Roy & Hugonnard, 1982), which ultimately reduces CO_2 emissions by providing easier access to public transport for residents. In the context of urban sustainability assessment, Munda (2006) has been one of the first to suggest using MCDA tools to compare cities on their sustainable development performance. MCDA tools were used for addressing the urban water management issues in the case of Berlin (Simon et al., 2004), Granada (Ruiz-Villaverde et al., 2012) and Athens (Kandilioti & Makropoulos, 2012). Waste management issues were explored with MCDA in the context of Beijing (Xi et al., 2010), Kampala (Oyoo et al., 2013), Dakar (Kapepula et al., 2007) and Barcelona (Bautista & Pereira, 2006).

The application of ecological-economic tools in the field of smart city analysis started in the early 2010s. Kim et al. (2016) apply input–output analysis to reveal the role played by the smart cities in South Korean economy, focusing on employment and economic impacts. Milan et al. (2015) used MCDA for smart city scenarios for Oslo. Shahrokni et al. (2015) and Beck et al. (2013) focus on material flow impacts of the smart city in Stockholm, and London and Atlanta respectively. The applications of optimization tools to the smart city analysis are abundant, a good example being the study by Papastamatiou (2017).

It would be very practical here to discuss the advantages and disadvantages of using the methods discussed above for the sustainability analysis in the urban context. MFA for water has the following advantages: it is relatively simple and focuses on one dimension only; it allows exploring issues of embodied water use through treating international trade in goods and services and could assist in exploring the different degree to which cities depend on imported water. At the same time, there are certain disadvantages: international trade flows are normally not available for cities and exist for countries only; water use figures for cities could only reflect direct use components by residents, industry and services. Potentially extremely useful, this method has only been applied to a handful of cities and is not a standard internationally applied at the city level.

MFA for resources allows analysis of material flows into and from the city. It assumes a large taxonomy of flows with a possibility of treating dynamics of resource use and allows exploring issues not possible with other methods (e.g. aluminium use in renewable energy, the use of rare metals in computer manufacturing). There is a possibility of analysing embodied flows occurring through trade, including unused components. On the other hand, cities are not required by law to compile material flows accounts; cross-city comparisons might be problematic for the lack of unifying international methodology, and input–output tables that are required to attribute resource flows at the urban level are often missing.

MFA for waste allows us to trace flows at a city scale including both municipal and industrial components; it helps to determine the degree of circularity and to identify problematic flows. Theoretically, it is possible to trace waste flows across borders. On the other hand, information on waste flows at a city scale is often incomplete; cities in the developing world do not have recycling facilities and composition of the waste stream in cities may need additional research.

MFA for CO_2 allows detailed consideration of urban CO_2 emissions, attributing them to climatic, economic and technological factors and policies; it also allows to transfer knowledge to developing countries on the effects of technological change, policies and lifestyle adjustments. On the other hand, urban CO_2 figures are estimated and frequently contain estimation errors, and not all factors influencing CO_2 emissions at the urban scale can be identified.

Input–output analysis allows us to consider urban intersectoral economic flows and makes it possible to analyse employment patterns, resource use and emissions on a sector-by-sector basis, making it possible to identify key sectors capable of above-average knock-on effects. At the same time, there are certain weaknesses: input–output tables for cities are not available in most cases, Singapore being a rare exception. Approximation from national tables risks oversimplifying the differences between the nation and the city, leading to under-representation of concentrated industries. At the very best, a monetary and not a physical table could be made available, which might lead to discrepancies in mapping resource use and pollution.

Optimization allows detailed analysis of systems change, e.g. in cases of waste and renewable energy, potentially highlighting trade-offs in case of multi-criteria optimization. At the same time, the method has extensive data requirements including spatial and temporal data; it requires a high level of computational complexity, especially for multi-criteria optimization.

MCDA has been selected in this book due to the following clear advantages in the urban context. First, due to its ability to compare multidimensional alternatives; second, the possibility of dealing with different types of information; third, for allowing analysis of performance under different policy priorities through changing weights, and, finally, for adopting a stronger sustainability perspective through outranking approaches that do not accept trade-offs so easily. At the same time, MCDA requires the full dataset covering all cities and all indicators in the pool, which implied that we had to start with cities for which those data were available. Recycling, biodiversity and income differentiation indicators might be difficult to obtain for cities in many developing countries; the basket of indicators must be balanced in terms of social, economic, environmental and smart dimensions, and finally, fine-tuning is required based on indifference and preference thresholds, cut-off points and weights.

Indicators for smart sustainable cities

There has been a strong interest in using indicator-based frameworks for sustainability analysis, manifested in a wide spectrum of studies: Valentin and Spangenberg (2000), Spangenberg (2002a,b, 2005), Monfaredzadeh and Berardi (2015), Hara et al. (2016), Manitiu and Pedrini (2016), Ahvenniemi et al. (2017), Garcia-Fuentes et al. (2017), Girardi and Temporelli (2017), Klopp and Petretta

(2017) and Pierce et al. (2017). This research has been driven by the United Nations Guidelines and Methodologies on Sustainable Development Indicators (UN, 2007), EU Sustainable Development Indicators (EC, 2009), Sustainable Development Indicators Framework (UNECE, 2013), new ISO 37120 standards on Sustainable Development of Communities (ISO, 2014), the Sustainable Development Goals Framework (UN, 2015) and the Smart Sustainable City Indicator Framework (UN ECOSOC, 2015).

Indicator-based sustainability assessments for cities have been conducted by many researchers in the past decade: Shmelev and Shmeleva (2009), Shen et al. (2011), Shen and Zhou (2014), Mori and Yamashita (2015), Wong (2015), Yigitcanlar et al. (2015) and Wei et al. (2015, 2016). Several assessment frameworks are widely accepted today: UN SDG indicators, ISO 37120 Sustainable Development of Communities and UNECE-ITU Smart Sustainable City Indicators, which we took into account in designing our methodology.

The UN SDG Indicator Framework comprises 232 indicators, which often becomes unmanageable for the sheer quantity of data. The ISO 37120 standard featuring 116 indicators exhibits indicators that are more precisely defined, although social and environmental aspects are given slightly greater prominence than economic and smart indicators. The UNECE-ITU Smart Sustainable Cities Indicator Framework is more balanced between different dimensions of sustainability and formulated with a lot more clarity and a forward-looking strategic vision in mind; it features 72 indicators, a much more manageable set. Selection of individual indicators for cities, chosen for this book (see Figure 1.1), was based on an earlier sustainable cities framework (Shmelev & Shmeleva, 2009), inspired by our dynamic sustainability assessments carried out for countries (Shmelev, 2011, 2017a) and adapted for the urban scale (Shmelev, 2017b), cross-checked with the indicators from the three frameworks described above and based on data availability across the database.

The final set of smart and sustainable indicators included a range of economic, environmental, social and smart cities indicators following an approach identified by the UNECE and ITU United for Smart and Sustainable Cities initiative (Figure 1.1).

Additional existing benchmarking frameworks (listed in Table 1.2) include:

- Economist Intelligence Unit's Global Liveability Index, which uses 30 indicators to rank 140 cities placing on top Vienna (1), Melbourne (2), Osaka (3), Calgary (4), Sydney (5), Vancouver (6), Toronto (7), Tokyo (8), Copenhagen (9), Adelaide (10);
- Rockefeller Foundation's Resilient Cities Index, employing 52 indicators to benchmark 100 cities;
- UN Habitat's Global Prosperity Index, focusing on 72 indicators and 295 cities highlighting the leading position of Oslo (1), Copenhagen (2), Stockholm (3), Helsinki (4), Paris (5), Vienna (6), Melbourne (7), Montreal (8), Toronto (9), Sydney (10);
- Arcadia's Sustainable Cities Index based on 32 indicators applied to 100 cities, selecting Zurich (1), Singapore (2), Stockholm (3), Vienna (4),

London (5), Frankfurt (6), Seoul (7), Hamburg (8), Prague (9), Munich (10) as most sustainable;

- Mori Foundation's Global Power City Index focusing on 70 indicators as applied to 44 cities emphasizing London (1), New York (2), Tokyo (3), Paris (4), Singapore (5), Seoul (6), Amsterdam (7), Berlin (8), Hong Kong (9), Sydney (10), and
- Monocle's Quality of Life Survey presenting the 25 cities with the best quality of life in the world, including Munich (1), Tokyo (2), Vienna (3), Zurich (4), Copenhagen (5), Berlin (6), Madrid (7), Hamburg (8), Melbourne (9), Helsinki (10).

Because the composition of the indicator sets varies tremendously as do the cities involved, different assessment systems place emphasis on different aspects of urban performance.

The cities chosen for our analysis include: New York, Los Angeles, Rio de Janeiro, São Paulo, London, Paris, Berlin, Moscow, Beijing, Tokyo, Hong Kong, Shanghai, Singapore and Sydney. Criteria for selecting the cities were economic importance and environmental impacts (all cities feature in world's top 30 cities by GDP, comprising over 8% of global GDP, employing nearly 130 million people; most cities are part of the C40 network focused on greenhouse gas emissions mitigation, producing around 3% of global CO_2 emissions, use more than 15 billion tonnes of water per year and generate over 55 million tonnes of municipal solid waste per year). Our study draws on a wide range of sources from Eurostat (2016), city governments (City of London, 2009; City of New York, 2012; City of Rio de Janeiro, 2011; Mairie de Paris, 2011; San Francisco Department for the Environment, 2004; Singapore, 2009), Siemens European Green City Index (Siemens, 2009), World Cities Culture Forum (Mayor of London, 2014), UN Habitat (UN Habitat, 2013) and World Bank publications (World Bank, 2013), LSE Going Green Report (LSE, 2013) as well as considerations on availability of data.

Below we will illustrate the diversity in sustainability performance of global cities on various individual dimensions (Figures 1.2–1.10). As can be seen from Figure 1.2, cities such as Beijing, New York, Los Angeles, Tokyo, Paris, Moscow and London exhibit the highest levels of Gross Regional Product at PPP. At the same time, unemployment has been largest in Africa and cities such as Madrid, Los Angeles, Berlin, Rome, New York, London, and lowest in Hong Kong, Moscow, Beijing and Singapore (Figure 1.3). The income differentiation (Figure 1.4) has been largest in Washington, DC, Rio de Janeiro, Hong Kong, New York, São Paolo, Paris, Beijing, Los Angeles, Moscow and lowest in Tokyo, Berlin, Stockholm. CO_2 emissions per capita (Figure 1.5) were very high in Melbourne and Sydney, and significant in Los Angeles, Shanghai, Washington, DC, Beijing, and somehow lower in Moscow, Berlin, Hong Kong, Paris, Tokyo and very low in Rio de Janeiro and São Paolo. PM_{10} concentrations (Figure 1.6), on the other hand are very high in cities of the Middle East and Africa, Delhi, Beijing, Shanghai, Rio de Janeiro, Hong Kong and much lower in Sydney, Toronto, Washington, DC, New York, Tokyo, Los Angeles and Berlin.

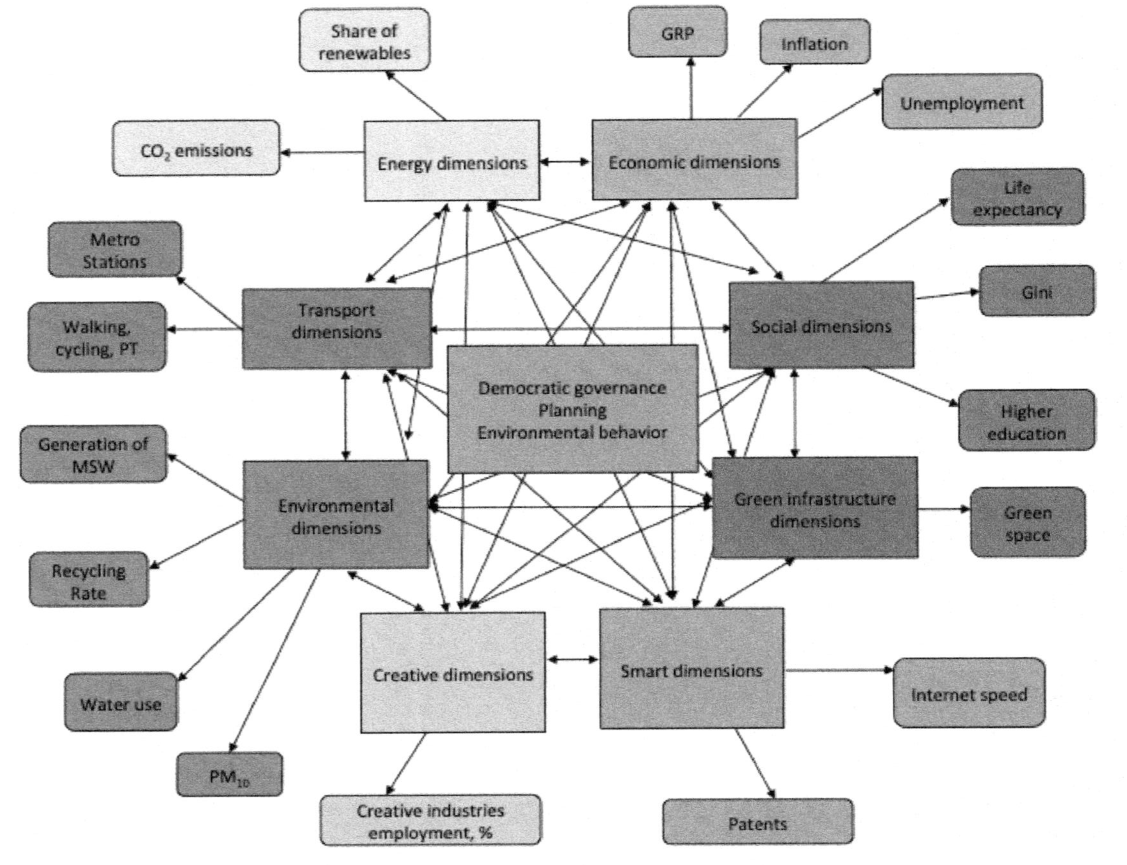

Figure 1.1 Conceptual diagram of a Smart and Sustainable City Assessment Methodology

Table 1.2 Alternative urban smart and sustainability indicator systems

Rating	*Year*	*Organization*	*Number of indicators*	*Indicators*	*Cities*	*Source*	*Criticism*	*Results*
Global Liveability Index	2018	Economist Intelligence Unit	30	Stability: 25%, 5; Healthcare: 20%, 6; Culture and the Environment: 25%, 9; Education: 10%, 3; Infrastructure: 20%, 7	140	EIU (2018)	No sustainability: renewables, recycling	Top: Vienna (1), Melbourne (2), Osaka (3), Calgary (4), Sydney (5), Vancouver (6), Toronto (7), Tokyo (8), Copenhagen (9), Adelaide (10); Bottom: Dakar, Algiers, Douala, Tripoli, Harare, Port Moresby, Karachi, Lagos, Dhaka, Damascus.
Resilient City Index	2016	Rockefeller Foundation	52	Health and well-being; Economy and society; Infrastructure and environment; Leadership and strategy; 52 indicators	100	Rockefeller Foundation (2016)	Indicators are often qualitative and not concrete, huge data gaps: 56% in Liverpool	Not available

(continued)

Table 1.2 (continued)

Rating	*Year*	*Organization*	*Number of indicators*	*Indicators*	*Cities*	*Source*	*Criticism*	*Results*
Global Prosperity Initiative	2015	UN Habitat	72	Productivity: city product per capita, unemployment rate; Infrastructure: Improved shelter, improved water, physical density, Internet access; traffic fatalities; Quality of life: life expectancy at birth, under-five mortality rate, literacy rate, mean years of schooling, homicide rate; Equity and Social Inclusion: Gini coefficient, poverty rate, slum households, youth unemployment, equitable secondary school enrolment; Environmental Sustainability: $PM_{2.5}$ concentration; solid waste collection; CO_2 emissions, share of renewable energy; Governance and Legislation: Voter turnout, days to start a business. 72 indicators	295	UN Habitat (2016)	No track of Inflation; FDI, patents, creative industry employment, No trips made by public transport, no metro; No Internet speed; No higher education or suicide rates; No water consumption, green space, waste generation, waste recycling, PM_{10}, NO_2, SO_2	Oslo (1), Copenhagen (2), Stockholm (3), Helsinki (4), Paris (5), Vienna (6), Melbourne (7), Montreal (8), Toronto (9), Sydney (10).
Sustainable Cities Index	2016	Arcadis	32	People Planet Profit. Education: Literacy rate, University rankings, Share of population with tertiary education; Health: Life expectancy, Obesity rate; Demographics: Dependency ratio; Income Inequality: Gini coefficient; Affordability: Consumer Price Index, Property Prices; Work Life Balance: Average annual hours worked; Crime: Homicide rate; Environmental risks: Natural catastrophes exposure; Green Spaces: Green Space as % of city area; Energy: Energy use, Renewables share, Energy consumption per GDP; Air Pollution: mean level of pollutants; GHG emission per capita; Waste Management: Landfill vs Recycling; Wastewater treatment; Drinking water and sanitation: Access to drinking water %, Access to improved sanitation; Transport infrastructure: Congestion, Rail infrastructure, Airport satisfaction; Economic Development: GDP per capita; Ease of Doing Business: Index; Tourism: International visitors per year (absolute and per capita); Connectivity: mobile connectivity; Broadband connectivity; Importance in Global Networks; Employment: Number of people employed %,	100	Arcadis (2016)	No track of inflation, patents, creative industry employment, public transport, metro, specific pollution metrics on PM_{10}, NO_2, SO_2, water consumption, waste generation.	Zurich (1), Singapore (2), Stockholm (3), Vienna (4), London (5), Frankfurt (6), Seoul (7), Hamburg (8), Prague (9), Munich (10).

Global Power City Index	2017	MORI Foundation	70	Economy: Nominal GDP, GDP per capita, GDP growth rate; Level of economic freedom; Total market value of listed shares on stock exchanges; World's to 500 companies; Total employment; Number of employees in service; Wage level; Ease of securing human resources; Office space per desk; Corporate Tax rate; Political, economic and business risk (13 all in all); Research and Development: Number of Researchers, Worlds top 200 universities; Academic Performance in Mathematics and Science; Readiness for Accepting Researchers; Research and Development Expenditure; Number of patents; Number of prize winners; Interaction opportunities between researchers (8 all in all); Cultural Interaction: Number of international conferences held; Number of world class cultural events held; Trade value of audio-visual and related services; Environment of creative activities; Number of world heritage sites; Opportunities for cultural, historical and traditional interaction; number of theatres and concert halls; number of museums; number of stadiums; number of luxury hotel guest rooms; number of hotels; attractiveness of shopping options; attractiveness of dining options; number of foreign residents; number of visitors from abroad, number of foreign students (16 all in all); Liveability: Total unemployment rate; Total working hours; Level of satisfaction of employees; Average housing rent; Price level; Murders per million; Economic risk of natural disaster; Life expectancy; Degree of social freedom, fairness and equality; Risk to mental health; Number of doctors per million people; ICT readiness; Variety of retail shops; Variety of restaurants (14 all in all); Environment: Companies with ISO 14000 certification; Renewables; Recycling; CO_2, PM, SO_2; NO_2; Water quality in rivers, Green coverage; Comfort level of temperature (9 all in all); Accessibility: Number of cities with direct international flights; International freight flows; Departing and arriving passengers; Runways; Density of railways stations; Punctuality and coverage of public transportation; Travel time between centre and international airports; Commuting convenience; Transportation fatalities per million; Taxi fare.	44	Mori Foundation (2017)	Most indicators reflected. Lack of coverage of health or creative economies, galleries.	London (1), New York (2), Tokyo (3), Paris (4), Singapore (5), Seoul (6), Amsterdam (7), Berlin (8), Hong Kong (9), Sydney (10).

(continued)

Table 1.2 (continued)

Rating	*Year*	*Organization*	*Number of indicators*	*Indicators*	*Cities*	*Source*	*Criticism*	*Results*
Quality of Life Survey	2018	MONOCLE	ND	Safety/crime, international connectivity, climate/sunshine, quality of architecture, public transportation, tolerance, environmental issues and access to nature, urban design, business conditions, pro-active policy developments and medical care	25	Monocle (2018)	Relative lack of transparency as to which specific indicators were used	Munich (1), Tokyo (2), Vienna (3), Zurich (4), Copenhagen (5), Berlin (6), Madrid (7), Hamburg (8), Melbourne (9), Helsinki (10), Stockholm (11), Lisbon (12), Sydney (13), Hong Kong (14), Vancouver (15), Amsterdam (16), Kyoto (17), Dusseldorf (18), Barcelona (19), Paris (20), Singapore (21), Fukoka (22), Auckland (23), Brisbane (24), Oslo (25)

Figure 1.2 Gross Regional Product per capita

Source: Environment Europe™ Sustainable Cities Database, 2018

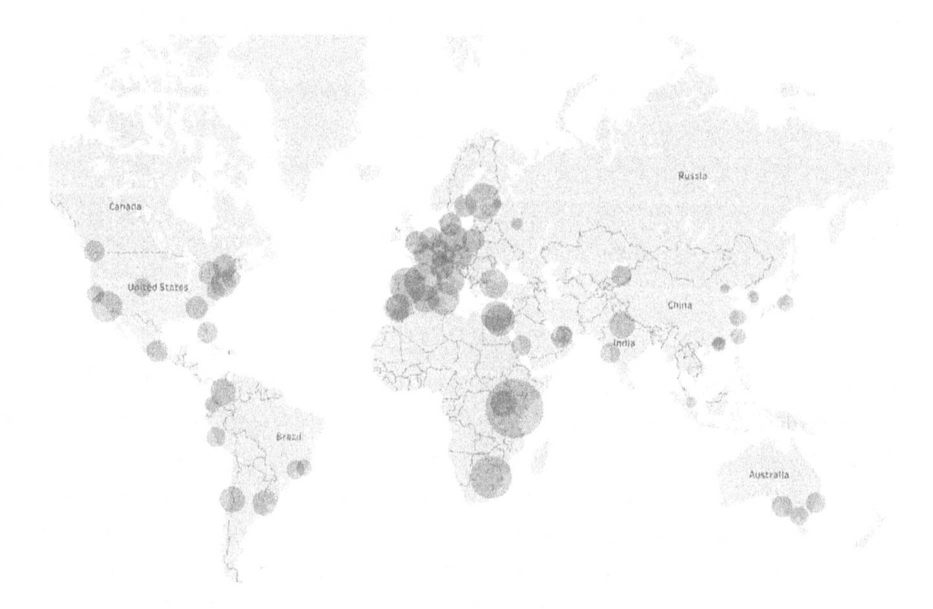

Figure 1.3 Unemployment

Source: Environment Europe™ Sustainable Cities Database, 2018

Figure 1.4 Gini Index of Income Inequality

Source: Environment Europe™ Sustainable Cities Database, 2018

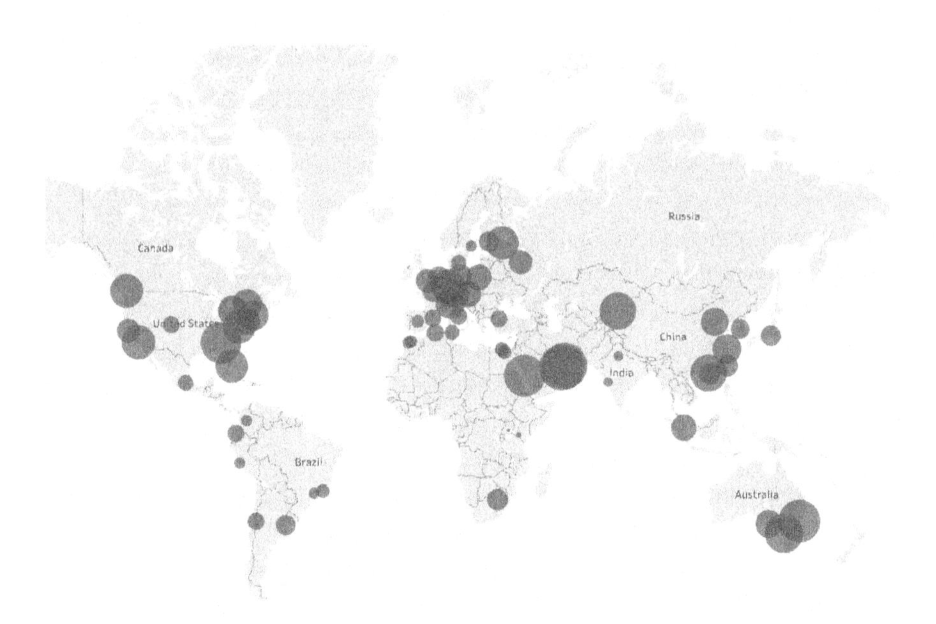

Figure 1.5 CO_2 emissions per capita

Source: Environment Europe™ Sustainable Cities Database, 2018

Figure 1.6 PM_{10} concentrations

Source: Environment Europe™ Sustainable Cities Database, 2018

Figure 1.7 Water use per capita

Source: Environment Europe™ Sustainable Cities Database, 2018

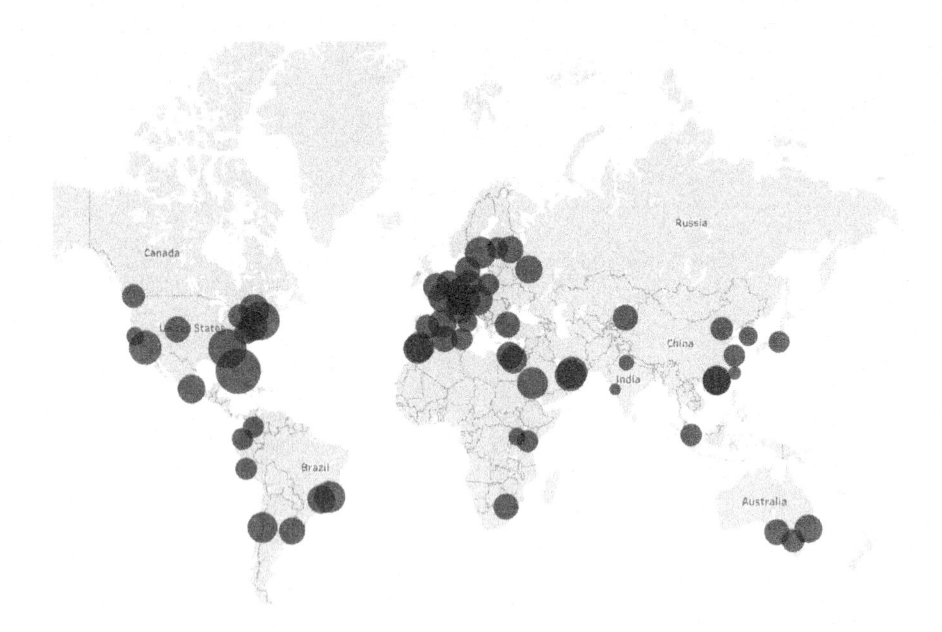

Figure 1.8 Waste generation per capita

Source: Environment Europe™ Sustainable Cities Database, 2018

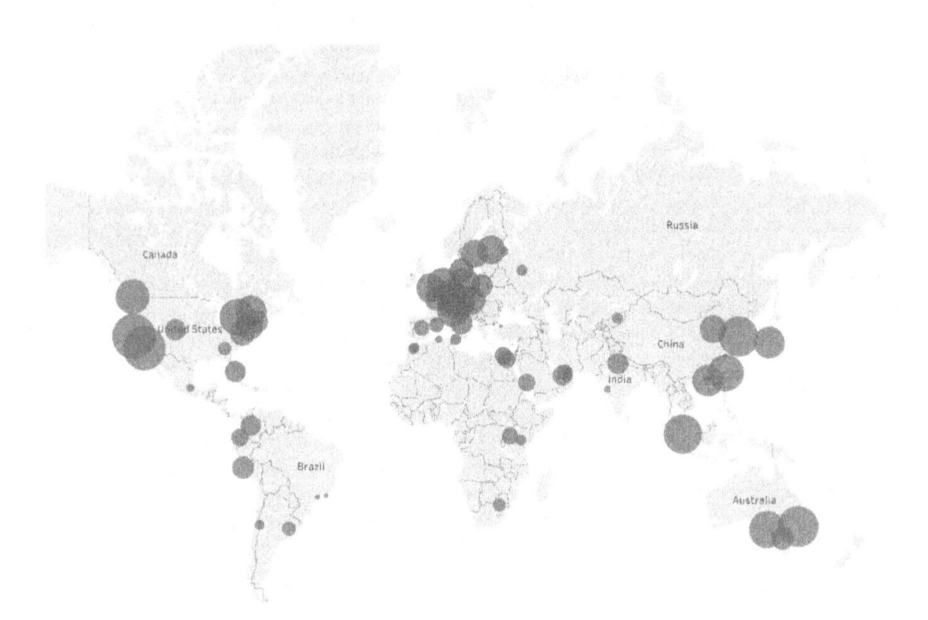

Figure 1.9 Recycling rate, %

Source: Environment Europe™ Sustainable Cities Database, 2018

Figure 1.10 Creative industries employment, %

Source: Environment Europe™ Sustainable Cities Database, 2018

Water use (Figure 1.7) has been highest in cities of the Middle East, Los Angeles, Washington, DC, Toronto, Shanghai, Moscow, Hong Kong and Tokyo and was lower in Copenhagen, Barcelona, Vienna, Berlin, London, Sydney, São Paolo and New York. More solid waste was generated per capita in Miami, Los Angeles, Rio de Janeiro, New York, Paris, Hong Kong, and less in London, Moscow, Sydney, Berlin, Shanghai, Beijing and Tokyo (Figure 1.8). Recycling was a major success in Los Angeles, Sydney and Singapore, with slightly more moderate rates of recycling observed in Berlin, Tokyo, Hong Kong, London, Beijing and New York, and much lower rates in Moscow, Rio de Janeiro, São Paolo and Shanghai (Figure 1.9).

Many of the world's top cities developed significant creative industries sectors, with leaders being Paris, New York, Tokyo, London, São Paolo, Rio de Janeiro and Berlin, and somehow lower levels of development of creative industries recorded in Los Angeles, Hong Kong, Moscow, Beijing and Singapore (Figure 1.10). We consider creative industries as a significant factor in stimulating smart economy and ultimately urban sustainability.

Figures 1.2–1.10 illustrate the heterogeneous performance of cities based on the level of economic development, geography, climatic conditions, lifestyles, policies, technological development and taxation regimes. There are no two cities that are alike and therefore multidimensional analysis is required to explain urban sustainability performance. For example, in Los Angeles we observe higher than average GRP, high CO_2 emissions, and high water use and waste generation. At the same time, the city is characterized by high unemployment, a high Gini index and success in recycling performance. Understanding the heterogeneity of their urban performance could allow cities to assess their policies and forge new strategies and innovations that deal with problem areas on a systemic level.

Multi-criteria decision aid methodology and applications

Multi-criteria decision aid (MCDA) methodology was chosen here for its ability to treat several dimensions of data simultaneously and its capacity to integrate such information via multi-criteria aggregation procedure (MCAP), with or without converting the data of different nature into a single composite index. MCDA requires the following components: alternatives that need to be compared; criteria used to assess performance of these alternatives; a multi-criteria aggregation procedure (MCAP) and policy recommendations resulting from the method's application. MCDA tools, namely outranking methods based on pair-wise comparisons of alternatives, allow strong sustainability assessment, understood here as a setting where less compensation is allowed among criteria accepted (Martinez-Alier et al., 1998). The MCDA method "ELimination Et Choix Traduisant la REalité" III (ELECTRE III, "ELimination and Choice Expressing Reality") is a discrete multi-criteria outranking method well suited for this task. For modelling various policy priorities focusing on environmental, economic, social and smart dimensions, we have chosen to use different sets of weights in the ELECTRE III method, allowing us to adopt various assessment perspectives.

The application of the ELECTRE III method for comparative sustainability benchmarking of the world's largest cities rests on the following assumptions. According to the set of 20 criteria selected for this assessment (Figure 1.1), a 10% difference in the value of each criterion is sufficient for domination and less than 5% presents an indifference. The results are presented in the form of the webs of domination relationships among the cities obtained through the pair-wise comparisons within the ELECTRE III tool under four different policy priorities: environmental, economic, social and smart (Figures 1.11–1.14). An arrow between two cities denotes a relationship of domination in the sense of the criteria chosen, but lack of such an arrow points to incomparability.

Under environmental priorities (Figure 1.11), Singapore outranks all the cities in the set, followed by Sydney and London, and then Tokyo. The worst performing cities in this setting were also Rio de Janeiro, Shanghai and Los Angeles. Mid-range, Moscow performed better than Beijing under the environmental policy priorities.

If economic priorities are chosen (Figure 1.12), then Tokyo clearly dominates in the assessment, followed by London and Beijing, Hong Kong and Sydney. At the bottom of the web of domination relationships in this setting are Rio de Janeiro, São Paolo, Los Angeles and Shanghai.

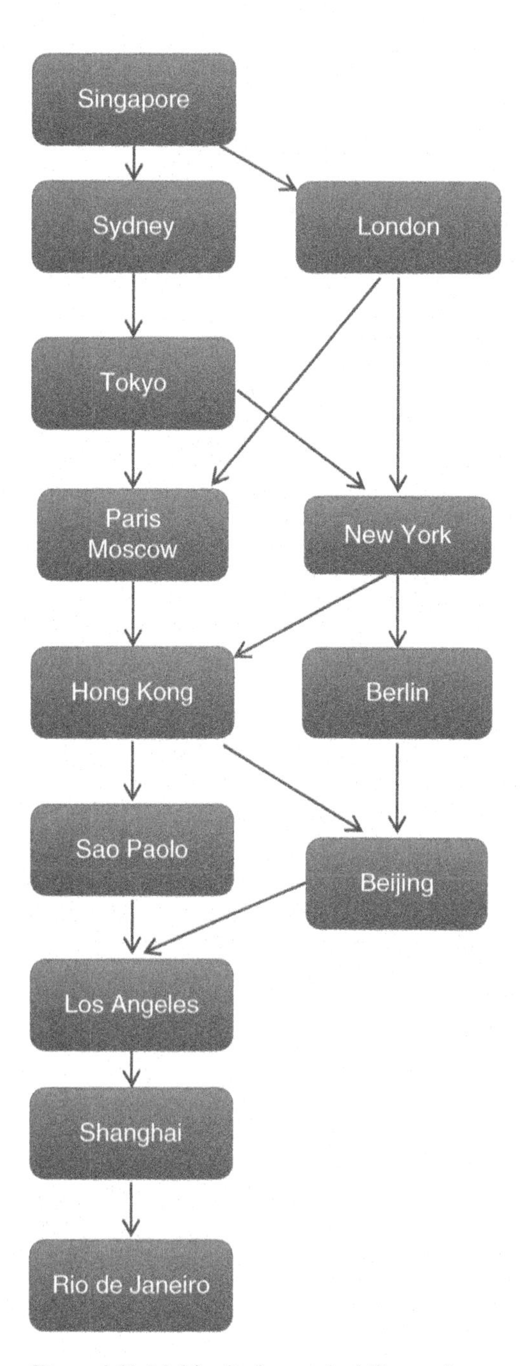

Figure 1.11 Multi-criteria sustainability performance assessment web of domination relationships among megacities, ELECTRE III: environmental priorities, A = 0.05 for indifference, A = 0.1 for preference

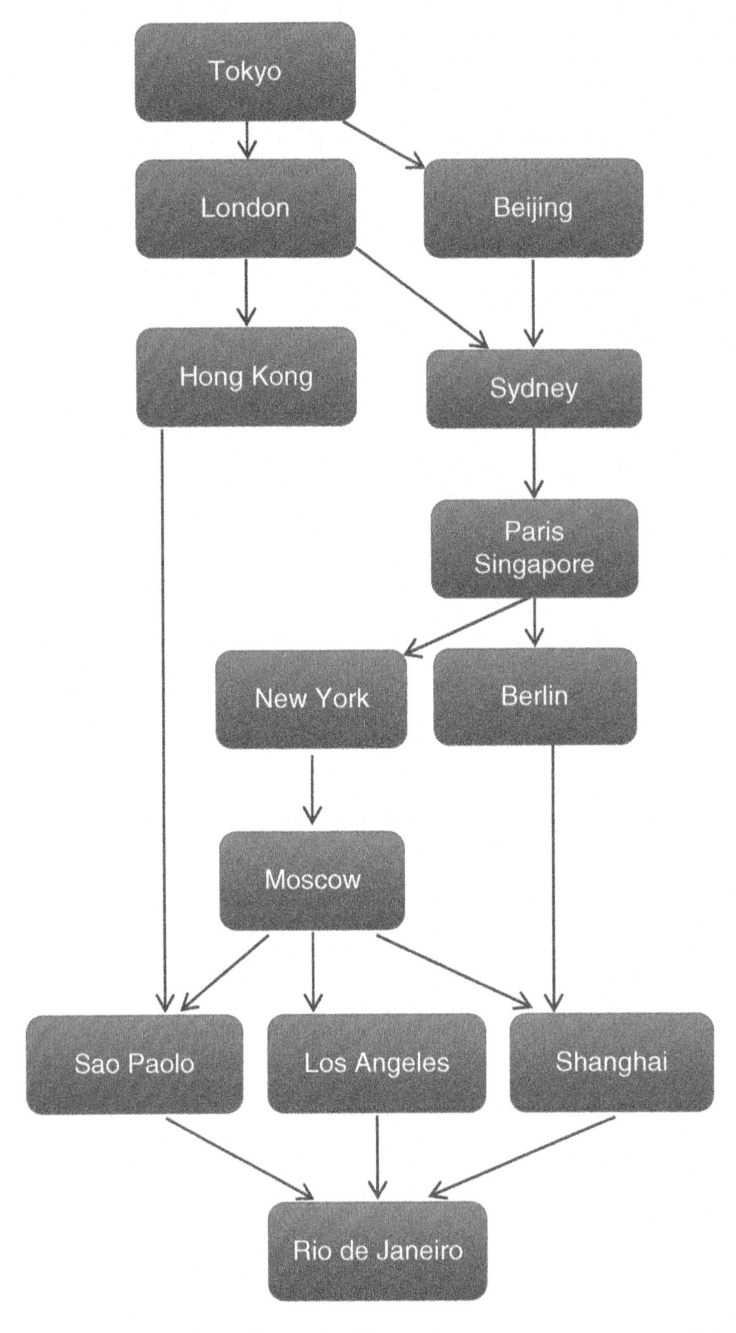

Figure 1.12 Multi-criteria sustainability performance assessment web of domination relationships among megacities, ELECTRE III: economic priorities, A = 0.05 for indifference, A = 0.1 for preference

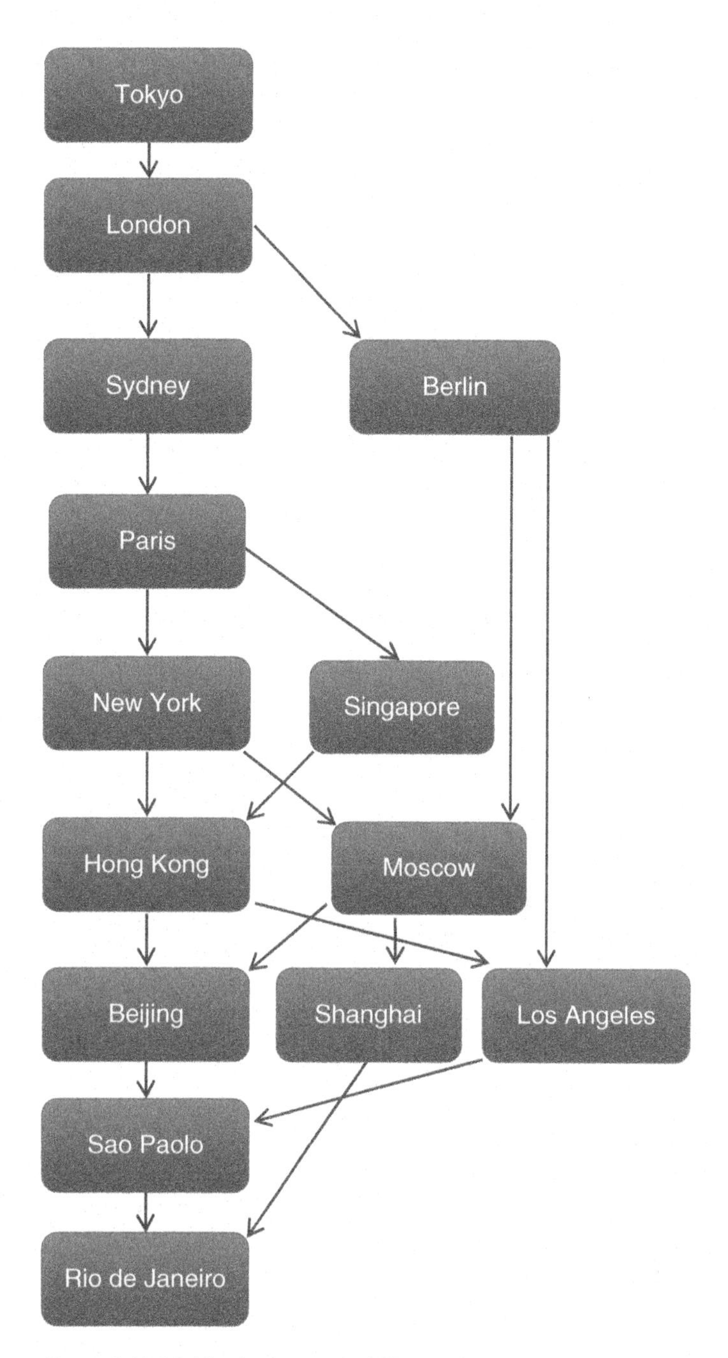

Figure 1.13 Multi-criteria sustainability performance assessment web of domination relationships among megacities, ELECTRE III: social priorities, A = 0.05 for indifference, A = 0.1 for preference

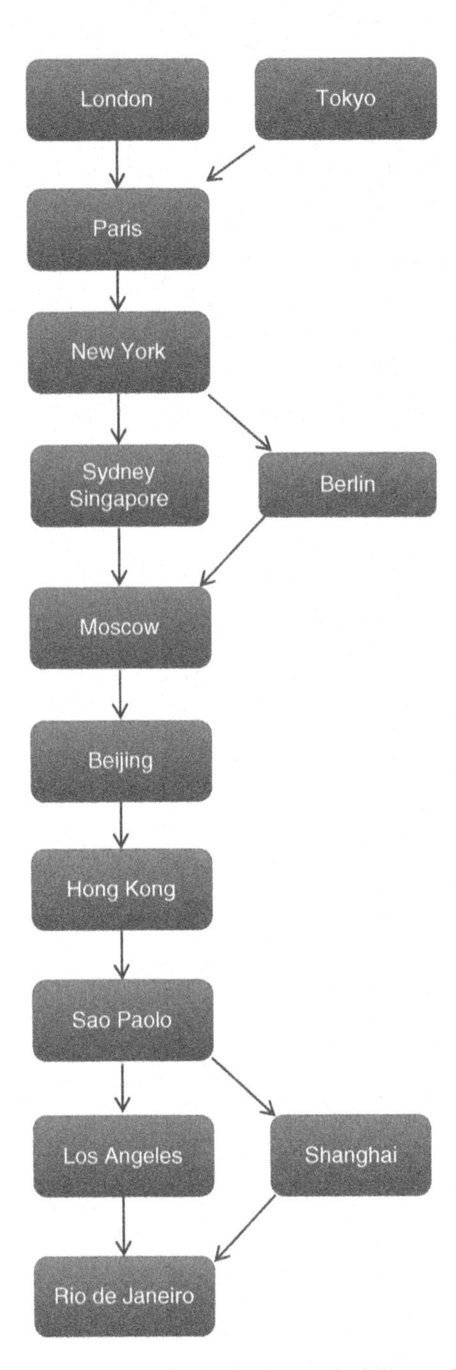

Figure 1.14 Multi-criteria sustainability performance assessment web of domination relationships among megacities, ELECTRE III: smart city priorities, A = 0.05 for indifference, A = 0.1 for preference

When social priorities are considered (Figure 1.13), the clear leaders are Tokyo, London, Sydney, Berlin and Paris. Somewhat lagging behind are Rio de Janeiro, São Paolo, Beijing, Shanghai and Los Angeles. In this assessment, performance of Moscow and Beijing are close, although Moscow performs somewhat better.

Based on the smart city performance (Figure 1.14), London and Tokyo lead, followed by Paris, New York, Sydney, Singapore and Berlin. Still lagging behind are Rio de Janeiro, Shanghai, Los Angeles, São Paolo and Hong Kong. The performance of Moscow and Beijing is close, although Moscow performed somewhat better.

It would be insightful to compare the present rankings with multiple existing urban rankings presented in Table 1.2. It should be mentioned that overall the rankings mentioned in Table 1.2 feature Sydney, Paris, Singapore, London, Tokyo, New York among the top five, which is broadly consistent with our ranking of top megacities, the obvious advantage of our outranking ELECTRE III approach being a stronger focus on sustainability. It should be added that the goals of the rankings presented in Table 1.2 are absolutely different: liveability in EIU (2018); resilience in Rockefeller Foundation (2016); prosperity in UN Habitat (2016); sustainability in Arcadis (2018); power in Mori Foundation (2017) and quality of life in Monocle (2018). In our analysis, this diversity equates to varying policy priorities, making our methodology a more general and overarching case. It appears that the cities featured at the top ten positions in our megacity rating, Singapore, Sydney, London, Tokyo, Paris, New York, Hong Kong, can be found at the very top of the Mori Foundation Global Power City Index, which clearly indicates the key characteristic of our sample. The sustainability-focused Arcadis index features Singapore, followed by London at the top of the global ranking, which closely corresponds to our setting with environmental priorities. According to social, economic and smart priorities, Tokyo in our assessment is performing better than other cities, which closely corresponds to the Monocle quality of life index. According to environmental priorities, Sydney features very high in our ranking and at the same time, Tokyo performs very well under social, economic and smart priorities, which corresponds to the EIU liveability index.

It should be noted that sensitivity analysis is often performed to understand the robustness of MCDA assessment results. Within the ELECTRE III method, a range of technical parameters, namely the indifference and preference thresholds, weights and the cut-off points are the essential dimensions of the method's sensitivity and are capable of modelling a stronger or weaker sustainability perspective. In this book, we were able to model sensitivity of the model's results to the particular stakeholder worldview and incorporated those by means of the economic, social, environmental and smart priorities. We have dealt with the sensitivity analysis at greater length in Shmelev (2017a), where one of the methods employed, APIS, allowed conducting a Monte Carlo analysis of weights from step one.

It would be highly beneficial to consider cities of Rio de Janeiro, Los Angeles and Shanghai that are consistently ranked lower in our set.

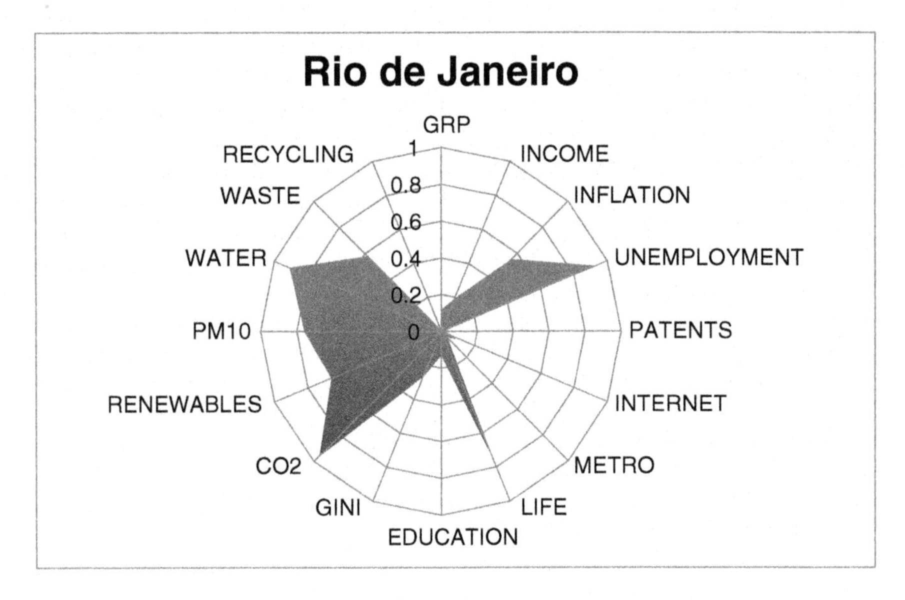

Figure 1.15 Rio de Janeiro sustainability performance, 2014

Rio de Janeiro (Figure 1.15) faces the following major challenges: relatively low economic potential and low incomes, insufficient development of innovation potential of the economy, undeveloped public transport infrastructure apart from several new metro lines, a low level of higher education, high discrepancy between rich and poor, and low recycling rates. In our opinion, to improve sustainability performance, Rio would clearly benefit from increased focus on innovation, stimulated by a faster Internet connection and stronger involvement into research and education, further stimulating regional economic development. The infrastructure would benefit from an expansion of the underground system and pioneering a robust waste recycling system, which is currently not in place. It would be advantageous to introduce some progressive policies to aim to reduce the gap between the rich and the poor.

Los Angeles (Figure 1.16) is characterized by wasteful consumption lifestyles, relatively high water consumption, low uptake of renewables, despite the city's sunny climate, quite high discrepancy between rich and poor, a relatively low level of public transport development and reasonably modest average incomes. It would be highly advisable for Los Angeles to look at ways to reduce unemployment. There is a strong potential to improve infrastructure by expanding the currently very small underground network and bringing in high-speed trams. Los Angeles is in a strong position to expand its solar electricity capacity by utilizing all existing roof and wall space that could be adapted for solar panels. There is a potential to rethink the consumption model and try minimizing water use. Additional green space, creative use of vertical greening and green roofs could be expanded to retain the water the city so badly needs.

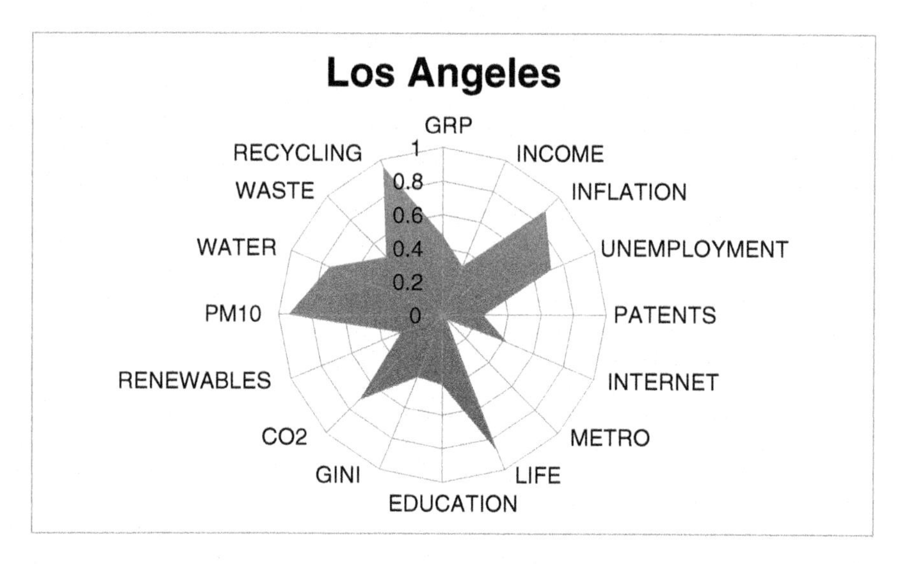

Figure 1.16 Los Angeles sustainability performance, 2014

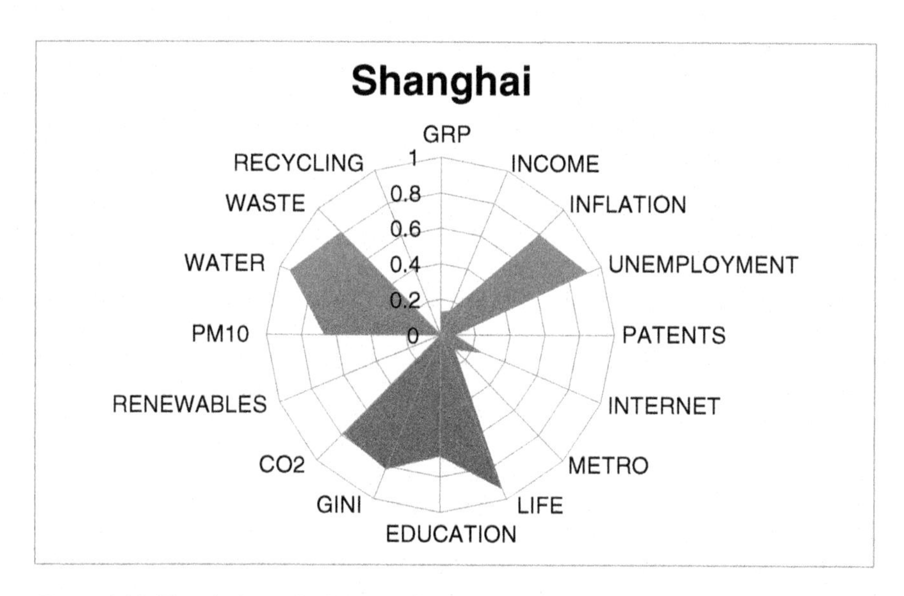

Figure 1.17 Shanghai sustainability performance, 2014

The main challenges faced by Shanghai are relatively low average incomes, a relatively low level of innovation in the economy, insufficient development of public transport infrastructure for a city of this size, relative low uptake of renewables and low recycling. Shanghai has a lot of potential to strengthen its strategic position on innovation that it currently loses to neighbouring innovation giants like Shenzhen and Taipei. At the same time, Shanghai could expand its public transport system, undertake additional efforts to expand renewables and build the capacity for recycling its waste. Doing this could strengthen its overall economic performance further.

Discussion

In the light of the information presented above, using four different weighing schemes under various sustainability priorities using ELECTRE III, there is a considerable degree of convergence in the results pointing towards Singapore as one of the most sustainable of all 14 cities in the pool on environmental performance, and Tokyo closely followed by London on economic and social performance.

In this regard it would be highly beneficial to turn to the experience of Singapore in implementing a sustainability strategy and using policy instruments to achieve sustainability at the city-state level, which is more significant than that of other world capitals.

Singapore created an Inter-Ministerial Committee on Sustainable Development (IMCSD) in January 2008 (Singapore, 2009). This body was set up to formulate a national strategy for Singapore's sustainable development. The IMCSD was co-chaired by the Minister for National Development and the Minister for the Environment and Water Resources. The members included the Ministers for Finance, Transport, and the Senior Minister of State for Trade and Industry. Setting very high aims of reaching a 70% recycling rate by 2030, achieving a 35% improvement in energy efficiency from 2005 levels by 2030 and reaching a level of domestic water consumption of 140l per person per day by 2030, the Strategy for Sustainable Growth formulated in 2009 presented a road map to the situation we observe today. The aim of the strategy for Singapore was to become the top city in Asia in terms of quality of life and to develop as a sustainable, high-density city that is clean and green, with excellent connectivity and a sense of space. The strategy set the aims to reduce the levels of $PM_{2.5}$ to 12mg/m^3 by 2020, to have 0.8ha of green space for every 1,000 residents and at the same time ensure that 70% of all journeys in the city are made by public transport. As of 2011, Singapore achieved a recycling level of 61%. In the process of designing the Blueprint 700, people from non-governmental organizations, businesses, grass-roots organizations, academia, media and mayors were consulted and over 1,300 suggestions were received from the public. Knowing that Singapore, being a small island, does not have a wide diversity of renewable energy resources available apart from solar (wind, geothermal or hydropower), the strategy focused on: (1) *raising energy efficiency* by pricing energy appropriately to reduce environmental impacts, providing information for better decisions, boosting energy-efficient industry designs, processes and technologies, building capabilities in renewable energy, promoting resource-efficient buildings, promoting public transport, expanding the water supply, improving water efficiency and minimizing waste. At the same time the decision was taken to stimulate facilitation of household recycling and enhance land use planning. The further priority of (2) *enhancing urban environment* was aimed at reviewing emission standards, adopting new technologies, pricing pollution, improving water quality, making the city cleaner, improving transport links, enhancing the city's greenery and conserving urban biodiversity. The priority of (3) *building capabilities* implied investment in R&D and facilitation of international sharing of knowledge. Finally, the goal of (4) *fostering community action* was focused on promoting community efforts, promoting industrial efficiency and stimulating development of the public sector (Singapore, 2009). Singapore keeps anticipating future change and in the recent foresight volume (Quah, 2016) a projection is made that by 2065 Singapore will generate 65TWh/year of electricity from renewable sources, which constitutes 50% of electricity demand. Among the technologies of the immediate future for Singapore, the authors mention solar photovoltaics (PV), biogas, marine energy, wind, biofuel from algae, co-generation and power from biomass as well as off-shore floating PV.

In the past ten years, Singapore estimated the potential damages from congestion to be in the range of $2–3 billion per annum. The city introduced a smart-card innovation for public transport, designed by IBM, which covered road tolls, bus travel, taxis, metro/subway usage and even shopping; the system was capable of registering 20 million transactions per day, and collected extensive traffic data, allowing city administration to constantly change routes to minimize congestion. The National Water Agency developed the Newater initiative, through which a new Siemens-designed desalination plant and a water recycling scheme provide up to 30% of Singapore's water needs, and two-thirds of the Singapore's land surface became a water catchment area. Between 2000 and 2007, the share of electricity produced by natural gas increased from 19% to 79% in Singapore, thereby reducing harmful CO_2 emissions. Since 2005, over 1,650 buildings in Singapore were made environmentally friendly. At present, around 80% of Singapore's residents are living in public housing provided by the Housing and Development Board. In a short space of time using a highly focused and strategic approach, Singapore achieved a great deal in the *economic* sphere by attracting 7,000 international companies and securing one of the highest per capita income levels in the world, in the *social* sphere by maintaining a very low unemployment rate of 1.8%, and achieving success in the *environmental* sphere by reducing the amount of waste generated per person and increasing recycling levels to 61%, keeping PM_{10} pollution at a relatively low level of 32 $\mu G/m^3$ through developing public transport and increasing the green space to 47% of its territory.

It is particularly reassuring to see such tremendous success achieved in Singapore through intensive interdepartmental and interdisciplinary collaboration (Singapore, 2009); the case for which was outlined in our earlier paper (Shmelev & Shmeleva, 2009).

Conclusion

In this chapter, we introduced the major sustainability analysis methods applied to urban areas and illustrated how multi-criteria decision aid could be applied to benchmark cities on their sustainability performance. The application focused on the most prominent megacities, which are global centres for economic activity and are responsible for a considerable share of global emissions of greenhouse gases, require considerable amounts of water and produce substantial volumes of waste. The application of multi-criteria decision aid allowed us to produce a multidimensional web of domination relationships among the top 14 world cities on 20 sustainability criteria with the help of the ELECTRE III tool. At the same time, varying indicator weights produced aggregate performance scores for megacities under four policy priorities: environmental, economic, social, and smart city criteria. The assessment carried out using the multi-criteria tool identified sustainability leaders (Singapore, Tokyo, London) and those that are lagging behind (Shanghai, Rio de Janeiro, Los Angeles). Application of environmental policy-focused ELECTRE III identified Singapore as the leader, Tokyo as the leader under economic and social priorities and Tokyo and London under smart city priorities. The result has put the performance of individual cities within the global context and presented the indicator-based sustainable development performance of individual cities within a coherent framework of multi-criteria decision aid. Learning from best practices and worst cases in this context provides an invaluable insight for policy reform to create smarter, greener, more compact, socially diverse, economically strong and less polluting cities around the world.

In the chapters that follow we will explore different alternative approaches to the assessment of urban sustainability from a large panel of cities compared using linear aggregation in Chapter 2, to the application of the outranking ELECTRE III method in Chapter 4, the outranking PROMETHEE method in Chapter 5, environmentally extended input–output analysis in Chapter 6, comparative analysis in Chapter 7 and econometric analysis in Chapter 9. The regions we explore range from Europe and North America (Chapter 4), cities of the Global South (Chapter 5), a city state, Singapore (Chapter 6), Asian cities Taipei and Almaty (Chapter 7) and global cities (Chapter 9). We will also explore alternative interdisciplinary ways of looking at a city, including the approach of urban planning and design (Chapter 2), participatory poverty analysis (Chapter 8) and renewable energy analysis (Chapter 9). Indicators of smart and sustainable performance will be dealt with in more depth in Chapters 3 and 11.

Notes

1 http://waterfootprint.org/.
2 The sources of relevant case studies are mentioned in subsequent sections further in the text.
3 Vanham & Bidoglio (2014).
4 Hoff et al. (2013).
5 Hoff et al. (2013).
6 Hoff et al. (2013).
7 Venkatesh & Brattebø (2012).
8 Jenerette et al. (2006).
9 Zhang et al. (2009).
10 Rosado et al. (2014).

11 Barles (2009).
12 Schulz (2007).
13 Hammer & Giljum (2006).
14 Hammer & Giljum (2006).
15 Hammer & Giljum (2006).
16 Zhang et al. (2009).
17 Newcombe et al. (1978).
18 San Francisco Department for the Environment (2004).
19 City of London (2009).
20 Mairie de Paris (2011).
21 City of New York (2012).
22 City of Rio de Janeiro (2011).
23 Shahrokni et al. (2015).
24 Beck et al (2013).
25 Zhang et al (2012).
26 Okadera et al. (2006).
27 Zhou et al. (2010).
28 Liang & Zhang (2012).
29 Leigh et al. (2012).
30 Hui et al. (2006).
31 Ala-Mantila et al. (2013).
32 Chen et al. (2013).
33 Hallegatte et al. (2011).
34 Lenzen, Dey & Foran (2004).
35 Kim et al. (2016).
36 Zarghami (2010).
37 Lü et al. (2012).
38 Mortazavi et al. (2012).
39 Santibañez-Aguilar et al. (2013).
40 Chang et al. (2012).
41 Dai et al. (2011).
42 Papastamatiou (2017).
43 Ruiz-Villaverde et al. (2012).
44 Kandilioti & Makropoulos (2012).
45 Simon et al. (2004).
46 Xi et al. (2010).
47 Oyoo, Leemans & Mol (2013).
48 Munda (2006).
49 Munda (2006).
50 Munda (2006).
51 Munda (2006).
52 Bautista & Pereira (2006).
53 Roy & Hugonnard (1982).
54 Milan et al. (2015).

References

Ahvenniemi, H., Huovila, A., Pinto-Seppä, I. & Airaksinen, M., 2017. What are the differences between sustainable and smart cities? *Cities*, 60(1 February), pp. 234–245.

Ala-Mantila, S., Heinonen, J. & Junnila, S., 2013. Greenhouse gas implications of urban sprawl in the Helsinki metropolitan area. *Sustainability*, 5(10), pp. 4461–4478.

Arcadis (2016) Sustainable Cities Index 2016, www.arcadis.com/en/global/our-perspectives/sustainable-cities-index-2016/

Barles, S., 2009. Urban metabolism of Paris and its region. *Journal of Industrial Ecology*, 13(6), pp. 898–913.

Bautista, J. & Pereira, J., 2006. Modeling the problem of locating collection areas for urban waste management: An application to the metropolitan area of Barcelona. *Reverse Production Systems*, 34(6), pp. 617–629.

Beck, M.B., Walker, R.V. & Thompson, M., 2013. Smarter urban metabolism: Earth systems re-engineering. *Proceedings of the Institution of Civil Engineers – Engineering Sustainability*, 166(5), pp. 229–241.

Chang, Y.J., Chu, C.-W. & Lin, M.-D., 2012. An economic evaluation and assessment of environmental impact of the municipal solid waste management system for Taichung City in Taiwan. *Journal of the Air & Waste Management Association*, 62(5), pp. 527–540.

Chen, G.Q. et al., 2013. Three-scale input-output modeling for urban economy: Carbon emission by Beijing 2007. *Communications in Nonlinear Science and Numerical Simulation*, 18(9), pp. 2493–2506.

City of London, 2009. City of London carbon footprint, www.conurbant.eu/file/1141-London_SUS_CarbonFootprintreport.pdf

City of New York, 2012. Inventory of New York City greenhouse gas emissions, http://s-media.nyc.gov/agencies/planyc2030/pdf/greenhousegas_2012.pdf

City of Rio de Janeiro, 2011. Greenhouse gas inventory and emissions scenarios of Rio de Janeiro, Brazil, www.rio.rj.gov.br/dlstatic/10112/1712030/4114527/CRJ_InventarioGEE2012_resumo_tecnicoINGLESFINAL1.pdf

Dai, C., Li, Y.P. & Huang, G.H., 2011. A two-stage support-vector-regression optimization model for municipal solid waste management: A case study of Beijing, China. *Journal of Environmental Management*, 92(12), pp. 3023–3037.

Dietzenbacher, E. & Velázquez, E., 2007. Analysing Andalusian virtual water trade in an input–output framework. *Regional Studies*, 41(2), pp. 185–196.

EC, 2009. Sustainable development indicators: An overview of relevant framework programme funded research and identification of further needs in view of EU and international activities, https://ieep.eu/archive_uploads/443/sdi_review.pdf

EIU, 2018. *The Global Liveability Index 2018*, https://pages.eiu.com/rs/753-RIQ-438/images/The_Global_Liveability_Index_2018.pdf

Eurostat, 2001. *Economy-wide Material Flow Accounts and Derived Indicators: A Methodological Guide*, Luxembourg.

Eurostat, 2016. Urban Europe: Statistics on cities, towns and suburbs, http://ec.europa.eu/eurostat/web/products-statistical-books/-/KS-01-16-691

Galante, G. et al., 2010. A multi-objective approach to solid waste management. *Waste Management*, 30(8–9), pp. 1720–1728.

García-Fuentes, M.Á., Quijano, A., De Torre, C., García, R., Compere, P., Degard, C. & Tomé, I., 2017. European cities characterization as basis towards the replication of a smart and sustainable urban regeneration model. *Energy Procedia*, 111(1 March), pp. 836–845.

Girardet, H., 1993. *The Gaia Atlas of Cities: New Directions for Sustainable Urban Living*, Gaia Books Ltd.

Girardet, H., 2004. *Cities People Planet: Liveable Cities for a Sustainable World*, John Wiley & Sons.

Girardet, H., 2014. *Creating Regenerative Cities*, Routledge.

Girardi, P. & Temporelli, A., 2017. Smartainability: A methodology for assessing the sustainability of the smart city. *Energy Procedia*, 111(1 March), pp. 810–816.

Hall, P., 2014. *Good Cities, Better Lives: How Europe Discovered the Lost Art of Urbanism* (Planning, History and Environment Series), Routledge.
Hall, P. & Pfeiffer, U., 2000. *Urban Future 21: A Global Agenda for Twenty-First Century Cities*, Routledge.
Hall, P., Buijs, S., Tan, W. & Tunas, D., 2010. *Megacities. Exploring a Sustainable Future*, nai010 Publishers.
Hallegatte, S. et al., 2011. Assessing climate change impacts, sea level rise and storm surge risk in port cities: A case study on Copenhagen. *Climatic Change*, 104(1), pp. 113–137.
Hammer, M. & Giljum, S., 2006. *Materialflussanalysen der Regionen Hamburg*, Wien und Leipzig.
Hara, M., Nagao, T., Hannoe, S. & Nakamura, J., 2016. New key performance indicators for a smart sustainable city. *Sustainability* (Switzerland), 8(3), Article number 206.
Hoff, H. et al., 2013. Water footprints of cities: Indicators for sustainable consumption and production. *Hydrology and Earth System Sciences Discussions*, 10(2), pp. 2601–2639.
Hui, Y. et al., 2006. Urban solid waste management in Chongqing: Challenges and opportunities. *Waste Management*, 26(9), pp. 1052–1062.
ISO, 2014. ISO 37120:2014(en) Sustainable development of communities: Indicators for city services and quality of life.
Jenerette, G.D. et al., 2006. Contrasting water footprints of cities in China and the United States. *Ecological Economics*, 57(3), pp. 346–358.
Kandilioti, G. & Makropoulos, C., 2012. Preliminary flood risk assessment: The case of Athens. *Natural Hazards*, 61(2), pp. 441–468.
Kapepula, K.-M. et al., 2007. A multiple criteria analysis for household solid waste management in the urban community of Dakar. *Waste Management*, 27(11), pp. 1690–1705.
Kim, K., Jung, J.K. & Choi, J.Y., 2016. Impact of the smart city industry on the Korean national economy: input–output analysis. *Sustainability*, 2016(8), p. 649.
Klopp, J.M. & Petretta, D.L., 2017. The urban sustainable development goal: Indicators, complexity and the politics of measuring cities. *Cities*, 63(1 March), pp. 92–97.
Leigh, N.G., Choi, T. & Hoelzel, N.Z., 2012. New insights into electronic waste recycling in metropolitan areas. *Journal of Industrial Ecology*, 16(6), pp. 940–950.
Lenzen, M., Dey, C. & Foran, B., 2004. Energy requirements of Sydney households. *Ecological Economics*, 49(3), pp. 375–399.
Leontief, W., 1936. Quantitative input and output relations in the economic systems of the United States. *Review of Economics and Statistics*, 18(3), pp. 105–125.
Leontief, W., 1970. Environmental repercussions and the economic structure: An input-output approach. *Review of Economics and Statistics*, 52(3), pp. 262–271.
Liang, S. & Zhang, T., 2012. Comparing urban solid waste recycling from the viewpoint of urban metabolism based on physical input-output model: A case of Suzhou in China. *Waste Management*, 32(1), pp. 220–225.
Liang, S. et al., 2012. Sustainable urban materials management for air pollutants mitigation based on urban physical input-output model. *8th World Energy System Conference, WESC 2010*, 42(1), pp. 387–392.
LSE, 2013. How cities are leading the next economy Going Green: A global survey and case studies of cities building the green economy. Final report.
Lü, Y.P. et al., 2012. Cost-effectiveness-based multi-criteria optimization for sustainable rainwater utilization: A case study in Shanghai. *Urban Water Journal*, 10(2), pp. 127–143.
Mairie de Paris, 2011. Le bilan carbone de Paris: bilan des émissions de gaz à effet de serre.

Manitiu, D.N. & Pedrini, G., 2016. Urban smartness and sustainability in Europe: An ex ante assessment of environmental, social and cultural domains. *European Planning Studies*, 24(10), 2 October, pp. 1766–1787.

Martinez-Alier, J., Munda, G. & O'Neill, J., 1998. Weak comparability of values as a foundation for ecological economics. *Ecological Economics*, 26, pp. 277–286.

Mayor of London, 2014. World cities culture report, www.worldcitiescultureforum.com/assets/others/World_Cities_Culture_Report_2014_hires.pdf

Milan, L., Kin, B., Verlinde, S. & Macharis, C., 2015. Multi-actor multi-criteria analysis for sustainable city distribution: A new assessment framework. *International Journal of Multicriteria Decision Making*, 5(4), pp. 334–354.

Minciardi, R. et al., 2008. Multi-objective optimization of solid waste flows: Environmentally sustainable strategies for municipalities. *Waste Management*, 28(11), pp. 2202–2212.

Monfaredzadeh, T. & Berardi, U. 2015. Beneath the smart city: Dichotomy between sustainability and competitiveness. *International Journal of Sustainable Building Technology and Urban Development*, 6(3), pp. 140–156.

Monocle, 2018. Quality of life survey, 2018, https://monocle.com/film/affairs/quality-of-life-survey-top-25-cities-2018/

Mori Foundation, 2017. Global power city index 2017, www.mori-m-foundation.or.jp/pdf/GPCI2017_en.pdf

Mori, K. & Yamashita T., 2015. Methodological framework of sustainability assessment in City Sustainability Index (CSI): A concept of constraint and maximisation indicators. *Habitat International*, 45(2015), pp. 10–14.

Mortazavi, M., Kuczera, G. & Cui, L., 2012. Multiobjective optimization of urban water resources: Moving toward more practical solutions. *Water Resources Research*, 48(3), p. 3514.

Munda, G., 1995. *Multicriteria Evaluation in a Fuzzy Environment: Theory and Applications in Ecological Economics*, Contributions to Economics Series, Physica-Verlag, Heidelberg.

Munda, G., 2005a. 'Measuring sustainability': A multi-criterion framework. *Environment, Development and Sustainability*, 7(1), pp. 117–134.

Munda, G., 2005b. Multiple criteria decision analysis and sustainable development. In: Figueira, J., Greco, S. & Ehrgott, M. (eds), *Multiple-criteria Decision Analysis: State of the Art Surveys*, Springer International Series in Operations Research and Management Science, New York, pp. 953–986.

Munda, G., 2006. Social multi-criteria evaluation for urban sustainability policies. *Resolving Environmental Conflicts: Combining Participation and Muli-Criteria Analysis*, 23(1), pp. 86–94.

Newcombe, K., Kalma, J.D. & Aston, A.R., 1978. The metabolism of a city: The case of Hong Kong. *Ambio*, 7(1), pp. 3–15.

Ni, J.R. et al., 2001. Total waste-load control and allocation based on input–output analysis for Shenzhen, South China. *Journal of Environmental Management*, 61(1), pp. 37–49.

Okadera, T., Watanabe, M. & Xu, K., 2006. Analysis of water demand and water pollutant discharge using a regional input–output table: An application to the City of Chongqing, upstream of the Three Gorges Dam in China. *Ecological Economics*, 58(2), pp. 221–237.

Ornetzeder, M. et al., 2008. The environmental effect of car-free housing: A case in Vienna. *Ecological Economics*, 65(3), pp. 516–530.

Oyoo, R., Leemans, R. & Mol, A.P., 2013. The determination of an optimal waste management scenario for Kampala, Uganda. *Waste Management & Research*, 31(12), pp. 1203–1216.

Quah, E.T.E. (ed.), 2016. *Singapore 2065: Leading Insights on Economy and Environment from 50 Singapore Icons and Beyond*, World Scientific.

Papastamatiou, I., Marinakis, V., Doukas, H. & Psarras, J., 2017. A decision support framework for smart cities energy assessment and optimization. *Energy Procedia*, 111(2017), pp. 800–809.

Peters, G.P. & Hertwich, E.G., 2006. Pollution embodied in trade: The Norwegian case. *Global Environmental Change*, 16(4), pp. 379–387.

Pierce, P., Ricciardi, F. & Zardini, A., 2017. Smart cities as organizational fields: A framework for mapping sustainability-enabling configurations. *Sustainability* (Switzerland), 9(9), 24 August, Article number 1506.

Rockefeller Foundation, 2016. City resilience index: Understanding and measuring city resilience, https://assets.rockefellerfoundation.org/app/uploads/20171206110244/170223_CRI-Brochure.pdf

Rosado, L., Niza, S. & Ferrão, P., 2014. A material flow accounting case study of the Lisbon metropolitan area using the urban metabolism analyst model. *Journal of Industrial Ecology*, 18(1), pp. 84–101.

Roy, B., 1985. *Méthodologie Multicritère d'Aide à la Décision*, Economica, Paris.

Roy, B., 1996. *Multicriteria Methodology for Decision Aiding*, Kluwer Academic Publishers.

Roy, B., 1991. The outranking approach and the foundations of electre methods. *Theory and Decision*, 31(1), pp. 49–73.

Roy, B. & Hugonnard, J.C., 1982. Ranking of suburban line extension projects on the Paris metro system by a multicriteria method. *Transportation Research Part A: General*, 16(4), pp. 301–312.

Ruiz-Villaverde, A., González-Gómez, F. & Picazo-Tadeo, A.J., 2012. Public choice of urban water service management: a multi-criteria approach. *International Journal of Water Resources Development*, 29(3), pp. 385–399.

San Francisco Department for the Environment, 2004. Climate action plan for San Francisco: Local actions to reduce greenhouse gas emissions, https://sfenvironment.org/download/2004-climate-action-plan-for-san-francisco

Santibañez-Aguilar, J.E. et al., 2013. Optimal planning for the sustainable utilization of municipal solid waste. *Waste Management*, 33(12), pp. 2607–2622.

Schulz, N.B. 2007. The direct material inputs into Singapore's development. *Journal of Industrial Ecology*, 11(2), pp. 117–131.

Sharokhni, H., Arman, L., Lazarevic, D., Nilsson, A. & Brand, N., 2015. Implementing smart urban metabolism in the Stockholm Royal Seaport. *Journal of Industrial Ecology*, 19(5), pp. 917–929.

Shen, L.-Y., Ochoa, J.J., Shah, M.N. & Zhang, X., 2011. The application of urban sustainability indicators: A comparison between various practices. *Habitat International*, 35(2011), pp. 17–29.

Shen, L. & Zhou J. 2014. Examining the effectiveness of indicators for guiding sustainable urbanization in China. *Habitat International*, 44(2014), pp. 111–120.

Shmelev, S. 2012. *Ecological Economics: Sustainability in Practice*, Springer.

Shmelev, S. (ed.) 2017a. *Green Economy Reader: Lectures in Ecological Economics and Sustainability*, Springer.

Shmelev, S. 2017b. Multidimensional Sustainability Assessment for Megacities. In: Shmelev, S. (ed.), *Green Economy Reader: Lectures in Ecological Economics and Sustainability*, Springer, pp. 205–236.

Shmelev, S.E., 2011. Dynamic sustainability assessment: The case of Russia in the period of transition (1985–2008). *Ecological Economics*, 70(11), pp. 2039–2049.

Shmelev, S.E. & Shmeleva, I.A., 2009. Sustainable cities: Problems of integrated interdisciplinary research. *International Journal of Sustainable Development*, 12(1/2009), pp. 4–23.

Shmelev, S.E. & Shmeleva I.A. 2018. Global urban sustainability assessment: A multidimensional approach. *Sustainable Development*, 26(6), pp. 904–920.

Siemens, 2009. European Green City Index: Assessing the environmental impact of Europe's major cities, www.siemens.com/entry/cc/features/greencityindex_international/all/en/pdf/report_en.pdf

Simon, U., Brüggemann, R. & Pudenz, S., 2004. Aspects of decision support in water management – example Berlin and Potsdam (Germany) I – spatially differentiated evaluation. *Water Research*, 38(7), pp. 1809–1816.

Singapore, 2009. A lively and liveable Singapore: Strategies for sustainable growth, www.greengrowthknowledge.org/sites/default/files/downloads/policy-database/SINGAPORE%29%20A%20Lively%20and%20Liveable%20Singapore%20-%20Strategies%20for%20sustainable%20growth.pdf

Spangenberg, J.H., 2002a. Environmental space and the prism of sustainability: Frameworks for indicators measuring sustainable development. *Ecological Indicators*, 2(2002), pp. 295–309.

Spangenberg, J.H., 2002b. Institutional sustainability indicators: An analysis of the institutions in Agenda 21 and a draft set of indicators for monitoring their effectivity. *Sustainable Development*, 10(2002), pp. 103–115.

Spangenberg, J.H., 2005. Economic sustainability of the economy: Concepts and indicators. *International Journal of Sustainable Development*, 8(1/2), pp. 47–64.

UN, 2007. *Indicators of Sustainable Development: Guidelines and Methodologies*, New York.

UN, 2015. Transforming Our World: The 2030 Agenda for Sustainable Development, Resolution adopted by the General Assembly on 25 September 2015, A/RES/70/1.

UNECE, 2013. Framework and suggested indicators to measure sustainable development, prepared by the Joint UNECE/Eurostat/OECD Task Force on Measuring Sustainable Development, 27 May.

UNECE & ITU, 2016. Rome Declaration Adopted by the participants of the Forum 'Shaping smarter and more sustainable cities: Striving for sustainable development goals', 19 May in Rome, www.itu.int/en/ITU-T/Workshops-and-Seminars/Documents/Forum-on-SSC-UNECE-ITU-18-19-May-2016/Rome-Declaration-19May2016.pdf

UN ECOSOC, 2015. The UNECE-ITU Smart Sustainable Cities Indicators, www.unece.org/fileadmin/DAM/hlm/documents/2015/ECE_HBP_2015_4.en.pdf

UNEP, 2011. *Towards a Green Economy: Pathways to Sustainable Development and Poverty Eradication*, https://sustainabledevelopment.un.org/index.php?page=view&type=400&nr=126&menu=35

UN Habitat, 2013. *Planning and Design for Sustainable Urban Mobility: Global Report on Human Settlements*, https://unhabitat.org/planning-and-design-for-sustainable-urban-mobility-global-report-on-human-settlements-2013/

UN Habitat, 2015. *Global City Report 2015*, https://unhabitat.org/wp-content/uploads/2016/02-old/CPI_2015%20Global%20City%20Report.compressed.pdf

UN Habitat, 2016. *Urbanisation and Development: Emerging Futures*, www.unhabitat.org/wp-content/uploads/2014/03/WCR-%20Full-Report-2016.pdf

Valentin, A. & Spangenberg, J.H., 2000. A guide to community sustainability indicators. *Environmental Impact Assessment Review*, 20(2000), pp. 381–392.

Vanham, D. & Bidoglio, G., 2014. The water footprint of Milan. *Water Science & Technology*, 69(4), pp. 789–795.

Venkatesh, G. & Brattebø, H., 2012. Assessment of environmental impacts of an aging and stagnating water supply pipeline network. *Journal of Industrial Ecology*, 16(5), pp. 722–734.

Wei, Y., Huang, C., Lam, P.T.I. & Yuan, Z., 2015. Sustainable urban development: A review on urban carrying capacity assessment. *Habitat International*, 46 (2015), pp. 64–71.

Wei, Y., Huang, C., Li, J. & Xie, L., 2016. An evaluation model for urban carrying capacity: A case study of China's mega-cities. *Habitat International*, 53(2016), pp. 87–96.

Wong, C., 2015. A framework for 'City Prosperity Index': Linking indicators, analysis and policy. *Habitat International*, 45(2015), pp. 3–9.

World Bank, 2013. Building sustainability in an urbanizing world: A Partnership Report, https://openknowledge.worldbank.org/handle/10986/18665

Xi, B.D. et al., 2010. An integrated optimization approach and multi-criteria decision analysis for supporting the waste-management system of the City of Beijing, China. *Engineering Applications of Artificial Intelligence*, 23(4), pp. 620–631.

Yigitcanlar, T., Dur, F. & Dizdaroglu, D., 2015. Towards prosperous sustainable cities: A multiscalar urban sustainability assessment approach. *Habitat International*, 45(2015), pp. 36–46.

Zarghami, M., 2010. Urban water management using fuzzy-probabilistic multi-objective programming with dynamic efficiency. *Water Resources Management*, 24(15), pp. 4491–4504.

Zhang, Y., Yang, Z. & Yu, X., 2009. Evaluation of urban metabolism based on emergy synthesis: A case study for Beijing (China). *Ecological Modelling*, 220(13–14), pp. 1690–1696.

Zhang, Z., Shi, M. & Yang, H., 2012. Understanding Beijing's water challenge: A decomposition analysis of changes in Beijing's water footprint between 1997 and 2007. *Environmental Science & Technology*, 46(22), pp. 12373–12380.

Zhou, S.Y., Chen, H. & Li, S.C., 2010. Resources use and greenhouse gas emissions in urban economy: Ecological input–output modeling for Beijing 2002. *Communications in Nonlinear Science and Numerical Simulation*, 15(10), pp. 3201–3231.

2 Global urban sustainability benchmarking

A multidimensional approach for smart and sustainable cities

Stanislav E. Shmelev and Irina A. Shmeleva

1. Introduction

UNEP Green Economy Report highlighted urban sustainability as one of its important dimensions (UNEP, 2011). This topic receives a lot of attention in the EU, the US and increasingly China and Latin America since the Rio Summit of 1992, the Rio+20 Summit in 2012 and, especially, in the light of the recent Habitat III forum held in Quito, Ecuador in 2016. The new UN Habitat World Cities Report firmly links the New Urban Agenda with Sustainable Development Goals (UN Habitat, 2016). SDG 11 'Sustainable Cities and Communities' aims to 'make cities and human settlements inclusive, safe, resilient and sustainable' (UN, 2015). UNECE and ITU have launched a new United for Smart and Sustainable Cities initiative in 2016.

Urban sustainability is defined as a multidimensional capacity of a city to operate successfully in economic, social and environmental domains simultaneously. Sustainable urban policy developments have been explored by Girardet (1993, 2004, 2014), Naess (1995), Hall and Pfeiffer (2000), Bithas and Christofakis (2006), Shmelev and Shmeleva (2009), Hall et al. (2010), Dassen et al. (2013), Hall (2014) and Martin and Rice (2014). The multidimensional nature of an urban system defines a central analytical approach for sustainability assessment of cities used in this paper, namely the methodology of Multi-Criteria Decision Aid (Roy, 1996), following an approach outlined in Shmelev (2017) and Shmelev and Shmeleva (2018, 2019).

The Rome declaration adopted at the UN Forum on 'Shaping smarter and more sustainable cities: striving for sustainable development goals' in May 2016 declared that 'cities need to become smarter, with technological solutions deployed to address a wide range of common urban challenges' of sustainable development (UNECE & ITU, 2016). The EU's European Economic and Social Committee considers smart sustainable cities to be a tremendous source of growth, productivity and employment. A smart sustainable city, according to UNECE, is an innovative city that uses information and communication technologies (ICTs) and other means to improve quality of life, efficiency of urban operation and services, and competitiveness, while ensuring that it meets the needs of present and future generations with respect to economic, social, environmental aspects, as well as cultural ones (UN ECOSOC, 2015).

Cities depend on a wide array of ecosystem processes, functions and ultimately ecosystem services (Bolund & Hunhammar, 1999; Gomez-Baggethun & Barton, 2013: Spangenberg & Settele, 2010. The latter are broadly defined as *economic* (provisioning: water, food, fibre, energy), *ecological* (regulation and maintenance: biogeochemical cycling, soil formation, photosynthesis, pollination, air quality regulation) and *social* (cultural: cultural diversity, educational values, inspiration, aesthetic values) (Shmelev, 2012). Urban economies rely on the natural world and the functioning of ecosystems on the territory of a much larger region than the city itself, making them important systems for ecological economics research (Girardet & Mendonca, 2009). The science of urban sustainability requires therefore the integration of the two approaches, namely a systemic description of the city and the analysis of city-ecosystem interactions. A new smart city paradigm could, if applied wisely, assist in reaching the goal of urban sustainability.

In this chapter, we start by exploring a large database of ninety global cities and search for meaningful relationships between various indicators in the global dataset. We then compare fifty-seven major global cities, for which twenty smart and sustainable indicators were available, to assess sustainability of their performance, identify the sustainability leaders as well as cities experiencing the strongest sustainability challenges. We use a linear aggregation multidimensional approach, characterised by full compensation among criteria. The chapter aims to test environmental, economic, social and smart policy priorities to assess the balance between sustainability dimensions and to provide guidance for policy makers. The assessment is based on a set of twenty urban sustainability indicators. We conclude with a description of sustainability strategies and policies adopted in the leading cities of our pool, which could help us to understand its success.

The chapter is organised as follows. Section 1 offers an introduction to the topic. Section 2 discusses data and indicators used. Section 3 presents the results of regression analysis of linkages among sustainability indicators. Section 4 discusses the application of linear aggregation under economic, social, environmental and smart sustainability priorities to 57 global cities. Section 5 explores the sustainability strategies and policies in the most sustainable cities identified in our research. Section 6 concludes.

2. Indicators for Smart Sustainable Cities

Existing smart and sustainable cities indicator frameworks include the United Nations Guidelines and Methodologies on Sustainable Development Indicators (UN, 2007), EU Sustainable Development Indicators (EC, 2009), a Sustainable Development Indicators Frameworks (UNECE, 2013), a new ISO 37120 standards on Sustainable Development of Communities (ISO, 2014), a Sustainable Development Goals framework (UN, 2015), a Smart Sustainable City Indicator Framework (UN ECOSOC, 2015). These frameworks are discussed extensively in a range of comparative reviews: Valentin and Spangenberg (2000), Spangenberg (2002a,b), Spangenberg (2005), Kierstead and Leach (2008), Monfaredzadeh and

Berardi (2015), Hara et al. (2016), Manitiu and Pedrini (2016), Ahvenniemi et al. (2017), Garcia-Fuentes et al. (2017), Girardi and Temporelli (2017), Spangenberg (2017), Klopp and Petretta (2017) and Pierce et al. (2017).

Recently there has been a growth of interest in indicator-based sustainability assessments for cities: Shen and Zhou (2014), Mori and Yamashita (2015), Wong (2015), Yigitcanlar et al. (2015), Wei et al. (2015, 2016). The indicators following the International Urban Sustainability Indicators List proposed in Shen et al. (2011) include the economic characteristics, such as income per capita; social and cultural dimensions, including unemployment rate, income differentiation rate in the form of a Gini coefficient and higher education level, and, finally, a wide range of ecological-economic or environmental dimensions, including the share of green space, CO_2 emissions, average PM_{10} concentrations, water use per capita per day, waste generation per capita per day and recycling rates.

Our comparative analysis of the three assessment frameworks (UN SDG indicators, ISO 37120 Sustainable Development of Communities and UNECE-ITU Smart Sustainable City Indicators) has shown a difference in focus, balance between economic, social and environment dimensions, and some inconsistencies. The UN SDG indicator framework is more focused on the problems of developing countries and with its 249 indicators that are often defined in an imprecise way could become unmanageable. The ISO 37120 standard shows a more precise definition of indicators, although social and environmental aspects are given slightly greater prominence at the expense of economic and smart indicators. On the contrary, the UNECE-ITU Smart Sustainable Cities Indicator framework is more balanced between different dimensions of sustainability and formulated with a lot of clarity and a forward-looking strategic vision in mind.

Selection of individual indicators for cities, chosen for the present chapter, was based on an earlier sustainable cities framework (Shmelev & Shmeleva, 2009), inspired by our dynamic sustainability assessments carried out for countries (Shmelev, 2011, 2012) and adapted for the urban scale (Shmelev, 2017a). The process of indicator selection for the study was performed in two parts. First, a large set of criteria was analysed, including economic indicators (income per capita at PPP, number of large companies headquartered in the city, creative industries employment), environmental indicators (CO_2 emissions per capita, share of nuclear energy, PM_{10} emissions, water use per capita, waste generation per capita, recycling rates) and socio-cultural indicators (unemployment rate, Gini index of income inequality, life expectancy). After performing a Principal Component Analysis (Shmelev, 2017b), identifying redundant variables and adding relevant dimensions, the set of criteria took its final shape numbering twenty criteria as a result of several iterations.

The cities chosen for our analysis include:

- in Europe: Amsterdam, Barcelona, Berlin, Copenhagen, Frankfurt, Edinburgh, London, Madrid, Milan, Moscow, Munich, Paris, Rome, Stockholm, St Petersburg, Vienna, Warsaw, Zurich;

- in North America: Atlanta, Austin, Boston, Denver, Los Angeles, Mexico City, Miami, Montreal, New York, Portland, San Francisco, Toronto, Vancouver, Washington, DC;
- in South America: Bogota, Buenos Aires, Lima, Quito, Rio de Janeiro, São Paulo, Santiago;
- in Asia: Almaty, Beijing, Delhi, Hong Kong, Istanbul, Mumbai, Seoul, Shanghai, Shenzhen, Singapore, Taipei, Tokyo;
- in Africa: Johannesburg, Kampala, and Nairobi;
- and Oceania: Adelaide, Melbourne and Sydney.

Criteria for selecting these cities from the database of ninety were economic importance, environmental impacts and, most importantly, availability of data on all characteristics of interest for us. Our study draws on a wide range of sources, from Eurostat (2016), city governments (City of London, 2009; City of New York, 2012; City of Rio de Janeiro, 2011; Mairie de Paris, 2011; San Francisco Department for the Environment, 2004; Singapore, 2009), Siemens European Green City Index (Siemens, 2009), World Cities Culture Forum (Mayor of London, 2015), UN Habitat (UN Habitat, 2013) and World Bank publications (World Bank, 2013), LSE Going Green Report (LSE, 2013), as well as considerations on availability of data.

Cities are characterised by multidimensional complexity, which we will illustrate by presenting three indicators: income per capita, unemployment and CO_2

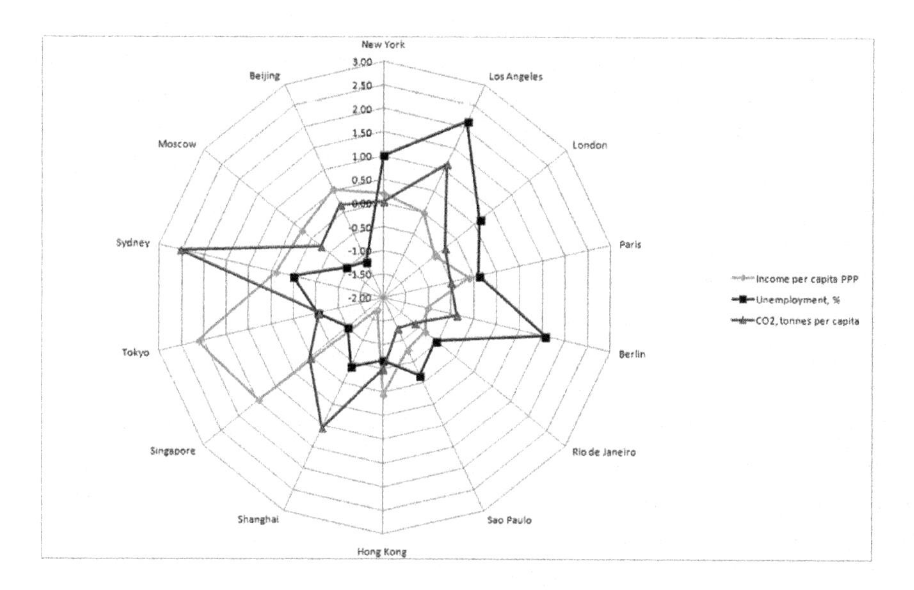

Figure 2.1 Comparative performance of 14 top megacities in economic, social and environmental dimensions: income per capita, unemployment and CO_2 emissions per capita, standardised data, 2013

Table 2.1 Urban smart and sustainable indicator weights

Type	*Abbreviation*	*Indicator*	*Equal*	*Environmental*	*Economic*	*Social*	*Smart*
Economic	GRP	Gross Regional Product, PPP, $million	0.05	0.040	0.080	0.040	0.040
Economic	COMPANIES	Number of Fortune 500 companies headquartered	0.05	0.040	0.080	0.040	0.040
Economic	INCOME	Disposable income per head, PPP, 2010 USD	0.05	0.040	0.080	0.040	0.040
Economic	INFLATION	Consumer price inflation rate, %	0.05	0.040	0.080	0.040	0.040
Economic	UNEM	Unemployment, %	0.05	0.040	0.080	0.040	0.040
Smart	PATENTS	Number of patents per 1,000 inhabitants	0.05	0.050	0.050	0.050	0.100
Smart	INTERNET	Average broadband Internet speed, Mb/c	0.05	0.050	0.050	0.050	0.100
Smart	METRO	Number of underground stations per 1,000,000 inhabitants	0.05	0.050	0.050	0.050	0.100
Smart	CREATIVE	Creative industries employment, %	0.05	0.050	0.050	0.050	0.100
Social	LIFE	Life expectancy at birth, years	0.05	0.067	0.067	0.133	0.067
Social	EDUC	Share of population aged 24–65 with a higher education	0.05	0.067	0.067	0.133	0.067
Social	GINI	Gini index of income inequality, %	0.05	0.067	0.067	0.133	0.067
Environmental	CO_2	CO_2 emissions per person per year	0.05	0.050	0.025	0.025	0.025
Environmental	NO_2	NO_2 average annual concentrations, mg/m^3	0.05	0.050	0.025	0.025	0.025
Environmental	SO_2	SO_2 average annual concentrations, mg/m^3	0.05	0.050	0.025	0.025	0.025
Environmental	PM_{10}	PM_{10} average annual concentrations, mg/m^3	0.05	0.050	0.025	0.025	0.025
Environmental	WATER	Domestic water consumption, m^3 per person per year	0.05	0.050	0.025	0.025	0.025
Environmental	WASTE	Municipal solid waste, kg per person per year	0.05	0.050	0.025	0.025	0.025
Environmental	RECYCL	Recycling rate, %	0.05	0.050	0.025	0.025	0.025
Environmental	GREEN	Green Space as % of total urban space	0.05	0.050	0.025	0.025	0.025
	TOTAL		**1.0000**	**1.0000**	**1.0000**	**1.0000**	**1.0000**

emissions for all cities in their standardised form, illustrating economic, social and environmental dimensions (with means subtracted from the raw figures and the results divided by standard deviations). As we can see in Figure 2.1, the cities differ substantially, e.g. Los Angeles has considerably higher unemployment and CO_2 emissions than New York, while being relatively close to it on the income scale. On the other hand, São Paolo and Rio de Janeiro exhibit considerably lower CO_2 emissions level due to the development of hydropower, have relatively low level of income but unemployment much lower than Berlin or London. Moscow and Beijing have relatively high income level at PPP and low unemployment, but still exhibit considerable potential in reducing CO_2 emissions.

The final set of smart and sustainable indicators included a range of economic, environmental, social and smart cities indicators following an approach identified by the UNECE and ITU United for Smart and Sustainable Cities initiative (Table 2.1). The table presents the indicators with varying weights reflecting different policy priorities to be tested in the multi-criteria assessment section. In each policy priority setting, more emphasis is placed on a particular dimension: economic, social, environmental or smart.

Below we will illustrate the diversity in sustainability performance of global cities on various individual dimensions.

As a matter of example, we would like to illustrate the geographical spread of the Environment Europe database with particulate matter (PM_{10}) concentrations data (Figure 2.2). PM_{10} concentrations are very high in Delhi, Beijing, Shanghai, Rio de Janeiro, Hong Kong and much lower in Sydney, Toronto, Washington, DC, New York, Tokyo, Los Angeles and Berlin.

Figure 2.2 PM_{10} concentrations

Source: Environment Europe™ Sustainable Cities Database, http://environmenteurope.org/, 90 global cities, 2017

3. Cross-section regression analysis

It would be highly beneficial to explore the world cities database from the point of view of interdependencies and trade-offs among various sustainability indicators, which will help us understand causes for certain performance across the whole pool of cities and the inherent trade-offs between indicators. Our goal in this section was to test several hypotheses regarding the inter-disciplinary links among urban sustainability dimensions. The hypotheses were derived from the assertion in the UN Guidelines on Sustainable Development Indicators (UN, 2007), which emphasised the interdisciplinary connections between sustainable development indicators. The exact formulation of the hypotheses is based on our previous research outlined in Shmelev and Shmeleva (2009), Shmelev (2017a) and Shmelev and Speck (2018). We explored a database of world cities, currently featuring 90+ cities from all inhabited continents and tried to see if there is a statistically significant relationship between pairs of indicator variables across the whole spectrum of cities.

The confirmation of our hypothesis of a highly significant correlation between the amount of CO_2 emissions and the share of coal, the most carbon-intensive technology at present in the energy mix, reinforces the need for an urgent transformation and decarbonisation of the energy sector. Such cities as Sydney, Warsaw, Hong Kong, Denver, Portland, Los Angeles, Washington, DC and Shenzhen have above-average levels of coal in the energy mix and exhibit high per capita CO_2 emission. On the other hand, cities such as São Paolo, Rio de Janeiro, Bogota, Quito, Madrid, Adelaide, Copenhagen and Rome have relatively low share of coal in the energy mix and lower levels of CO_2 emissions per capita.

A significant correlation between CO_2 emissions and the share of trips made by walking, cycling and public transport has been confirmed, which enriches our understanding of this wonderful urban planning tool for improving air quality and making the cities greener.

Such cities as Stockholm, Mumbai, Bogota, Delhi, Mexico City, Paris, Amsterdam, Seoul, Barcelona, São Paolo, Berlin, Singapore and Moscow have a significant percentage of trips made by walking, cycling and using public transport and are associated with lower per capita CO_2 emissions. On the other hand, cities such as Sydney, Shenzhen, Almaty, Los Angeles, Miami, Kuala Lumpur, Boston, Vancouver and Toronto rely on private car transport in a much more pronounced way and therefore have significantly higher CO_2 emissions per capita.

The role of renewable energy in reducing CO_2 emissions in global cities has been confirmed at a very high level of statistical significance. This clearly reinstates the tendency in cities such as São Paolo, Bogota, Montreal, Stockholm, Rio de Janeiro, Zurich and Copenhagen, that are largely powered by hydro energy to have lower per capita CO_2 emissions. At the same time cities such as Sydney, Atlanta, Almaty, Frankfurt, Miami, St Petersburg, Shanghai, Boston, Los Angeles, Vancouver and Shenzhen that tend to have lower levels of renewables in the energy mix, tend to exhibit higher per capita CO_2 emissions.

The hypothesis of a strong water-energy nexus, whereby larger CO_2 emissions tend to go hand-in-hand with higher water consumption has been confirmed. Cities such as Los Angeles, Almaty, Atlanta, Miami, Toronto and Kuala Lumpur use larger amounts of water with higher per capita CO_2 emissions. At the same time, cities such as Bogota, Lima, Lagos, Madrid, Adelaide, Barcelona, Copenhagen, Seoul and Rome exhibit lower levels of per capita CO_2 emissions accompanied by lower water consumption.

The hypothesis of a strong statistical link between life expectancy and PM_{10} concentrations echoes the recent WHO report on ambient air pollution and diseases it causes (WHO, 2016). Cities with lower PM_{10} concentrations have significantly higher life expectancy, which confirms the WHO estimates. On average, 10 extra micrograms of PM_{10} per cubic meter of air means a lowering of one's life expectancy by 0.7 of a year. Such cities like Delhi, Kampala, Mumbai, Cairo and Johannesburg exhibit considerably lower levels of life expectancy on the background of higher PM_{10} concentrations. On the positive end of spectrum, Tokyo, Madrid, Stockholm and Copenhagen have higher life expectancy and lower levels of PM_{10}.

The correlation between PM_{10} concentrations and the availability of underground stations illustrates one possible way of tackling high PM_{10} pollutions in cities like Bogota, which do not currently have an underground network, as well as Delhi, Xian, Cairo, Kampala, Mumbai, or Kolkata. Our hypothesis on the existence of such a relationship has been confirmed at a high level of statistical significance. In this regard, cities such as Washington, DC, Paris, Barcelona, Lille, Frankfurt and Madrid show the way in offering their residents a diversified and reliable underground system, which could be responsible for avoiding unnecessary PM_{10} emissions associated with private transportation.

On the other hand, such often neglected phenomenon as inflation could have a profound effect on life expectancy through stress. Our hypothesis about such a statistical link has been confirmed. Cities such as Lagos, Kinshasa, Moscow, St Petersburg, Buenos Aires and Cairo exhibit high level of inflation and lower levels of life expectancy. On the other hand, Tokyo, Milan, Madrid, Barcelona, Paris, Seoul, Toronto, Copenhagen and Vienna show low levels of inflation and higher life expectancy.

The research outlined above forms an important step in creating multivariate regression models explaining variation in key urban sustainability indicators, such as CO_2 or PM_{10}. Such models are used for out-of-sample forecasting.

4. Sustainability assessment: linear aggregation

As a first step in the multidimensional sustainability assessment, we have carried out a linear aggregation with different weights representing varying policy priorities. The total number of cities, for which enough data on the respected sustainability indicators was available is 57, which was determined largely by data availability. This set includes A+++, A++, A+ most important global cities representing Europe, Africa, Asia, North and South America and Oceania. The cities in question cover 6.7% of the European population, 3.2% of the population of Asia, 5.4% of the North American population, 10.5% of the South American population, 26.1% of the population of Oceania and 0.7% of the African population, which indicates some imbalance, which we are planning to address in the future by increasing the share of Asian and African cities in the database. The cities in the Environment Europe database include both A++ cities, London and New York, most A+ cities, including Singapore, Shanghai, Tokyo, Hong Kong, Beijing and Paris, and most significant A–, B+, B and B– cities from different regions.

The linear aggregation assumed perfect substitutability among sustainability criteria and represented a weak sustainability case. Several different policy priorities were applied, placing the emphasis on economic, social, environmental, or smart dimensions through varying weights. The results of the assessment applied to 57 global cities can be seen in Table 2.2.

The results clearly show that under environmental priorities the top five cities have been San Francisco, Stockholm, Seoul, Copenhagen and Zurich, while under smart policy priorities the leading cities are Stockholm, San Francisco, Paris, Tokyo and Boston. San Francisco leads under environmental and economic priorities and Stockholm under social and smart priorities. Seoul is at the 3rd place globally under environmental priorities, 2nd place under economic, 5th under social and 8th under smart, which is extremely strong performance overall. Copenhagen occupies the 4th place in the world under economic priorities, 11th position under economic, 2nd under social and 7th under smart policy priorities. London can be found at 25th place under environmental priorities, 26th under economic, 21st under social and 30th under smart. Washington, DC occupies 29th place under environmental, 8th under economic, 29th under social and 11th place under smart priorities.

The worst performing cities in our database are Johannesburg, Almaty, Delhi, Buenos Aires and Nairobi under environmental priorities; Nairobi, Buenos Aires, Johannesburg, Delhi and Kampala under economic priorities; Johannesburg, Nairobi, Mumbai, Delhi and Rio de Janeiro under social priorities, and Johannesburg, Nairobi, Delhi and Kampala under smart priorities.

5. Most sustainable global cities

Below we will explore some of the most sustainable and smart cities globally trying to explain how they achieved their remarkable success. Among the most successful cities are San Francisco, the US high tech and sustainability hub in California, the US's most economically successful state, and equivalent to the economy of France in size, as well as two national capitals: Stockholm and Seoul (Figure 2.3). Two other national capitals, Copenhagen and Tokyo, are following close behind.

San Francisco

San Francisco leads our ranking in Economic and Environmental Dimensions worldwide. It has been featured as a top global city in Global Cities Index 2017 by A.T. Kearney. The World Economic Forum places San Francisco second in

Table 2.2 Multidimensional sustainability assessment of global cities under environmental, economic, social and smart priorities, linear aggregation

	Equal		*Environmental*		*Economic*		*Social*		*Smart*	
	City	*Score*	*City*	*Score*	*City*	*Score*	*City*	*Score*	*City*	*Score*
1	San Francisco	0.728837829	San Francisco	0.745884997	San Francisco	0.708565601	Stockholm	0.800145459	Stockholm	0.734842073
2	Stockholm	0.724200576	Stockholm	0.735417008	Seoul	0.696053238	Copenhagen	0.739615123	San Francisco	0.659201186
3	Seoul	0.682860509	Seoul	0.685249975	Tokyo	0.694588071	Zurich	0.725674633	Paris	0.619604064
4	Tokyo	0.669353853	Copenhagen	0.6842292	Stockholm	0.674812206	Munich	0.716716327	Tokyo	0.605628111
5	Copenhagen	0.662977225	Zurich	0.67217179	Munich	0.620223872	Seoul	0.711239543	Boston	0.602802376
6	Zurich	0.653376419	Tokyo	0.66247683	Zurich	0.619852675	San Francisco	0.710184418	Frankfurt	0.593562987
7	Munich	0.647959494	Munich	0.66018521	Paris	0.618142335	Madrid	0.708767143	Copenhagen	0.58570181
8	Frankfurt	0.619244104	Montreal	0.639838834	Washington, DC	0.617623044	Tokyo	0.701973518	Seoul	0.578860239
9	Madrid	0.616861292	Madrid	0.637499828	New York	0.613968237	Vancouver	0.688488867	Taipei	0.569712512
10	Paris	0.611380641	Frankfurt	0.624050907	Boston	0.612578192	Toronto	0.683852649	Munich	0.56890388
11	Montreal	0.609328494	Amsterdam	0.618056991	Copenhagen	0.610265388	Amsterdam	0.682670552	Washington, DC	0.553811694
12	Vancouver	0.600465053	Toronto	0.617863979	Frankfurt	0.600977045	Barcelona	0.679793758	Barcelona	0.540994193
13	Singapore	0.600070917	Singapore	0.611425822	Beijing	0.595973084	Frankfurt	0.670926054	Singapore	0.53526509
14	Toronto	0.599721965	Vancouver	0.611411576	Singapore	0.595959409	Edinburgh	0.667160688	Zurich	0.528210603
15	Amsterdam	0.598851736	Paris	0.607608958	Taipei	0.586658173	Vienna	0.662481827	Madrid	0.527752497
16	Boston	0.595889689	Adelaide	0.603481526	Austin	0.58455946	Montreal	0.660955734	Vancouver	0.50542939
17	Taipei	0.593502435	Edinburgh	0.598758802	Portland	0.576649332	Taipei	0.6569491	Shenzhen	0.50272304
18	Vienna	0.583862992	Vienna	0.598268302	Vancouver	0.575087561	Berlin	0.634732411	Amsterdam	0.498058688

(continued)

Table 2.2 (continued)

	Equal		*Environmental*		*Economic*		*Social*		*Smart*	
	City	*Score*	*City*	*Score*	*City*	*Score*	*City*	*Score*	*City*	*Score*
19	Barcelona	0.583177689	Milan	0.597465354	Toronto	0.570393749	Warsaw	0.633965376	Austin	0.483968156
20	Milan	0.58038166	Barcelona	0.593772608	Madrid	0.56480666	Sydney	0.622556009	Vienna	0.481498643
21	Washington, DC	0.578732932	Taipei	0.592371942	Montreal	0.559134644	London	0.620502525	New York	0.477920684
22	Sydney	0.574032499	Sydney	0.59008489	Denver	0.558332608	Milan	0.617602012	Milan	0.469122942
23	Edinburgh	0.573834067	Berlin	0.589412927	Milan	0.556397937	Paris	0.614817019	Berlin	0.464880591
24	Adelaide	0.571711582	Boston	0.588554207	Amsterdam	0.555313391	Shanghai	0.600154385	Sydney	0.464391224
25	New York	0.571231864	London	0.578254515	Vienna	0.554011758	Boston	0.596544585	Hong Kong	0.455196953
26	Berlin	0.569256833	Portland	0.576392478	London	0.550082315	Adelaide	0.591568055	Portland	0.454998553
27	Portland	0.568699732	Rome	0.569579517	Sydney	0.549396592	Austin	0.589318111	Montreal	0.454249488
28	London	0.564710891	Warsaw	0.565584215	Atlanta	0.547492229	Rome	0.585334325	Toronto	0.452383872
29	Austin	0.559870401	Washington DC	0.565520018	Shenzhen	0.544864677	Washington, DC	0.581151193	Atlanta	0.450998826
30	Warsaw	0.547939401	New York	0.565035794	Barcelona	0.543320505	Portland	0.574510999	London	0.432838965
31	Beijing	0.546580373	Austin	0.55472764	Adelaide	0.538468696	Shenzhen	0.573266165	Los Angeles	0.410511808
32	Shenzhen	0.545416223	Shenzhen	0.547207327	Edinburgh	0.533860112	Singapore	0.568662475	Beijing	0.404308052
33	Rome	0.542919547	Beijing	0.543157107	Los Angeles	0.532351182	New York	0.556414778	Denver	0.39812501
34	Denver	0.532147353	Denver	0.533315352	Berlin	0.528807585	Melbourne	0.555410767	Warsaw	0.393872008
35	Los Angeles	0.524475876	Los Angeles	0.53326136	Warsaw	0.521041397	Denver	0.550529334	Edinburgh	0.391991023
36	Hong Kong	0.511830226	São Paulo	0.531032486	Melbourne	0.519633675	Beijing	0.526295598	Shanghai	0.379785135

37	Shanghai	0.499676319	Bogota	0.53062545	Rome	0.50736467	Atlanta	0.526069018	Rome	0.376887078
38	Atlanta	0.495355013	Hong Kong	0.521778257	Hong Kong	0.506517338	Los Angeles	0.519841336	Miami	0.37510133
39	Melbourne	0.493648583	Quito	0.50368792	Shanghai	0.497131851	Moscow	0.504114777	Adelaide	0.367893436
40	São Paulo	0.49145537	Shanghai	0.502713837	Miami	0.483726207	St Petersburg	0.502147951	Melbourne	0.36583582
41	Bogota	0.481960749	Lima	0.493246607	São Paulo	0.448258864	Hong Kong	0.49099525	Moscow	0.345222003
42	Quito	0.461863238	Melbourne	0.489935071	Moscow	0.437122687	Almaty	0.460732654	St Petersburg	0.321464479
43	Lima	0.461767099	Kampala	0.479298573	Lima	0.435144816	Miami	0.451267446	São Paulo	0.319247299
44	Moscow	0.461742958	Moscow	0.47907333	Mexico City	0.430736523	São Paulo	0.441283691	Mexico City	0.297696202
45	Miami	0.438184988	Atlanta	0.475945113	Bogota	0.429110194	Lima	0.437054842	Santiago	0.29578621
46	Mexico City	0.434856726	St Petersburg	0.461413807	Quito	0.421072373	Santiago	0.436593786	Lima	0.278749801
47	St Petersburg	0.434508227	Rio de Janeiro	0.454332021	Santiago	0.41062972	Istanbul	0.423974389	Bogota	0.278007389
48	Kampala	0.429601593	Mexico City	0.452530326	Rio de Janeiro	0.4003587	Mexico City	0.412340806	Istanbul	0.273777439
49	Santiago	0.425280725	Istanbul	0.449706691	Almaty	0.380221247	Bogota	0.411217219	Almaty	0.271746822
50	Rio de Janeiro	0.424220642	Santiago	0.443018098	St Petersburg	0.378754202	Quito	0.402072178	Rio de Janeiro	0.267282964
51	Istanbul	0.419113227	Miami	0.425037308	Istanbul	0.377885619	Buenos Aires	0.377719961	Quito	0.267106988
52	Mumbai	0.388648046	Mumbai	0.421730586	Mumbai	0.367693566	Kampala	0.377711632	Buenos Aires	0.258732775
53	Buenos Aires	0.387823252	Nairobi	0.417263843	Kampala	0.364726571	Rio de Janeiro	0.376011815	Kampala	0.244507359
54	Almaty	0.376338391	Buenos Aires	0.414663003	Delhi	0.361456021	Delhi	0.319621638	Mumbai	0.225173887
55	Delhi	0.375118567	Delhi	0.401285845	Buenos Aires	0.35564851	Mumbai	0.305548895	Delhi	0.224037724
56	Nairobi	0.34437642	Almaty	0.376863919	Johannesburg	0.274162102	Nairobi	0.234899074	Nairobi	0.176755155
57	Johannesburg	0.290983502	Johannesburg	0.323259721	Nairobi	0.231506827	Johannesburg	0.168994392	Johannesburg	0.157769727

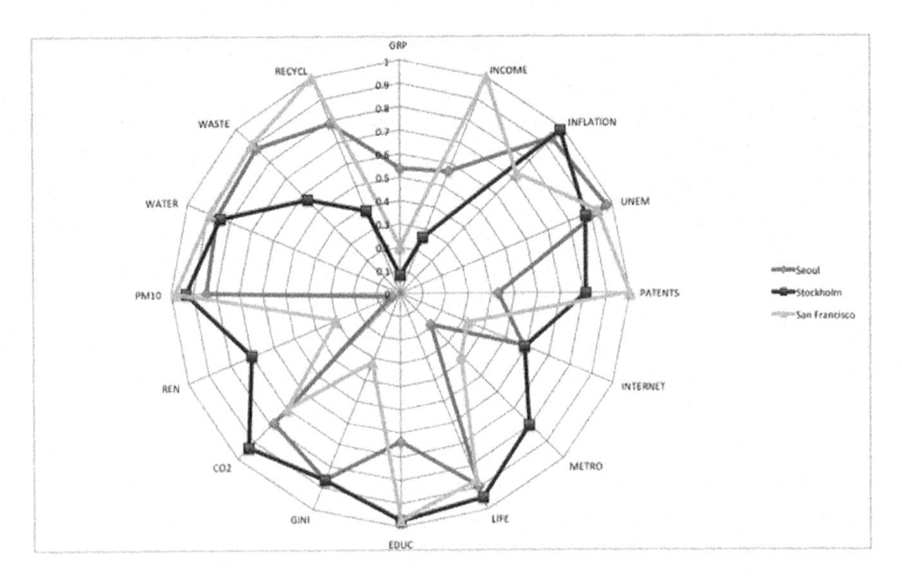

Figure 2.3 Comparison of the most successful cities globally: San Francisco, Stockholm and Seoul. Further away means better performance on each of the indicators

the world for tech in 2017. The Strategic Plan of San Francisco for 2016–2020 has a mission 'to provide solutions that advance climate protection and enhance quality of life for all San Franciscans'. The Strategic Plan has five goals: 1) Promoting Healthy Communities and Ecosystems; 2) Leading on Climate Action; 3) Strengthening Community Resilience; 4) Eliminating Waste; 5) Amplifying Community Action.

Goal 1 has a target to foster healthy and sustainable communities through science, with an emphasis on supporting San Francisco's most vulnerable populations and is further subdivided into subgoals: increase adoption of safer alternatives to harmful products and materials; support residents, businesses and city staff in limiting use of toxic and hazardous products, practices and materials; partner with key stakeholders to ensure sustainability initiatives are equitable and accessible; leverage the purchasing power of municipal operations to advance markets for green products and services; lead and leverage inter-agency efforts to green San Francisco's built and natural environments; maximise carbon sequestration through natural ecosystems; support sustainable and healthy food options for all individuals and families in San Francisco, especially the food insecure.

Goal 2 uses an active target to reduce greenhouse gas emissions by 40% by 2025 and has the following subgoals: maximise energy efficiency in existing buildings; reduce dependency on single-occupancy vehicles by improving access to sustainable and affordable modes of transportation; commit to ambitious carbon reduction targets across city agencies; continue to share San Francisco's practices and lessons to show the world what is possible; decarbonise the energy used for heating and cooling buildings; accelerate shift to 100% renewable grid electricity by 2030 and maximise local on-site generation of renewable electricity through policy development and investment; decarbonise the transport sector by facilitating deployment of electric and zero-emissions vehicles.

Goal 3 focuses on supporting economically resilient communities and has the following subgoals: keep small businesses and community-based organisations in San Francisco by minimising costs associated with energy use, water use and waste generation; reduce cost of living in San Francisco by ensuring cost-effective energy efficiency upgrades in all housing, with a special focus on affordable housing; create jobs and economic opportunity by keeping sustainability investments in our communities; increase equitable distribution and installation of local renewable energy and battery storage; connect communities most adversely impacted by environmental injustices with resources that enable them to become more resilient to the impacts of climate change; make San Francisco's infrastructure, landscapes and neighbourhoods resilient to climate change and coordinate with other city agencies and jurisdictions on adaptation planning and community engagement to ensure everyone has a seat at the table.

Goal 4 aims achieve zero waste and work towards closing landfills serving San Francisco. It has the following sub-goals: increase participation in recycling and composting programmes, expand accessibility and structure of programmes for collecting hazardous products, modernise San Francisco's refuse collection and process infrastructure, increase reuse and recycling of construction and demolition, prevent food waste, reduce consumption of single-use items and expand use of sustainable packaging materials.

Goal 5 aims to build a shared culture of environmental stewardship across San Francisco. It aims to reflect stakeholders' values, needs and everyday lives in Department programme offerings and environmental action; challenge businesses and local influencers to commit to meaningful action on climate strategy; align programme services with partners across city departments to maximise impact and reduce confusion; support and grow local environmental leaders, particularly in communities that have historically been under-represented in the environmental movement; provide grants and resources to a wide range of organisations in order to increase reach and collaboration; increase personal actions that reduce our impacts while preparing for climate change, and increase the funding pool available to community groups for neighbourhood environmental work by expanding the carbon fund grant programme.

San Francisco is one of the world leaders on recycling (80%), generating very small amounts of municipal solid waste per person (195.4 kg per year). Forty-nine per cent of the trips made by citizens are carried out by walking, cycling, or using public transport. San Francisco generates 6.2 tonnes of CO_2 per person per year and generates 30% of its energy through renewable sources. In the field of air quality, San Francisco exhibits low levels of PM_{10} pollution at 15.77 μg/m^3, which is within the WHO limit of 20μg/m^3. It has a reasonably diverse systems of underground public transport.

Economically, San Francisco is one of the most vibrant places in the world. With high per capita income of US$88,518 at PPP in 2010 prices, inflation is low at 3.8%, unemployment at 4.4%, which is three times lower than Los Angeles. San Francisco is a world innovation hub with 3.24 patents registered per 1,000 inhabitants, which is higher than Boston. Income differentiation in San Francisco is high, illustrated by a Gini index of income inequality of 0.51. Such relatively high income inequality could limit San Francisco's performance in the social dimension.

Stockholm

Analysis of Stockholm's performance as a sustainable city has been the focus of our recent publication (Tonks & Shmelev, 2018). Our research as well as several other metrics, including the European Green City Index compiled by Siemens in 2009, highlight the very strong position of Stockholm compared to other capital cities. Stockholm has received a prestigious prize of a 'European Green Capital' in 2010, awarded by experts following the detailed assessment of cities' performance. The city of Stockholm adopted an Environmental Programme for 2016–2019, based on complementarity between environmental protection and human needs. The six priority areas of this programme include: sustainable energy use, environmentally friendly transport; sustainable land and water use, resource efficient recycling, a non-toxic Stockholm, and a healthy indoor environment.

Our research shows that along with a serious concern about the environment, Stockholm exhibits extremely strong economic performance. Sweden is consistently ranked high in the World Economic Forum's Global Competitiveness Index. Sweden is a very open economy and outperforms the US, Japan and Brazil by attracting approximately 4.7% of GDP in foreign direct investment per annum. At the same time, it invests in the range of 3.7% of GDP in research and development, which is considerably higher than the EU average of 1.8%. Sweden and Stockholm managed to decouple economic development from the growth in CO_2 emissions as a result of technological modernisation in the 1970s with the extensive use of hydropower and nuclear energy as well as the successful application of environmental taxes since 1991 (Shmelev & Speck, 2018). Stockholm aims to be fossil fuel free by 2050 and is actively involved in new programmes on green urban transport.

Despite rather modest per capita disposable income at US$23,456 at PPP in 2010 prices – which is higher than those of London, Berlin, Madrid, Rome, Copenhagen and slightly higher than Vienna, and lower than Paris, Frankfurt, Zurich, Munich and Moscow – Stockholm has been capable of focusing on the qualitative aspects of development. Stockholm's economy is largely innovations-based with the number of new patents registered (2.62 per 1,000 inhabitants) higher than all other regional European centres, including technological giants like Copenhagen, Munich and Zurich. Stockholm outperforms Tokyo, but is at a lower level than Shenzhen, Taipei, Boston and San Francisco. At the same time, Stockholm is characterised by very low inflation – in fact, deflation at 0.04%. Unemployment in Stockholm has been recorded at 7.09%, which is lower than Amsterdam, London, Rome, Berlin and Madrid. Unemployment lower than

Stockholm's is observed in Copenhagen, Frankfurt, Zurich, Munich, Boston and San Francisco. The level of higher education in Stockholm is 58% of all residents aged 25–64. This is higher than Berlin, Rome, London, Amsterdam, Paris and Munich. In Asia, Stockholm compares favourably on education level with Seoul, Shanghai, Beijing. In the US, Stockholm outperforms San Francisco, Washington, DC and Boston.

Stockholm, representing the Nordic governance model, is characterised by a high level of taxation as a percentage of GDP and a reasonably low Gini index of income inequality (0.3). This is the level very similar to that of Barcelona, Amsterdam and Seoul, slightly higher than Tokyo and Berlin, and considerably lower than that of Hong Kong, Singapore, Beijing, San Francisco, Washington, DC and New York.

In the environmental dimension, Stockholm is characterised by a very distinct position among world cities on certain issues: for example, its leading CO_2 emissions at 1.44 tonnes (t) per capita, Stockholm outperforms such world capitals as Madrid, Copenhagen, Barcelona, Rome, Berlin, London and Moscow. A comparison with Asian cities reconfirms the leading position of Stockholm, with CO_2 emissions approximately three times lower than Seoul and over three times lower than Tokyo, Hong Kong, and considerably lower than in Singapore, Beijing and Shanghai. Among American cities, San Francisco, the regional leader, exhibits CO_2 emissions over four times higher than Stockholm, while New York, Washington, DC and Boston are left far behind.

One of the possible reasons for such low CO_2 emissions in Stockholm could be its active reliance on renewable energy. Stockholm occupies one of the leading positions in Europe on the share of renewables in the energy mix (70%), following Zurich. Stockholm's performance on renewables is considerably better than other European cities – Copenhagen, Edinburgh, Madrid, Rome, Moscow, Vienna, Paris, London and Amsterdam. On the other hand, according to the data on the share of all trips made by walking, cycling and using public transport, Stockholm is unfortunately not in the lead, following Vienna, Madrid, Moscow, Amsterdam and London at a modest level of 53%. For comparison, Asian cities like Singapore, Seoul, Beijing, Shanghai, Tokyo and Hong Kong perform better, while American cities such as San Francisco and New York are at a similar level and Los Angeles and Washington, DC perform much worse.

Another important parameter for 'explaining' low CO_2 emission levels is infrastructure, which gives affordance to public transport use by residents. In this regard, Stockholm is characterised by a highly diversified system of underground networks with 108 underground stations per 100,000 inhabitants. This is better than most European cities, including Madrid, Amsterdam, London, Rome and Berlin, the exception being Paris. Compared to Stockholm, Asian cities such as Kuala Lumpur, Seoul, Singapore, Shanghai, Beijing and Tokyo have considerably less stations per 100,000 inhabitants. For comparison, in the US, the leader is Washington, DC (the only city outperforming Stockholm), followed by Boston, San Francisco, New York and Los Angeles, which has a very small underground system.

Air quality in Stockholm is at a good European level with an average annual concentration of PM_{10} at 26μg/m^3, which is nevertheless higher than the maximum recommended by the World Health Organization of 20μg/m^3. Better air quality is observed in such European cities as Edinburgh, Madrid, Zurich, Amsterdam and Vienna; it is worse in London and Paris. In Asia, air quality is better in Tokyo, and worse in Singapore, Hong Kong, Shanghai and Beijing. In American cities air quality tends to be better: San Francisco, Washington, New York and Los Angeles.

In the field of circular economy, Stockholm generates rather large amounts of municipal solid waste of 597 kg/person per year, 31% of which is recycled. Other European cities, such as Madrid, Amsterdam, Berlin, London, Paris and Vienna, practice less resource-intensive lifestyles. In Asia, Tokyo generates less than half of municipal solid waste person per year compared to Stockholm; Seoul, Shanghai, Singapore, Beijing and Hong Kong generate much less, but Kuala Lumpur generates considerably more. Of American cities, San Francisco generates almost three times less than Stockholm, Washington, DC nearly half, New York a similar amount, Los Angeles slightly more, and Boston considerably more. Recycling rates are lower than Stockholm in Madrid, Rome, Paris and Copenhagen, and higher in Vienna, London, Berlin and Amsterdam. In Asia, recycling in Beijing is slightly lower (30), and in Hong Kong (39), Singapore (61), Seoul (63.50) considerably higher than in Stockholm. In the US, New York recycles 27% of its municipal solid waste, Washington 26%; however Los Angeles recycles 76.4% and San Francisco 80%.

Seoul

In November 2017, the Metropolitan Government of Seoul adopted 17 Sustainable Development Goals and 96 targets. Seoul Plan 2030, the city's urban planning document, covers three central dimensions: environment, society and culture, and the economy, and includes 30 urban development indicators. Among Seoul's strategic priorities are the reduction of Seoul's reliance on nuclear power, an energy efficiency and sustainable energy action plan, and increasing female participation in economic activities. Already in 2013, the International Telecommunication Union issued a Smart Cities report devoted to Seoul's achievements. The Seoul's Smart City programme includes fast optical wire and wireless network, Seoul began distributing second-hand smart devices to low-income families, established a U-Seoul net in 2003, which connected major public buildings, offices and municipalities via fibre-optic cables arranged along Seoul's underground tunnels. The Smart Work Center was established to allow government employees to work closer to home and 30% of staff were covered by this initiative in 2015. Seoul's open governance model implies a strong system of community mapping, through which citizens could raise concerns about their neighbourhoods and communities. Seoul's smart metering project aims to reduce electricity consumption by 10% and in 2012 a pilot project issued a thousand families with smart meters. Addressing SDG Goal 11 'Sustainable cities and communities' aimed at making cities inclusive, safe, resilient and sustainable, the U-Seoul safety service was established to assist vulnerable groups: the elderly, children, people with Alzheimer's; when the holder leaves the safety zone or pushes the emergency button, the alert is sent to police, guardians, fire departments and CCTV control centres. The administrative information is made available to citizens through the Open Governance 2.0 programme. The Open Data Square covers information on general administrative work, welfare, culture and tourism, city management, environment, safety, education, health, industry, economy and transportation. Smart solutions are used in Seoul to optimise citizens' personal travel, with opportunities for planning routes, choosing green transport solutions and reducing carbon emissions.

Seoul's metropolitan area maintained a significant share of the Korean economy, approaching 50% in 2013; at the same time, the Seoul metropolitan area provided employment for 50% of the country's population. Seoul's unemployment rate of 2.3% in 2014 was at the level of regional leaders like Beijing and Singapore, but lower than that of Tokyo. Unemployment in Seoul is comparable to that of Munich, but is lower than that in Stockholm and San Francisco. At the same time, it is considerably lower than unemployment in Vienna, London, Berlin, New York, Boston and Washington, DC. Seoul has a significant rate of residents with higher education (40.6), which is slightly lower than Singapore but higher than regional centres Beijing, Hong Kong, Shanghai and Shenzhen, and is higher than similar levels in Berlin, Vienna and Rome. The reasonably low Gini index of 0.3 underlines the values of equality in Korean society and is considerably lower than that of regional leaders such as Singapore, Beijing and Hong Kong. Compared with other global cities, the Gini index of Seoul is comparable to that of Canadian cities: Vancouver, Montreal; of European cities: Stockholm, Vienna, Amsterdam and Munich. At the same time, Seoul's Gini index is significantly lower than in New York and San Francisco, and lower than Paris, Moscow and London; however, the accuracy of the latter the authors find problematic due to the high number of super-rich individuals that have property in London and reside there for part of the year. Inflation in Seoul is low at 0.71%, which is comparable only to Copenhagen, not mentioning deflation in Stockholm. Compared to the regional centres, Seoul's inflation is lower than Beijing, Singapore, Shanghai and Hong Kong. The cities with low inflation rates, such as Toronto, New York, Berlin and Munich, are exhibiting higher rates of inflation than Seoul, to say nothing of other European cities such as Vienna, Amsterdam and London, and cities with high levels of inflation such as Moscow and St Petersburg.

According to our model, which uses the Smart and Sustainable Urban Development Indicator Framework, the number of patents registered per 1,000 inhabitants in Seoul is at a very respectable level of 1.4 per year. It is higher than the levels of the regional centres of Shanghai, Hong Kong, Singapore and Beijing, but lower than the innovation powerhouses of Tokyo, Shenzhen and Taipei. Compared with European cities, patent registration is higher than in Vienna, London, Barcelona, Amsterdam and Berlin, and lower than European leaders Munich, Copenhagen and Stockholm. In North America, Seoul outperforms Montreal, Washington, DC, Vancouver and New York, and follows the global leaders of Boston and San Francisco.

In the environmental field, CO_2 emissions per capita measured in Seoul on an annual basis at 4.5 are lower than regional leaders Tokyo, Hong Kong, Singapore, Beijing, Shanghai and Shenzhen. Compared with European cities, Seoul is outdone by Scandinavian cities which traditionally exhibit very high performance – Stockholm and Copenhagen, but is performing better than Paris, Amsterdam, Berlin, London, Vienna and Munich. Seoul also outperforms San Francisco, New York, Montreal, Boston, Washington, DC and Los Angeles.

Seoul is characterised by very low water consumption per capita of 69.10 m^3/person per year, which compares favourably to Beijing, Shanghai, Shenzhen, Singapore, Tokyo and Hong Kong. Copenhagen, London and St Petersburg outperform Seoul, but Stockholm, Paris, Moscow, New York, Washington, DC and Los Angeles perform worse than Seoul.

In the sphere of waste management, Seoul generates 226.4 kg of municipal solid waste per person per year, which is less than Tokyo, Shanghai, Singapore, Beijing, Hong Kong and Shenzhen. Seoul compares favourably with Copenhagen, Amsterdam, Berlin, London, Paris, Stockholm and Vienna. In the US, Seoul outperforms Washington, DC, New York, Los Angeles. San Francisco is one of the few cities generating less municipal solid waste than Seoul.

Recycling is definitely one of the main strengths of Seoul, with 63.5% of all collected municipal solid waste being recycled. In this regard, Seoul outperforms regional centres such as Singapore, Tokyo, Hong Kong and Beijing. European cities perform worse than Seoul on recycling: Vienna, London, Berlin, Munich and Amsterdam. In the US, Seoul performs better than New York but worse than Los Angeles and San Francisco.

Green space in Seoul is not particularly abundant at 1.39 m^2 per person, which is lower than Tokyo, Beijing, Shanghai, Barcelona, London, Paris, Stockholm, Berlin, Rome and Copenhagen.

6. Conclusion

In this chapter, we focused on global cities; the centres for economic activity and the cities that are responsible for a considerable share in global CO_2 emissions and produce substantial volumes of waste. The application of multi-criteria analysis allowed us to produce a multidimensional rankings of the fifty-seven world cities on twenty sustainability criteria. At the same time, varying indicator weights produced aggregate performance scores of global cities under four policy priorities: economic, social, environmental and smart city criteria. The assessment carried out identified sustainability leaders – San Francisco, Stockholm and Seoul – and cities that are lagging behind – Johannesburg, Nairobi, Delhi, Mumbai, Almaty, Buenos Aires. It is important to note that there is no absolute global leader that outperforms all other cities on all dimensions. San Francisco dominates the global rankings on economic and environmental criteria, Stockholm on social and smart priorities. The result has put the performance of individual cities within the global context and presented the indicator-based sustainable development performance of individual cities within a coherent framework of multi-criteria decision aid.

Learning from best practices and worst cases in this context provides an invaluable insight for policy reform to create smarter, greener, more compact, socially diverse, economically strong and less polluting cities around the world.

References

Ahvenniemi, H., Huovila, A., Pinto-Seppä, I., Airaksinen, M., 2017. What are the differences between sustainable and smart cities? *Cities*, 60 (1 February 2017), pp. 234–245.

Bithas K.P., Christofakis, M., 2006. Environmentally sustainable cities: Critical review and operational conditions. *Sustainable Development*, 14, pp. 177–189.

Bolund, P., Hunhammar, S., 1999. Ecosystem services in urban areas. *Ecological Economics*, 29 (1999), pp. 293–301

City of London, 2009. City of London Carbon Footprint, www.conurbant.eu/file/1141-London_SUS_CarbonFootprintreport.pdf

City of New York, 2012. Inventory of New York City Greenhouse Gas Emissions. DOI: 10.13140/2.1.2801.0240, http://s-media.nyc.gov/agencies/planyc2030/pdf/greenhouse gas_2012.pdf

City of Rio de Janeiro, 2011. Greenhouse Gas Inventory and Emissions Scenarios of Rio de Janeiro, Brazil, www.rio.rj.gov.br/dlstatic/10112/1712030/4114527/CRJ_InventarioGEE2012_resumo_tecnicoINGLESFINAL1.pdf

Dassen, T., Kunseler, E. van Kessenich, L.M., 2013. The sustainable city: An analytical-deliberative approach to assess policy in the context of sustainable urban development. *Sustainable Development*, 21, pp. 193–205.

EC, 2009. Sustainable development indicators: An overview of relevant Framework Programme funded research and identification of further needs in view of EU and international activities, https://ieep.eu/archive_uploads/443/sdi_review.pdf

Eurostat, 2016. Urban Europe: Statistics on cities, towns and suburbs, http://ec.europa.eu/eurostat/web/products-statistical-books/-/KS-01-16-691

García-Fuentes, M.Á., Quijano, A., De Torre, C., García, R., Compere, P., Degard, C., Tomé, I., 2017. European cities characterization as basis towards the replication of a smart and sustainable urban regeneration model. *Energy Procedia*, 111 (1 March), pp. 836–845.

Girardet, H., 1993. *The Gaia Atlas of Cities: New Directions for Sustainable Urban Living*, Gaia Books Ltd.

Girardet, H., 2004. *Cities People Planet: Liveable Cities for a Sustainable World*, John Wiley & Sons.

Girardet, H., 2014. *Creating Regenerative Cities*, Routledge.

Girardet, H., Mendonca, M., 2009. A renewable world: Energy, ecology, equality. A report to the World Future Council, http://www.stiftung-drittes-millennium.com/en/aktivitaeten/documents/WFC-ARenewableWorld_en.pdf

Girardi, P., Temporelli, A., 2017. Smartainability: A methodology for assessing the sustainability of the smart city. *Energy Procedia*, 111 (1 March), pp. 810–816.

GLA, 2013. Smart London Plan: Using the creative power of new technologies to serve London and improve Londoners' lives, www.london.gov.uk/sites/default/files/smart_london_plan.pdf

Gomez-Baggethun E., Barton D. N. 2013. Classifying and valuing ecosystem services for urban planning. *Ecological Economics*, 86, pp. 235–245.

Hall, P., 2014. *Good Cities, Better Lives: How Europe Discovered the Lost Art of Urbanism* (Planning, History and Environment Series), Routledge.

Hall, P., Pfeiffer, U., 2000. *Urban Future 21: A Global Agenda for Twenty-First Century Cities*, Routledge.

Hall, P., Buijs, S., Tan, W., Tunas, D., 2010. *Megacities: Exploring a Sustainable Future*, nai010 Publishers.

Hara, M., Nagao, T., Hannoe, S., Nakamura, J., 2016. New key performance indicators for a smart sustainable city. *Sustainability* (Switzerland), 8(3), Article number 206.

ISO, 2014. ISO 37120:2014(en) Sustainable development of communities: Indicators for city services and quality of life, www.iso.org/standard/62436.htm

Kierstead, J., Leach, M., 2008. Bridging the gaps between theory and practice: A service niche approach to urban sustainability indicators. *Sustainable Development*, 16, pp. 329–340

Klopp, J.M., Petretta, D.L., 2017. The urban sustainable development goal: Indicators, complexity and the politics of measuring cities. *Cities*, 63 (1 March), pp. 92–97.

LSE, 2013. Going Green: How cities are leading the next economy. A global survey and case studies of cities building the green economy, Final report, https://lsecities.net/.../reports/going-green-how-cities-are-leading-the-next-economy/

Mairie de Paris, 2011. *Le bilan carbone de Paris. Bilan des émissions de gaz à effet de serre*.

Manitiu, D.N., Pedrini, G., 2016. Urban smartness and sustainability in Europe: An ex ante assessment of environmental, social and cultural domains. *European Planning Studies*, 24 (10), pp. 1766–1787.

Martin, N., Rice, J., 2014. Sustainable development pathways: determining socially constructed visions for cities. *Sustainable Development*, 22, pp. 391–403.

Mayor of London, 2015. World Cities Culture Report, www.london.gov.uk/sites/default/files/wccf_report_interior_151102.pdf

Monfaredzadeh, T., Berardi, U. 2015. Beneath the smart city: Dichotomy between sustainability and competitiveness. *International Journal of Sustainable Building Technology and Urban Development*, 6(3), pp. 140–156.

Mori, K., Yamashita T., 2015. Methodological framework of sustainability assessment in City Sustainability Index (CSI): A concept of constraint and maximisation indicators. *Habitat International*, 45 (2015), pp. 10–14.

Naess, P., 1995. Central dimensions in a sustainable urban development. *Sustainable Development*, 3, pp. 120–129.

Pierce, P., Ricciardi, F., Zardini, A., 2017. Smart cities as organizational fields: A framework for mapping sustainability-enabling configurations. *Sustainability* (Switzerland), 9(9), Article number 1506.

Roy, B., 1996. *Multicriteria Methodology for Decision Aiding*, Kluwer Academic Publishers.

San Francisco Department for the Environment, 2004. Climate action plan for San Francisco. Local actions to reduce greenhouse gas emissions, https://sfenvironment.org/download/2004-climate-action-plan-for-san-francisco

Shen, L.-Y., Ochoa, J.J., Shah, M.N., Zhang, X., 2011. The application of urban sustainability indicators: a comparison between various practices. *Habitat International*, 35 (2011), pp. 17–29.

Shen, L., Zhou, J., 2014. Examining the effectiveness of indicators for guiding sustainable urbanization in China. *Habitat International*, 44 (2014), pp. 111–120.

Shmelev, S., 2012. *Ecological Economics: Sustainability in Practice*, Springer.

Shmelev, S., ed., 2017a. *Green Economy Reader: Lectures in Ecological Economics and Sustainability*, Dordrecht and New York: Springer.

Shmelev, S., 2017b. Multidimensional sustainability assessment for megacities, in Shmelev, S., ed. *Green Economy Reader: Lectures in Ecological Economics and Sustainability*, Springer, pp. 205–236.

Shmelev, S.E., 2011. Dynamic sustainability assessment: The case of Russia in the period of transition (1985–2008). *Ecological Economics*, 70(11), pp. 2039–2049.

Shmelev, S.E., Shmeleva, I.A., 2009. Sustainable cities: Problems of integrated interdisciplinary research. *International Journal of Sustainable Development*, 12 (1/2009), pp. 4–23.

Shmelev, S.E., Shmeleva, I.A., eds, 2012. *Sustainability Analysis: An Interdisciplinary Approach*, Palgrave.

Shmelev S.E., Shmeleva, I.A., 2018. Global urban sustainability assessment: A multidimensional approach. *Sustainable Development*, 26(6), pp. 904–920.

Shmelev, S.E., Shmeleva, I.A., 2019. Multidimensional sustainability benchmarking for smart megacities. *Cities*, 92, pp. 134–163.

Shmelev, S.E., Speck, S.U., 2018. Green fiscal reform in Sweden: Econometric assessment of the carbon and energy taxation scheme, *Renewable & Sustainable Energy Review*, 90, pp. 969–981.

Siemens, 2009. European Green City Index: Assessing the environmental impact of Europe's major cities, www.siemens.com/entry/cc/features/greencityindex_international/all/en/pdf/report_en.pdf

Singapore, 2009. A lively and liveable Singapore: Strategies for sustainable growth, www.greengrowthknowledge.org/sites/default/files/downloads/policy-database/SINGAPORE%29%20A%20Lively%20and%20Liveable%20Singapore%20-%20Strategies%20for%20sustainable%20growth.pdf

Spangenberg, J.H., 2002a. Environmental space and the prism of sustainability: Frameworks for indicators measuring sustainable development. *Ecological Indicators*, 2 (2002), pp. 295–309.

Spangenberg, J.H., 2002b. Institutional sustainability indicators: An analysis of the institutions in Agenda 21 and a draft set of indicators for monitoring their effectivity. *Sustainable Development*, 10, pp. 103–115.

Spangenberg, J.H., 2005. Economic sustainability of the economy: Concepts and indicators. *International Journal of Sustainable Development*, 8 (1/2), pp. 47–64.

Spangenberg, J.H., 2017. Hot air or comprehensive progress? A critical assessment of the SDGs. *Sustainable Development*, 25, pp. 311–321.

Spangenberg, J. H., Settele, J., 2010. Precisely incorrect? Monetising the value of ecosystem services, *Ecological Complexity*, 7 (2010), pp. 327–337.

Tonks, E., Shmelev, S.E., 2018. A multi-criteria approach for indicator-based urban sustainability assessment under varying policy priorities. *Ecological Indicators*, (in press).

UN, 2007. *Indicators of Sustainable Development: Guidelines and Methodologies*, New York.

UN, 2015. Transforming Our World: The 2030 Agenda for Sustainable Development, Resolution adopted by the General Assembly on 25 September 2015, A/RES/70/1

UNECE, 2013. Framework and suggested indicators to measure sustainable development, prepared by the Joint UNECE/Eurostat/OECD Task Force on Measuring Sustainable Development, 27 May.

UNECE & ITU, 2016. Rome Declaration Adopted by the participants of the Forum 'Shaping smarter and more sustainable cities: Striving for sustainable development goals', on 19 May in Rome, www.itu.int/en/ITU-T/Workshops-and-Seminars/Documents/Forum-on-SSC-UNECE-ITU-18-19-May-2016/Rome-Declaration-19May2016.pdf

UN ECOSOC, 2015. The UNECE–ITU Smart Sustainable Cities Indicators, www.unece.org/fileadmin/DAM/hlm/documents/2015/ECE_HBP_2015_4.en.pdf

UNEP, 2011. *Towards a Green Economy: Pathways to Sustainable Development and Poverty Eradication*, https://sustainabledevelopment.un.org/index.php?page=view&type=400&nr=126&menu=35

UN Habitat, 2013. *Planning and Design for Sustainable Urban Mobility: Global Report on Human Settlements*, https://unhabitat.org/planning-and-design-for-sustainable-urban-mobility-global-report-on-human-settlements-2013/

UN Habitat, 2016. *Urbanisation and Development: Emerging Futures*, www.unhabitat.org/wp-content/uploads/2014/03/WCR-%20Full-Report-2016.pdf

Valentin, A., Spangenberg, J.H., 2000. A guide to community sustainability indicators. *Environmental Impact Assessment Review*, 20 (2000), pp. 381–392.

Wei, Y., Huang, C., Lam, P.T.I., Yuan, Z., 2015. Sustainable urban development: A review on urban carrying capacity assessment. *Habitat International*, 46 (2015), pp. 64–71.

Wei, Y., Huang, C., Li, J., Xie, L., 2016. An evaluation model for urban carrying capacity: A case study of China's mega-cities. *Habitat International*, 53 (2016), pp. 87–96.

WHO, 2016. Ambient air pollution: A global assessment of exposure and burden of disease, www.who.int/phe/publications/air-pollution-global-assessment/en/

Wong, C., 2015. A framework for 'City Prosperity Index': Linking indicators, analysis and policy. *Habitat International*, 45 (2015), pp. 3–9.

World Bank, 2013. Building sustainability in an urbanizing world: A Partnership Report, https://openknowledge.worldbank.org/handle/10986/18665

Yigitcanlar, T., Dur, F., Dizdaroglu, D., 2015. Towards prosperous sustainable cities: A multiscalar urban sustainability assessment approach. *Habitat International*, 45 (2015), pp. 36–46.

3 Comparative analysis of indicator-based urban sustainability assessment frameworks

Ellie Tonks and Stanislav E. Shmelev

1. Introduction

There is no doubt that the world is becoming increasingly urban. This trend will only continue into the foreseeable future (WHO, 2016). Cities, however, generate 80% of greenhouse gas emissions and claim 75% of the world's energy production (Lazaroiu & Roscia, 2012). Therefore, as urbanisation continues, the sustainability of cities must take central stage in both political and scientific spheres (Huang et al., 2015).

In the face of inherent pluralism, sustainable development is commonly regarded as the quest for developing and sustaining qualities of life for the current generation, without limiting the possibility for others, elsewhere or in the future, to achieve a desirable quality of life (World Commission on Environment and Development, 1987). The concept of the sustainable city, on the other hand, has been subject to a number of definitions (Michael et al., 2014). Publications on the topic of sustainable cities often refer to the three pillars of sustainability – economy, environment and society – as well as how to balance them in order to develop and sustain qualities of life for city dwellers (Bagstad & Shammin, 2012; Dizdaroglu, 2015; Huang et al., 2015; Mori & Christodoulou, 2012). Increasingly, the three pillars are underpinned by a fourth dimension: the institutional and governance structures required to make sustainability successful (Michael et al., 2014). According to the United Nations Human Settlement Program (UN-Habitat), for example, a sustainable city is one where urban planning addresses the environmental, demographic, economic, socio-spatial and institutional urban challenges (United Nations Human Settlements Programme, 2009). Echoing the assessment of UN-Habitat, the European Commission views a sustainable city as one where urban development integrates physical urban renewal, education, economic development, social inclusion and environmental protection, whilst fostering strong partnerships between citizens and civil society (European Commission, 2016). Ibrahim, Omar and Mohamad (2015) describe the sustainable city as one that can meet the basic needs of its inhabitants, including civic amenities, education, employment, health and medical care, infrastructure, transportation and good governance. This review takes the stance that a three-pillar definition – one that does not take governance into consideration – is a static definition. Sustainability and the 'sustainable city' concept should therefore be viewed and assessed against the four pillars of sustainability.

Table 3.1 Similar concepts to the sustainable city

Term	*Definition*	*Source*
Low Carbon City	A city in which low-carbon progress and a high level of energy efficiency are adopted in the development plan to ensure an equitable contribution to stabilise global concentrations of CO_2. Low-carbon behaviours are fortified, whilst ensuring that the needs of all members within society are met.	(Skea & Nishioka, 2008)
Resilient City	A city in which individuals, communities, institutions, businesses and systems, survive, adapt and grow regardless of the natures of chronic stress and acute shocks they experience.	(Spaans & Waterhout, 2016)
Eco City	A city that enhances the well-being of its citizens and society through integrated planning and management that wholly capture the benefits of ecological systems, and protects these assets for future generations.	(Suzuki et al., 2009)
Smart City	A city which utilises data, information and Information Technologies to provide more efficient services to citizens, to embrace concepts of sustainability, to optimise existing infrastructure and to increase collaboration between economic actors whilst encouraging innovative business models.	(Marsal-Llacuna et al., 2015)
Urban Ecosystem	A dynamic urban system of the natural environment, the built environment and socio-economic environment.	(Dizdaroglu, 2015)
Urban Sustainability	An urban system that fosters a balance between economic, environmental and social aspects of sustainability.	(Azami et al., 2015)

A variety of terms exist that contribute to the broader sustainable city discourse, as shown in Table 3.1. Some may be used effectively as a synonym for the 'sustainable city' concept, whilst others may overlap in meaning but give differing emphasis in terms of aspirations or the pathways through which such aspirations can be met (Tan et al., 2016). This chapter includes papers that use any of the terms in Table 3.1 on the condition that, as a minimum, they overlap in meaning with the 'sustainable city' concept and that the publication employs indicators to assess or discuss the term.

1.1 Sustainability indicators

Indicators are measures that represent the state of a system or process within defined boundaries. An index is the resulting measure of indicator aggregation, and provides a simplified, coherent multidimensional understanding of a system (Mayer, 2008). Sustainability indicators structure and communicate information about sustainable development issues and their trends (Rametsteiner et al., 2011). Appropriately selected indicators result in a reduced number of measurements needed to gain a picture of a system's performance (Guy & Kibert, 1998). Additionally, indicators provide insight on the nature and strength of linkages connecting different system components and offer an enhanced understanding of the consequences of human actions on sustainability dimensions (Rametsteiner et al., 2011). The selection, measurement and application of sustainability indicators, however, remain major challenges to policy makers, scientists and citizens tasked with sustainability (McCool & Stankey, 2004).

1.2 Indicator-based sustainability assessments

Sustainability assessment is a generic term for a methodology that applies the broad principles of sustainability to ascertain whether or not, and to what extent, various actions (such as policy changes, new business models, or building developments) advance or hinder the cause of sustainability (Shmelev, 2017; Shmelev & Shmeleva, 2018; Hiremath et al., 2013). Four main uses of sustainability assessments have been identified: (1) acting as a decision-making, project design and strategic planning tool for governmental, international and non-governmental organisations; (2) providing evidence for monitoring, evaluation and impact analysis; (3) functioning as a source for reporting on international conventions, on the state of the environment and on specific sustainability themes, and (4) providing a platform for raising awareness about sustainable development issues (Guijt et al., 2001). Many urban sustainability assessment methods have been identified (Hiremath et al., 2013), and can be categorised into three general areas: indicators and indices, product-related assessment tools and integrated assessments. Indicators and indices are further subdivided into non-integrated and integrated. Product-related tools focus, from a life-cycle perspective, on material and/or energy flows of a product or service. Integrated assessments utilise an assortment of tools, usually focused on policy change or project implementation; examples include multi-criteria analysis, risk and vulnerability analysis, cost-benefit analysis or impact assessments (Ness et al., 2007). This literature review is focused on publications that address the first category of sustainability assessments: indicators and indices, however, where publications in category (3) have employed city sustainability indicators, they are also included.

Within the first category of sustainability assessments presented by Ness and colleagues (2007) – i.e. indicators and indices – the tools are either non-integrated (they do not incorporate multiple pillars of sustainability) or integrated (they aggregate the different sustainability dimensions into one index). The Ecological Footprint and the Human Development Index are examples of integrated tools. There is a sub-category of non-integrated tools that focuses on regional flow indicators, for example, Input–output Analysis or Economy-wide Material Flow Analysis. The publications presented in this literature review, with some exceptions, fit into the non-integrated and integrated categories. In these cases, however, non-integrated usually refers to a framework of non-integrated indicators from a number of domains, to capture the dimensions of sustainability, rather than one non-integrated index.

This chapter presents a systematic literature review of indicator-based city sustainability assessments, seeking to address the following questions: which indicator categories (sustainability pillars) are commonly applied to assess city sustainability? What limitations exist with regard to indicator categories? What governance indicators are currently applied? As far as the available evidence suggest, no prior systematic literature reviews on indicator-based city sustainability assessments have been undertaken. Hiremath and colleagues (2013) present a review on indicator-based urban sustainability; however, the review does not provide a systematic methodology and defines differing methodologies according to whether or not they focus on the transport sector. Additionally, many of the reviewed publications in this chapter have been published since 2013; it is therefore suggested that the field has advanced. Consequently, the aims of this chapter are twofold: first, to address the non-systematic nature of Hiremath and colleagues' review, and second, to provide a response to the three questions defined above. This chapter later describes a number of other non-systematic reviews on the topic; these, however, focus on reviewing specified real-world sustainable city indicators (SCI) frameworks, rather than literature reviews.

In the next section, this chapter goes on to offer an outline of the systematic literature review research methodology. Thereafter, the literature is categorised into two groups (case studies and non-case studies) and then summarised. Section 5 presents the results of the review, with an emphasis on governance indicators. Finally, the findings are discussed in the context of the initial three research questions.

2. Research methodology

The aim of this study was to implement an exhaustive literature search which would therefore be representative of publications on the topic of indicator-based city sustainability assessments (Siddaway, 2014). The research questions were translated into the following keywords: 'sustainable city indicators' and 'urban sustainability indicators'. The keywords were used as selection criteria for the 'abstract', 'title' and 'keywords' (Science Direct) and the 'topic' and 'title' (Web of Knowledge) fields in each paper for the two respective databases. Document type was specified as 'articles' and 'review articles' and no time limits were established.

Objective inclusion and exclusion criteria were then established before the review and applied consistently throughout the review process. Inclusion criteria were defined by the research design, whereby publications were included if they employed indicators to assess city/urban sustainability, and research scope, whereby publications were included that focused on city (metropolitan) or urban sustainability. The exclusion criterion was defined by the research design; if a study did not employ or discuss the application of indicators, it was excluded.

For the next step, selection of publications, searches were conducted within two databases: 'Web of Science' and 'Science Direct'. Figure 3.1 presents a PRISMA flow diagram of the selection process, during which secondary inclusion criteria were employed to select publications that directly related to the three questions of this study. Publications were included only if the study did one of two things. First, if they aimed to assess overall city sustainability (addressing more than one pillar of sustainability), i.e. if a study focused only on waste management or transport, for example, it was excluded. Second, the centrally discussed city sustainability methods had to employ indicators. This resulted in a final inclusion of 55 publications. Eight papers were later excluded during the assessment of the full-text articles; this could be because the publication focused only on assessing environmental sustainability, assessed only the sustainability of an urban development, or focused on urban metabolism but did not include SCI and instead focused on energy flows. Citations were exported and stored in Mendeley Desktop reference manager.

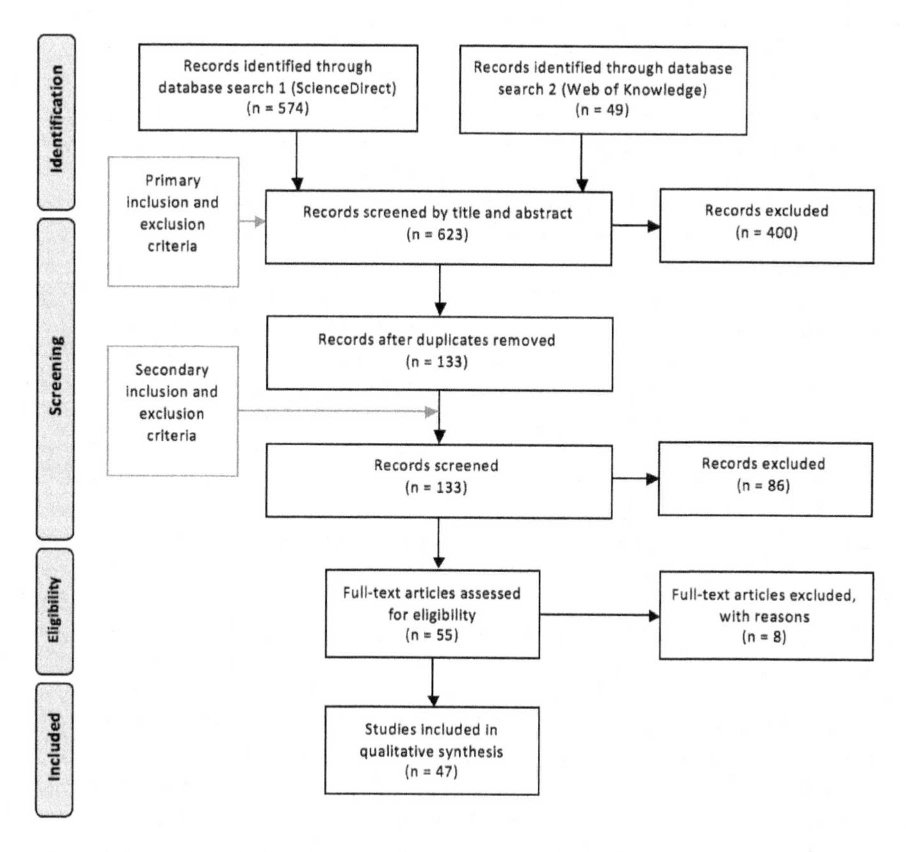

Figure 3.1 A PRISMA flow diagram of the steps taken in the literature selection process

Source: Moher et al. (2009)

3. Literature classification

The 47 publications have been organised into two categories – either 'empirical case study' or 'non-case study' – according to the research method highlighted in the abstract:

1 *Empirical case studies* make up 45% of the reviewed papers. Empirical studies utilise data to test if relationships hold in the real world (Wacker, 1998). In this literature review, empirical case studies consist of papers that investigate the sustainable city phenomena within its real-life context, through case studies regarding sustainability assessments of either one or multiple cities.
2 *Non-case studies* form 55% of the reviewed studies and include both empirical studies that are not city-specific sustainability assessments, and theoretical studies. Theoretical research includes papers using the following methods: conceptual definitions, domain limitations, predications and relationship-building (Wacker, 1998).

The literature is then further subdivided between the two main categories:

1 Empirical case studies

  i Single-city specific sustainability assessments (integrated or non-integrated)
  ii Multi-city sustainability assessment comparisons (integrated or non-integrated)
  iii Alternative city sustainability assessments

2 Non-case studies

  iv Appraisals of singular real-world sustainability assessment methods
  v Reviews of a number of real-world sustainability assessment methods
  vi Proposed city sustainability assessment methods
  vii Applied methods for sustainable city indicators (SCI) selection

The next section focuses on each of the various categories identified in this classification framework and identifies to what extent the literature captures the four pillars of sustainability. This chapter terms the four pillars as economic, environmental, social and governance. Publications may categorise their indicator sets into four or more themes; this review, however, re-categorises them based on the four pillars. (See Figure 3.2 Section 5, for a breakdown of the four pillars into broad indicator topics.)

3.1 Empirical case studies

Researchers have applied an expansive number of indicators, within a range of indicator categories, to assess the sustainability of real-world cities. The term 'unsubstantial' is introduced in this review to describe an indicator framework that drastically under-represents a pillar through the use of one, or comparatively few indicators compared to the overall indicator framework.

3.1.1 Single-city specific sustainability assessments

Of the eight publications using single-city specific sustainability assessment, three papers utilise a four-pillar indicator framework and five adopt a three-pillar method. No publications assessing singular cities sustainability within this literature review employed a two-pillar indicator set.

Turning first to the four-pillar approach, Lee and Huang (2007) are among the three papers applying a four-pillar framework of 51 SCIs to Taipei City, Taiwan, a method developed through discussions with experts, researchers and government departments. Their indicators are classified into economic, social, environmental and institutional (governance) dimensions and are used to identify sustainability trends. Huang and colleagues (1998) also assess Taipei and present a four-pillar procedure, including an 80 SCI system developed through participation with non-governmental organisations, to discuss the ecological economic dimensions of urban sustainability. Yuan and James (2002) present an unsubstantial four-pillar historical performance analysis of the evolution of the Shanghai city region from 1978 to 1998. The city is assessed against ten economic, environmental, social and governance indicators; however, the social and governance pillars are each only represented by one indicator and the governance indicator is methodically non-transparent. The framework consequently has limited value to assessing overall social and governance sustainability performance of Shanghai. The four-pillar framework applied by Lee and Huang is therefore concluded to be the most comprehensive of the three in capturing sustainability.

A three-pillar approach is more widely adopted by researchers assessing singular-city sustainability. In one of the most recent studies, Azami and colleagues (2015) employ a three-pillar indicator-based approach to determine areas of unsustainability in Sanandaj City, Iran. Over a 10-year period (2001–11), a framework of 15 SCIs is analysed, and classified under economic sustainability, social (including cultural) sustainability, physical sustainability (for example, public services) or ecosystem sustainability. Similarly, Li and colleagues (2009) develop a three-pillar Full Permutation Polygon Synthetic Indicator method to evaluate current and future sustainability of Jining City, China. Future evaluations are based upon predicated indicator values for the following two decades. The method includes a comprehensive framework of 52 SCIs that address four categories: economic growth and efficiency, ecological and infrastructural construction, environmental protection, and social and welfare progress. Rosales (2011) also introduces a three-pillar set of indicators for quantifying urban sustainability performance and integrating it into planning processes. Indicators are operationalised into planning tools by moving from an ex-post evaluation of cities to an ex-ante position. The identified SCIs can be categorised according to four 'modules': the urban safety indicator (social), the urban health indicator (environmental), the visually and culturally appealing city indicator, and the urban efficiency indicator (economic). Mexico City is presented as a case study from which the urban health module is applied. Seifollahi and Faryadi (2011) present another three-pillar indicator-based model, utilising indicator importance coefficients to compare Tehran's sustainability performance in 2006 to the city's ideal sustainability performance. The model employs a SCIs framework structured into four groups of needs: basic, socio-economic, build environmental needs, and cultural and recreational. Finally, Yigitcanlar, Dur, and Dizdaroglu (2015) develop a very unsubstantial three-pillar multi-scalar urban sustainability approach, linking two existing sustainability assessment models (MUZIX and ILTIM), and test the model in Gold Coast (Australia). The multi-scalar approach, however, includes only a singular economic SCI (jobs) and a singular social SCI (transport). When researchers adopt a three-pillar approach, it is most common to assess economic, environmental and social sustainability, as documented here.

3.1.2 Multi-city sustainability assessment comparisons

Ten multi-city sustainability comparisons are presented in this review. Three of these employ to some extent, sustainability assessments capturing sustainability's four pillars. Five publications utilised a three-pillar SCIs framework in their multi-city comparisons. Two papers failed to capture three pillars of sustainability.

Papers encapsulating the four-pillar dimension of sustainability include Lazaroiu and Roscia (2012) who propose and apply a four-pillar model for computing 'the smart city' indices. The framework of 18 non-homogenous SCIs captures five of the six characteristics of 'the smart city', including smart economy, smart mobility, smart environment, smart people and smart governance; but the model excludes smart living. The approach, based on fuzzy logic, is used to assess eleven Italian cities against 'the smart city' model. Secondly, Zoeteman, Mommaas, and Dagevos (2016) describe, and then implement an unsubstantial four-pillar method previously developed by Telos, Tilburg University, for integrated sustainability monitoring of municipalities. Data for all 403 municipalities in the Netherlands on 90 SCIs describing three essential subsystems of the total city social system – termed ecological, socio-cultural and economic 'capital' – is retrieved and compared. Unsubstantial is used to describe the method for its inclusion of only a singular governance indicator topic: turnout in municipal and national elections. The study also presents the key factors determining municipal sustainability scores, including municipality size, city typology and additional factors, such as average income and geographical position. Thirdly, Kılkış (2016) compares twelve South-east European cities on an unsubstantial four-pillar composite indicator, the Sustainable Development of Energy, Water and Environment Systems (SDEWES) City Sustainability Index. The index consist of 35 main SCIs across seven dimensions: energy consumption and climate; penetration of energy and carbon dioxide-saving measures; renewable energy potential and utilisation; water and environmental quality; carbon dioxide emissions and industrial profile; city planning and social welfare; and research, development, innovation and sustainability policy. The method employs a min-max normalisation practice to then aggregate and rank the cities. Unsubstantial is employed to describe the singular SCIs representing economic and governance sustainability: GDP per capita and reductions target for CO_2 respectively. Of the three papers that initially appear to employ all-encompassing methods across all four pillars, two pose limited scope in capturing one or two of these pillars of sustainability.

Three-pillar SCIs frameworks were used in five multi-city comparisons. Firstly, Bagstad and Shammin (2012) use a three-pillar 26-component indicator framework of economic, social and environmental sustainability to measure the Genuine Progress Indicator (an integrated index) – a proposed alternative to Gross Domestic Product (GDP) – of the State of Ohio, of Akron and Cleveland cities, and of 17 North-east Ohio counties for the years 1990 to 2005. The GPI is also calculated for two locally relevant scenarios of policy change on renewable energy and urban agriculture. It is argued that component indicators of GPI include the fundamentals of the three pillars of sustainability that are likely contributors to quality of life. Secondly, Huang, Yan and Wu (2016) evaluate the sustainability of ten megacities in China based on time-series data from 1978 to 2012. The study uses a three-pillar definition of sustainability and therefore employs a set of SCIs capturing environmental, economic and social sustainability. Sun and colleagues (2015) developed a three-pillar 22-indicator system to assess the sustainability of 277 Chinese cities of differing sizes (from megalopolises to small-sized cities). A Full Permutation Polygon Synthetic Indicator method is employed to assess economic development, social progress and ecological infrastructure. In their analysis of Low Carbon Cities (LCC), Tan and colleagues (2016) also develop a three-pillar evaluation framework, as well as a subsequent certification benchmark level. The framework consists of SCIs in seven categories (energy pattern, water, social and living, carbon and environment, solid waste, urban mobility and economic) and is used to assess ten global cities. Conversely to the above statement that three-pillar methods address only economic, environmental and social sustainability, Li and Qiu (2015) present a paper that considers economic, environmental and governance aspects. Li and Qiu review the development of the eco-city framework in China, and perform an unsubstantial three-pillar comparative analysis investigating the quality of Chinese eco-cities. Suzhou (China) and Kitakyushu (Japan), both acknowledged best practice eco-cities, are compared against 19 SCIs from the Chinese national eco-city framework. The indicators include economic, environmental and social aspects of sustainability. The social pillar, as defined by this review, however, is not addressed, whereas the governance pillar is captured unsubstantial through the inclusion of an indicator on environmental protection investment share in GDP. Of the five multi-city publications employing three-pillar SCIs frameworks only a singular study employed an unsubstantial model, with the first four providing comprehensive three-pillar frameworks.

Two papers providing multi-city comparisons did not cover three pillars of sustainability. One of these is a study by Egilmez, Gumus and Kucukvar (2015) who develop and apply an unsubstantial two-pillar four-step fuzzy multi-criteria decision aid (MCDA) to assess the sustainability performance of 27 US and Canadian metropoles. Sixteen SCIs, representing seven main impact areas (air, energy use, water consumption, buildings, land use, transport and waste), are used to rank the cities. Consequently, only one topic of social sustainability, transport, is addressed.

The authors acknowledge the lack of socio-economic metrics within the study and plan to integrate such indicators into future MCDA. The second study, conducted by Castellani and Sala (2013), applies a single-pillar indicator based Strategic Environmental Assessment (SEA) to urban master plans in four Italian municipalities, which covers the following categories: air and climate, water, land use, agriculture, biodiversity and landscape, waste and energy. The SEA captures three pillars through the inclusion of a cost-benefit analysis and two additional measures capturing aspects of consumption patterns: ecological footprint and carbon balance. Rather than tackling overall city sustainability assessments, these final two papers instead focus on environmental sustainability of cities.

3.1.3 Alternative method city assessments

In addition to the sustainability assessments that can be generalised into Ness and colleagues' (2007) category 1 (see section 1.2. on indicator-based sustainability assessments), three papers are identified that employ SCIs in alternative procedures. Of these, one study applied a four-pillar method and two address three pillars. No publications assessing city sustainability with an alternative procedure employed a two-pillar indicator set.

Shen and Guo (2014) spatially quantify and weigh urban sustainability with an unsubstantial four-pillar SCIs model for the city of Saskatoon, Canada. Unsubstantial refers to the singular indicators of governance: electoral turnout. The spatial distribution of urban sustainability is demonstrated across neighbourhoods, with respect to main indicator domains: urbanisation and quality of life. The study captures economic, social and environmental qualities of sustainability but is limited by its lack of governance and built-environment indicators.

Two of the alternative procedure publications adopt a three-pillar approach. Both employ urban metabolism methods and can therefore be categorised as regional flow indicator tools (see section 1.2.). Metabolism studies are based on material and physical flows through systems; when applied to the urban context, the system is that of the city system. Chrysoulakis and colleagues (2013) utilise environmental, social and economic SCIs in a three-pillar comparative urban metabolism study of Helsinki, Athens, London, Florence, and Gliwice. The study employs the BRIDGE Decision Support System, which combines *in situ* measurements of physical flows, numerical models simulating flows, and indicators. BRIDGE links biophysical processes with socio-economic parameters, with the aim of aiding the evaluation of the sustainability of urban planning interventions. In the second such alternative procedure addressing three pillars, Moles and colleagues (2008) examine the relationship between settlement size, functionality, location and sustainability through an analysis of 79 Irish settlements. Two methods are employed: first, Metabolism Accounting and Modelling of Material and Energy Flows (MA), and second, Sustainable Development Index Modelling (SDIM). A set of 40 SCIs, representing three pillars of sustainability and classified into four domains (environment, quality of life, socio-economics, and transport), is identified for use in the SDIM. Transport is given its own domain due to the importance it holds in addressing urban sustainability issues.

It should be noted that additional urban metabolism studies were identified within the first 623 identified papers; however, most were excluded on the basis that they did note employ indicators, but rather measures of physical flows (see for example; Kennedy et al., 2014).

3.2 Non-case study

In addition to real-world sustainability assessments, the literature on indicator-based city sustainability includes theoretical papers, reviews on existing indicator frameworks, proposed methodologies and indicator sets, and applied methods for SCIs selection. Indicator categories and types can be drawn out from these publications, either from presented indicator frameworks or from theoretical discussions on indicators.

3.2.1 Appraisals of singular real-world sustainability assessment methods

A four-pillar approach is adopted by three of the papers apprising singular real-world city assessment methods. The remaining three of the six singular real-world city sustainability method appraisals employed a three-pillar approach, with no publications employing a two- or singular-pillar framework when discussing the sustainable city.

Although none of the three studies covering all four pillars of sustainability state specific SCIs, they can be defined as four-pillar approaches given their definitions of and discussions on domains of sustainability. In one such study, Block, Van Assche, and Goeminne (2013) argue that the City Monitor for Sustainable Urban Development in the Flanders (Belgium) is relevant to the 'governance' pillar in urban sustainable development: 200 four-pillar SCIs contain actor-exceeding and policy-exogenous information. The City Monitor is developed on a 'vision matrix' of four core principles of sustainability (economic, social, physical-ecological, institutional) and eight 'activity domains', including learning and education, safety and protection, nature and the environment. In another similar study, Van Assche, Block, and Reynaert (2010) focus on the innovative characteristics of the participatory methodology of the aforementioned four-pillar City Monitor, used to assess sustainability of Flemish cities. The participatory approach of the City Monitor is concluded to foster the implementation of the most desirable community indicators. An earlier study conducted by Holden (2006) reports on the theory, methods and design of the four-pillar Regional Vancouver Urban Observatory (RVu) project: the first member of the UN-Habitat Global Urban Observatory network established in a developed nation. RVu is discussed in terms of the following three areas: a process-based position on city sustainability; indicators and values in decision making, and moving toward consensus, advocacy and action through indicator projects. Values are integrated through RVu study groups clustered in an alternative manner in order to prevent segregation simply along 'economic', 'environmental', and 'social' lines. They can be approximately categorised into the following domains: community, city as a living organism, poverty, arts and culture, governance, mobility, food, and regional growth. The groups worked on identifying SCIs, though the indicators themselves are not discussed in the paper. Despite not stating specific SCIs, these four publications are defined as four-pillar approaches; showing that the fourth pillar is indeed discussed in the literature as a domain of sustainability.

Three singular real-world city sustainability method appraisals utilised a three-pillar methodology in their studies. Ibrahim, Omar, and Mohamad (2015) review the theoretical structure of 55 SCIs in the three-pillar Malaysian Urban Indicators Network (MURNInet). MURNInet, based on a computer network design, is used to analyse present city conditions and the effects of scenario changes. The impact of MURNInet on the three primary city stakeholders – federal and state government, local authority and local community – is also discussed. The indicators provide an indication of city sustainability in the following domains: population, housing, economy, utility and infrastructure, public facilities and environment, sociology and social impact, land use, tourism and heritage, transportation, and management and finance. In an earlier study, Keirstead and Leach (2008) state that SCIs often portray a restricted, easily quantifiable, picture of urban systems, and therefore fail to address many fundamental issues on how to define, and achieve, a sustainable city. The study examines London's SCIs using a three-pillar indicator system, and confirms this theory-practice gap. Keirstead and Leach propose a service-niche method, founded upon the notion that urban services (such as energy, water or waste) can often be associated with tangible goals to improve the effectiveness of indicators. Finally, Pires, Fidélis, and Ramos (2014) review the efforts of European institutions and research projects towards the coordination of local SCIs, in light of the arguments against common sets of indicators. Evidence from two case studies of an unsubstantial three-pillar Portuguese initiative using common sustainability indicators across cities, ECOXXI, is presented. ECOXXI includes both environmental and governance indicators but is deemed unsubstantial due to its inclusion of only a singular social indicator (transport). In the absence of national or European guidelines, the review concludes on the major advantages of developing and delivering guidelines of a top-down but adaptable SCI framework. Of the three singular real-world city sustainability method appraisals, Ibrahim, Omar and Mohamad (2015) and Keirstead and Leach (2008) present comprehensive three-pillar methodologies.

3.2.2 Reviews of a number of real-world sustainability assessment methods

Nine larger review papers on indicator-based city sustainability assessments are now considered, six of which take into account, to some extent, a four-pillar view of sustainability. The other three papers adopt a three-pillar approach; thus none of the multi-paper reviews portray sustainability as a concept consisting of two or less pillars.

In one such four-pillar conception of sustainability, Srivastava (2016) evaluates the plans that the city of Fargo, North Dakota, has to assess its sustainable growth, against two established indicator-based models of urban sustainability. The three models are assessed on the following four-pillar sustainability domains: environmental, economic, social, governance, state, policy, water, neighbourhoods, infill and new development, arts and culture, transportation, health, safety and education. Shen and colleagues (2011) conduct a comparative study of nine SCI-based city sustainability assessment procedures. The procedures are compared to the International Urban Sustainability Indicators List and are benchmarked against sustainability's four pillars. The study emphasises the need for consistent indicator selection methods, obtained from best practices. Shen and Zhou (2014) compare nine existing SCI systems in China against the four pillars of sustainability. The study concludes that the differences between the existing systems, with no official Chinese department supervising them, limit the effectiveness of their application. Tanguay and colleagues (2010) analyse 17 publications of SCI applications in developed Western countries, concluding that there is a lack of consensus on both the conceptual framework and indicator approach. Based on the prior analyses, which catalogued 188 indicators, a condensed set of 29 indicators is proposed, taking into account the four pillars of sustainable development and their constituent categories. Michael, Noor, and Figueroa (2014) review the processes, methodologies and resulting sets of three- and four-pillar SCIs on urban sustainability in Malaysia, Taiwan, and China. The operationalisation of the indicators, whether or not they are integrated into the policy process, is also assessed. Of the methods compared, both Taiwan and Malaysia included indicators on governance. Meijering, Kern, and Tobi (2014), based on a literature review, develop a methodology for identifying the methodological characteristics of city rankings. The developed method is applied to six city rankings, aimed at measuring the environmental sustainability of European cities, and finds a great diversity of methodological characteristics, but also methodological weaknesses across the six. Meijering, Kern and Tobi focus on SCI methods, rather than coverage. Nonetheless, of the six rankings assessed, two (the European Green Capital Award, and European Green City Index) address all four pillars; two (Urban Ecosystem Europe, and the European energy award) address three pillars (environmental, social and governance); one (European Soot-free City Ranking) addresses environmental and social pillars; and one (the Renewable Energy Systems Champions League) addresses a singular pillar (energy).

Three of the review publications address sustainability's three principal pillars. Hiremath and colleagues (2013) present a literature review and suggest the need for benchmarking indicator-based sustainability approaches in enhancing long-term city sustainability. The review concludes that SCIs should contribute to the following: making urban sustainable development more visible and transparent; aiding in comparison, evaluation and prediction; helping construct data banks; providing decision makers with relevant information, and promoting citizen participation. Of the papers reviewed, only three included indicators of governance; therefore the review is here regarded as presenting a three-pillar sustainability definition. Huang, Wu, and Yan (2015) present a review on the key concepts of urban sustainability and assess ten existing composite sustainability indicators and three SCI sets. Of the reviewed methods, eight address three pillars of sustainability (environment, economy and society) and five cover two of the three pillars. The study also assesses the methods in terms of 'weak' and 'strong' sustainability. Mori and Christodoulou (2012) discuss and present new conceptual requirements for a City Sustainability Index. Existing major sustainability indices and indicators are reviewed firstly in terms of their ability to capture the three pillars of sustainability, secondly depending on whether or not their sustainability is 'weak' or 'strong', and finally based on their ability to capture leakage effects (external impacts). None of the multi-paper reviews considered sustainability as consisting of two or less pillars.

3.2.3 Proposed city sustainability assessment methods

Of the proposed city assessment methods, three studies suggested approaches that took a four-pillar approach to sustainability, whilst three proposed methods addressing three pillars of sustainability. In this particular case, not all studies can be said to follow the 'pillar' strategy and the last paragraph in this section will thus address one such paper.

Under the comprehensive four-pillar approach, Braulio-Gonzalo, Bovea, and Rua (2015) present a review of SCIs in 13 national and internationally developed urban sustainability assessment tools. A four-pillar indicator system is proposed – composed of 14 categories (including management and institution), 63 sub-categories, indicators and measurement methods – to assess the urban sustainability of Mediterranean cities. Similarly, Egger (2006) proposes a four-pillar model of SCIs through which the city can be assessed at both the global city network level and the 'self' level, whereby a city must meet the needs of its own inhabitants. The model addresses the sustainable city from three vantage points: first, global capital (economic, climate change, air quality, water, resource efficiency, society); second, city capacity (innovation, development, infrastructure, social capital, planning, economic, political); and third, city condition (economic, social, culture and leisure). Finally, Sharifi and Yamagata (2014) develop a broad four-pillar set of criteria, based on a review of literature on urban resilience and sustainability. The criteria cover the following themes: infrastructure, security, environment, economy, institutions, and social and demographics. Sharifi and Yamagata suggest that the criteria should be used to guide the selection of indicators to assess urban resilience and sustainability. All three proposed city sustainability methods applying four-pillar approaches present comprehensive methodologies, addressing each pillar in a substantial manner.

Three publications proposed methods addressing three pillars of sustainability. Boyko and colleagues (2012) utilise a toolkit of 120 SCIs, developed as part of the Urban Futures research project, to assess the impact of uncertain futures in the form of four scenarios on urban sustainability. The toolkit represents nine sustainability topics: biodiversity; air quality; water and wastewater; sub-surface built environment, infrastructure and utility services; surface built environment; density and design decision making; organisational behaviour and innovation, and social needs and planning policy. Moussiopoulos and colleagues (2010) develop a framework of 88 SCIs, in 13 discrete categories, as a dynamic tool for assessing the three pillars of city sustainability. A case study of the Greater Thessaloniki Area, Greece is used to demonstrate the framework's applicability and to make recommendations on local stakeholder involvement. Wiek and Binder (2005) have developed a theoretical tool to define Sustainability Solution Spaces for Decision-making, which provides a consistent set of targets concerning the systematic relations between SCIs of city sustainability from a three-pillar perspective. Boyko and colleagues (2012), Moussiopoulos and colleagues (2010) and Wiek and Binder (2005) propose comprehensive methods addressing three pillars of sustainability.

In this case, not all publications use a 'pillar' approach to assessing city sustainability. Marsal-Llacuna, Colomer-Llinàs, and Meléndez-Frigola (2015) present a review of current Smart City initiatives and also of past assessment tools. According to the definition outlined in this review, the latter of these – Local Agenda 21 – actually presents a two-pillar approach, addressing only environment and society. Two improvements for monitoring the initiative are offered. The first of these is the construction of synthetic indices to form a final index for visualising an initiative's achievement, specifically by using principle component analysis. Second, is the need for real-time data that is currently lacking in the Smart City initiative. The publication can therefore not be categorised into a particular pillar category; though it reviews a two-pillar approach, it goes on to discuss a number of three- or four-pillar systems.

3.2.4 Applied methods for indicator selection

With regard to SCI selection methods, two publications out of four employ sustainability's four pillars in their analysis, whilst one focuses on three pillars and the remaining study on only two.

In one such four-pillar approach, Turcu (2012) develops a set of 170 SCIs, drawing on five existing lists. Based on a three-filter process of selection and on expert opinions, the list is reduced to 26 indicators organised under the four pillars of sustainability: economic, social, environmental and institutional. Turcu discusses the indicator set with 130 residents, concluding that local perspective of urban sustainability is more complex and less 'predictable' than its top-end 'expert' view. In the other study, Zhou and colleagues (2015) present a responsibility-based method for SCI selection, named Strategic Goal-responsibility Department Response (SRR). Indicators are selected from the perspective of relevant departments' responsibilities. The SRR model is used for the city of Jinan, China, and is then compared to the SCIs of the United Nation Commission on Sustainable Development (CSD), finding that 31 of the 38 indicators defined by CSD are composed in the list produced for Jinan. Both the selected set for Jinan and CSD are based on a four-pillar sustainability framework.

Munier (2011) is the single indicator selection method publication that employs a three-pillar sustainability approach. Munier proposes a methodology to determine the best framework of core SCIs from an initial three-pillar dataset of indicators. The method provides a platform for (1) identifying the components of an initial indicator dataset, (2) selecting indicators through the application of linear programming, and (3) employing the entropy concept for maximising indicator information content.

Tran (2016) captures two pillars of sustainability in a method of SCIs selection using an integrated objective quantitative analysis, combining participatory dialogue and input from stakeholders. The method is based on variable clustering and enables the selection of a small subset of indicators accounting for a large amount of information. The SCIs presented in the study represent two pillars of sustainability (environmental and social), capturing the following areas of sustainable development: biodiversity conservation, clean air, clean and plentiful water, climate stabilisation, food, fuel and materials, natural hazard mitigation, and recreation, culture and aesthetics.

All four applied methods of SCI selection present substantial models that address sustainability domains with a number of indicators. Only two however, Turcu (2012) and Zhou and colleagues (2015), employ sustainability's four pillars in their analysis, and are therefore recommended as the most comprehensive indicator selection methods.

4. Results

The reviewed publications mostly favoured a three- or four-pillar approach to sustainability, with an almost equal split between the two: 22 publications included three pillars in their conception of city sustainability and 21 publications used all four (see Table 3.2). Similarly, the number of unsubstantial four-pillar and three-pillar studies is found to be in the same region, at four and three respectively. 'Non-case study' publications account for two-times the number of four-pillar publications (see Table 3.2). Within this literature review, studies that focus on city case study sustainability assessments are more likely to employ a three-pillar methodology. A similar conclusion can be observed in Figure 3.2, depicting the number of publications in the sustainable city indicator-based literature review employing specific indicator types or categories. Both environmental and economic pillars are dominant areas of focus but there is less inclusion of social, and minimal inclusion of governance indicators. Specifically, issues of safety and culture are lacking within indicator frameworks. Citizen participation and sustainability focus are the two indicator types most commonly employed to represent governance in SCI frameworks.

Table 3.2 Number of sustainable development pillars included in reviewed publications*

Publication category	*Pillars*			
	4	*3*	*2*	*1*
Case study	7	12	1	1
Non-case study	14	10	1	0

*n = 46 as the publication by Marsal-Llacuna, Colomer-Llinàs, and Meléndez-Frigola (2015) could not be classified into a pillar category.

4.1 Governance indicators

Table 3.3, based on the systematic literature review, presents the specific governance indicators of 13 publications that refer to precise indicators. For a number of the publications, indicators of governance are not included in a 'governance' domain; they are instead allocated into other domains, most commonly 'social'. The indicators are classified under six topic headings: (1) citizen participation, (2) local authority policy and targets, (3) international cooperation, (4) management, (5) sustainability focus and (6) transparency. Across these topics, there was most consistency within citizen participation indicators: publications referred to either/both election turnout or public participation in policy. Equally, the topic of sustainability focus was often quantified using indicators measuring government expenditure on either the environment or society. With the exception of these two topics, there is little convergence on how to measure the other governance topics. It must also be noted that, though many publications name indicators, they do not explain how they have calculated, or they would advise to calculate, said indicators. In addition to the ambiguity surrounding the number of pillars of sustainability, the unclear and inconsistent methodology underpinning many governance indicators could further explain the lack of governance indicators implemented within SCI frameworks.

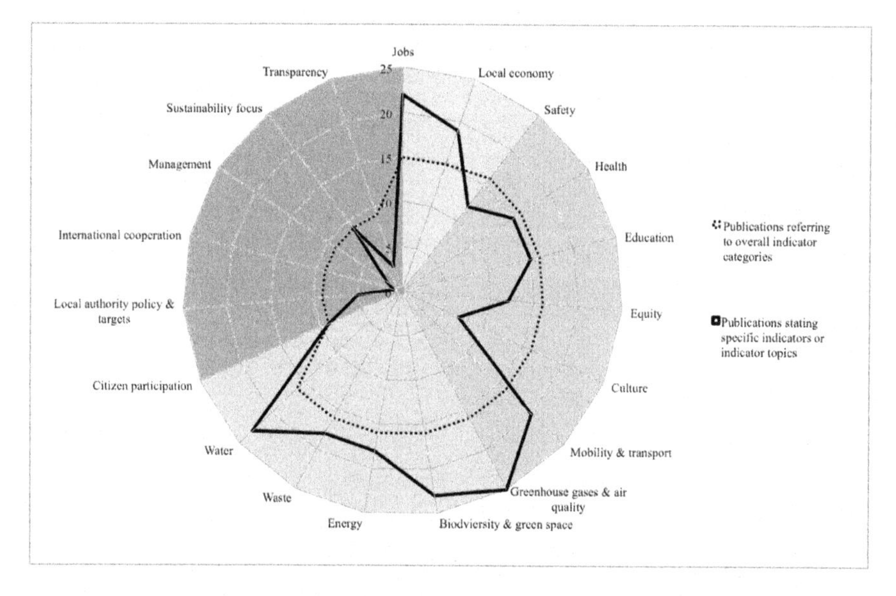

Figure 3.2 Number of publications, in the sustainable city indicator-based literature review, employing specific indicator types or categories (n = 47, tinted segments represent the four pillars of sustainability: economy, social, environmental, and governance. Based on a systematic literature review conducted July 2016)

Table 3.3 Summary of governance indicators employed or discussed by publications in the literature review

Indicator topic	*Indicator*	*Type*	*Measurement*	*Categories in studies*	*Number of publications*	*Sources*
Citizen participation	Election turnout	Qn	% (votes cast/eligible voters for municipal and national elections)	Social and cultural capital	3	(Zoeteman, Mommaas, & Dagevos, 2016)
				Socio-political index		(Shen & Guo, 2014)
		NS		Governance		(Srivastava, 2016)[*]
	Public participation in decision making	Qn	No. consultations No. adopted	Social aspect	5	(Braulio-Gonzalo, Bovea, & Rua, 2015)
		NS	decisions	Institutional dimension		(Lee & Huang, 2007)[*]
		NS		Smart governance		(Lazaroiu & Roscia, 2012)[*]
		NS		Governance		(Srivastava, 2016)[*,a]
		NS		Institutions		(Pires et al., 2014)[*]
Local authority policy targets	Reduction target for CO_2 emissions	D		R&D, innovation and sustainability policy	1	(Kılkış, 2016)[*]
	Strategy implementation of SD	NS		Institutional	1	(Zhou et al., 2015)[*]
	Political strategies and perspectives	NS		Smart governance	1	(Lazaroiu & Roscia, 2012)[*]
	Enforcement of local environmental plans	NS		Institutional dimension	1	(Lee & Huang, 2007)[*]
	Ratio of completed sustainability assessments to initiated sustainability assessments	NS		Institutional dimension	1	(Lee & Huang, 2007)[*]

	Regulations to improve sustainability (incorporating public parking rates into city centres/in corporation of discounts and bonuses to use public transport)	Qn	% of users of public transport	Management and institution	1	(Braulio-Gonzalo, Bovea, & Rua, 2015)
International cooperation	Joint international cooperation regarding sustainable development	NS		Institutional Dimension	1	(Lee & Huang, 2007)*
Management	Sustainable management of the authorities	NS		Governance		(Srivastava, 2016)*
	Cooperation among government administrations	Qn	No. of workshops held	Management and institution	i	(Braulio-Gonzalo, Bovea, & Rua, 2015)
	Proportion of companies and institutions with an implemented management system. Administrative	Qn	No. of well-managed companies/ No. of companies	Management and institution		(Braulio-Gonzalo, Bovea, & Rua, 2015)
Sustainability focus	Public expenditure on environment	Qn	Ratio (environmental and ecological budget: total budget)	Institutional dimension	4	(Lee & Huang, 2007)
		Qn	% of public expenditure on environmental protection	Environmental management		(Huang et al., 1998)
			Environmental protection investment share in GDP	Environmental		(Li & Qiu, 2015)
		NS				(Yuan & James, 2002)*
	Government expenditure on pollution prevention and resource recycling	NS		Institutional dimension	1	(Lee & Huang, 2007)*
	Disaster preparedness and response	NS		Institutional		(Zhou et al., 2015)*

(continued)

Table 3.3 (continued)

Indicator topic	*Indicator*	*Type*	*Measurement*	*Categories in studies*	*Number of publications*	*Sources*
	Non-governmental effort on environmental protection	Qn	No. of NGOs	Environmental management	1	(Huang et al., 1998)
	Development of information material on environmental matters	Qn	No. of campaigns	Management and institution	1	(Braulio-Gonzalo, Bovea, & Rua, 2015)
	Development of information material with official data and technical reports	Qn	No. of campaigns	Knowledge and information management	1	(Braulio-Gonzalo, Bovea, & Rua, 2015)
	Public expenditure on activities for society	Qn	Public expenditure for society/ total expenditure	Management and institution	2	(Braulio-Gonzalo, Bovea, & Rua, 2015)
	Public and social services	NS		Smart governance		(Lazaroiu & Roscia, 2012)*
		NS		Institutional sustainability		(Turcu, 2012)*
	Public expenditure on research & development (R&D)	Qn	Gross domestic expenditure on R&D	R&D and innovation policy orientation	1	(Kılkış, 2016)
Transparency	Transparent, accountable and efficient governance	NS		Governance	2	(Srivastava, 2016)*
		NS		Smart governance		(Lazaroiu & Roscia, 2012) *
	Administrative transparency integrating Agenda 21 into urban planning	Q1	Yes/No	Management and institution	1	(Braulio-Gonzalo, Bovea, & Rua, 2015)
	Government corruption	NS		Governance	1	(Srivastava, 2016)*

"Type" responses represent the following: Qn = Quantitative, Ql = Qualitative, N = Not specified, and D = Dimensionless (D was specified by the authors in the publication)
*The publication did not state a method for calculating the proposed or applied indicator
[a]The publication specifically focused on public participation in energy policies

Note: indicators are included if they are either employed or mentioned (discussed as potential indicators) in publications.

5. Discussion

This chapter presents a systematic literature review of existing indicator-based city sustainability assessments. The results of the review show that a three-pillar approach is most commonly employed in indicator frameworks developed for city sustainability assessments. Considering all literature categories, however, both three- and four-pillar sustainability definitions are used equally across publications. This lack of consensus on and commitment to a definition of sustainability, and therefore of the sustainable city, leads to ambiguity and inconsistency when using sets of sustainable city indicators (SCI). Tanguay and colleagues (2010) take a similar stance on the lack of agreement surrounding the creation of sustainable development indicators, which they see as resulting from uncertain definitions of sustainable development and as a consequence of ill-defined objectives for such indicators. On the issue of indicator development, Turcu (2012) similarly concludes that sustainability indicators are as much value-based and political as they are scientific: until there is consensus on what exactly should be sustained – in other words, on which dimensions of sustainability are considered important – it is impossible to identify or interpret any indicators. Turcu also emphasises the divergences in definitions of sustainability between a city's inhabitants and those with 'expert' knowledge of said city. Within the 'expert' category, Cartwright (2000) reports on inconsistencies across local authorities in England when it comes to the importance of different sustainability indicator. It is therefore advised that, in order to move forward with indicator-based city sustainability assessments, clear definitions of sustainability must be established in both political and scientific fields, as well as at both local and expert level.

The SCI categories (economic, environmental, social and governance) and types (such as jobs, local economy, safety and health) used in this review are tied to two assumptions in their illustration and assessment of the literature. The first of these is the fluid location of transport indicators within frameworks: for example, the placement of transport indicators within the social category in this review saw a greater number of publications labelled as 'three-pillar' through the inclusion of social indicators. For example, Moles and colleagues (2008), guided by the centrality of transport issues to urban sustainability, set transport as one of their four main domains (environment, quality of life, socio-economics and transport). Meanwhile, Seifollahi and Faryadi (2011), in their four-pillar needs-based indicator model, place the topic of transport under built environmental needs. Other studies solve this issue by disregarding the use overarching domains or 'pillar', instead employing a number of categories (usually more than eight), enabling transport and mobility to stand alone (see for example; Block et al., 2013; Braulio-Gonzalo et al., 2015; Ibrahim et al., 2015; Moussiopoulos et al., 2010). The second such assumption is that the cohort SCI types used in this review to group indicators is a highly subjective one. Subjectivity, however, underpins most publications in which an indicator framework is proposed or applied, as frameworks reflect the issues of greatest concern to researchers. Sustainability indicators should be largely objective, 'measurable', comprehensible and reflect local circumstances (Cartwright, 2000). They do not need to be purely objective, however; indeed, in reality, few are (Turcu, 2012). Despite the subjectivity at the root of indicator categories and types, this chapter presents a comprehensive systematic literature review, from which a number of interesting observations have been drawn.

Across the four possible 'pillars' of sustainability, this review finds that any hierarchy tends to place the most emphasis on the environmental pillar, followed by economic, social and governance pillars in that order. A principal limitation of SCI frameworks is the lack of representation of the governance pillar of sustainable development. Shen and Zhou (2014) report a similar finding from their analysis of eight SCI systems employed in China: of the 105 indicators used, 29 relate to economic issues, 35 to social, 36 to environmental and only five to governance. Here, it is argued that the fourth pillar, governance, is in fact pivotal in implementing and achieving the sustainable city concept. The concept of 'governance' focuses on the complex network of government, authorities and civil society: institutions that structure society with a view to providing solutions to problems, which in turn facilitate sustainability (Leal Filho et al., 2015). Governments and urban actors are partners in realising city sustainability. This relates to the quality of civil culture, of the institutions themselves and of the processes that combine to govern and steer society at the city level (Block et al., 2013). These qualities, as stated by Block and colleagues, include public trust and legitimacy, collective responsibility, facilitation of participation, accountability and transparency, efficiency (of organisations, policy and targets), and integration and coordination (within and between authorities). Good governance is necessary to ensure objectivity during the process of diagnosing, determining, drafting and approving city strategies and thus essential to integrating sustainable city development priorities (Braulio-Gonzalo et al., 2015; Huang et al., 1998). In addition to achieving sustainable development strategies, governance is important for citizen quality of life (Marans, 2015). A city's citizens clearly occupy a central position in the sustainable city concept if we consider sustainable development to be the quest for developing and sustaining quality of life, not only for the current generation but also for those living elsewhere and in the future (World Commission on Environment and Development, 1987). Good governance is a fundamental means of instilling confidence in citizens and other city stakeholders (Cruz & Marques, 2014). Having confidence in the capabilities of local government is central to improving quality of life (Marans, 2015). Turcu (2012) reported that over 50% of local residents viewed institutional sustainability, and its corresponding indicators, as very important to them (and thus to their quality of life). Governance can therefore be viewed as important at two levels: first, on a top-down scale, given its capability in ensuring sustainable development strategies, and second, at the bottom-up scale, given its capacity to ensure and improve residents' quality of life.

The set of governance indicators reported in this systematic literature review are limited by a lack of consistency across SCI frameworks, and by a lack of methodological transparency across publications. In order to progress with four-pillar indicator-based city sustainability assessments, sets of comprehensive governance indicators must be identified. Of the publications reviewed, Braulio-Gonzalo and colleagues (2015) present a comprehensive set of management and institutional SCIs, along with their respective measurements. Cruz and Marques (2014), not included in the review, present a review of existing assessment tools for the performance, measurement and benchmarking of the sustainability of local governments. From this, Cruz and Marques produce a conceptual four-pillar model for benchmarking local authority sustainability, in the form of the municipal scorecard, which is applied to the local government of the city of Lisbon. Leal Filho and colleagues (2015) present a framework of 'good governance' indicators, which they apply to six Baltic countries. Their framework is based on six governance performance categories: (1) voice and accountability, (2) political stability and absence of violence/terrorism, (3) government effectiveness, (4) regulatory quality, (5) rule of law and (6) control and corruption. It is also recommended that researchers looking to include governance indicators in sustainability indicator sets should refer to the resilient city literature (see for example, Milman & Short, 2008; Spaans & Waterhout, 2016) where large emphasis is placed on governance. Further research is therefore required in order to not only identify current practices but also to develop future methodology for appropriate and effective governance indicators.

6. Conclusion

Sustainability indicators are widely used to assess city sustainability. While there is no lack of sustainable city indicators (SCIs), their effective application is hindered by ambiguous definitions of sustainable development. As a result of this lack of clarity, the representation of governance in sustainable city indicator frameworks is also lacking, despite the vital role that this so-called fourth pillar plays in the development and implementation of the sustainable city concept. To move forward with indicator-based city sustainability assessments, the following suggestions are proposed: (1) a clear definition of city sustainable development should be decided upon; (2) this definition should consider the fourth pillar of sustainable development, governance and (3) further research should be conducted into the definition of governance and the method through which it can be represented by indicators.

References

Azami, M., Mirzaee, E., & Mohammadi, A. (2015). Recognition of urban unsustainability in Iran (case study: Sanandaj City). *Cities*, *49*, 159–168. http://doi.org/10.1016/j.cities.2015.08.005

Bagstad, K. J., & Shammin, M. R. (2012). Can the Genuine Progress Indicator better inform sustainable regional progress? A case study for Northeast Ohio. *Ecological Indicators*, *18*, 330–341. http://doi.org/10.1016/j.ecolind.2011.11.026

Block, T., Van Assche, J., & Goeminne, G. (2013). Unravelling urban sustainability: How the Flemish City Monitor acknowledges complexities. *Ecological Informatics*, *17*, 104–110. http://doi.org/10.1016/j.ecoinf.2011.04.001

Boyko, C. T., Gaterell, M. R., Barber, A. R. G., Brown, J., Bryson, J. R., Butler, D., & Rogers, C. D. F. (2012). Benchmarking sustainability in cities: The role of indicators and future scenarios. *Global Environmental Change*, *22*(1), 245–254. http://doi.org/10.1016/j.gloenvcha.2011.10.004

Braulio-Gonzalo, M., Bovea, M. D., & Rua, M. J. (2015). Sustainability on the urban scale: Proposal of a structure of indicators for the Spanish context. *Environmental Impact Assessment Review*, *53*, 16–30. http://doi.org/10.1016/j.eiar.2015.03.002

Cartwright, L. E. (2000). Selecting local sustainable development indicators: Does consensus exist in their choice and purpose? *Planning Practice and Research*, *15*(1–2), 65–78. http://doi.org/10.1080/713691874

Castellani, V., & Sala, S. (2013). Sustainability indicators integrating consumption patterns in strategic environmental assessment for urban planning. *Sustainability*, *5*(8), 3426–3446. http://doi.org/10.3390/su5083426

Chrysoulakis, N., Lopes, M., San José, R., Grimmond, C. S. B., Jones, M. B., Magliulo, V., & Cartalis, C. (2013). Sustainable urban metabolism as a link between bio-physical sciences and urban planning: The BRIDGE project. *Landscape and Urban Planning*, *112*(1), 100–117. http://doi.org/10.1016/j.landurbplan.2012.12.005

Cruz, N. F. da, & Marques, R. C. (2014). Scorecards for sustainable local governments. *Cities*, *39*, 165–170. http://doi.org/10.1016/j.cities.2014.01.001

Dizdaroglu, D. (2015). Developing micro-level urban ecosystem indicators for sustainability assessment. *Environmental Impact Assessment Review*, *54*, 119–124. http://doi.org/10.1016/j.eiar.2015.06.004

Egger, S. (2006). Determining a sustainable city model. *Environmental Modelling & Software*, *21*(9), 1235–1246. http://doi.org/10.1016/j.envsoft.2005.04.012

Egilmez, G., Gumus, S., & Kucukvar, M. (2015). Environmental sustainability benchmarking of the U.S. and Canada metropoles: An expert judgment-based multi-criteria decision making approach. *Cities*, *42*, 31–41. http://doi.org/10.1016/j.cities.2014.08.006

European Commission. (2016). The EU Emissions Trading System (EU ETS). Retrieved February 15, 2016, from http://ec.europa.eu/clima/policies/ets/index_en.htm

Guijt, I., Moiseev, A., & Prescott, A. (2001). *IUCN Resource Kit for Sustainability Assessment, Part B: Facilitator's Materials. Evaluation*. Gland, Switzerland. http://cmsdata.iucn.org/downloads/resource_kit_b_eng.pdf

Guy, G. B., & Kibert, C. (1998). Developing indicators of sustainability: US experience. *Building Research & Information*, *26*(1), 39–45. http://doi.org/10.1080/096132198370092

Hiremath, R. B., Balachandra, P., Kumar, B., Bansode, S. S., & Murali, J. (2013). Indicator-based urban sustainability: A review. *Energy for Sustainable Development*, *17*(6), 555–563. http://doi.org/10.1016/j.esd.2013.08.004

Holden, M. (2006). Urban indicators and the integrative ideals of cities. *Cities*, *23*(3), 170–183. http://doi.org/10.1016/j.cities.2006.03.001

Huang, L., Wu, J., & Yan, L. (2015). Defining and measuring urban sustainability: A review of indicators. *Landscape Ecology*, *30*(7), 1175–1193. http://doi.org/10.1007/s10980-015-0208-2

Huang, L., Yan, L., & Wu, J. (2016). Assessing urban sustainability of Chinese megacities: 35 years after the economic reform and open-door policy. *Landscape and Urban Planning*, *145*, 57–70. http://doi.org/10.1016/j.landurbplan.2015.09.005

Huang, S. L., Wong, J. H., & Chen, T. C. (1998). A framework of indicator system for measuring Taipei's urban sustainability. *Landscape and Urban Planning*, *42*(1), 15–27. http://doi.org/10.1016/S0169-2046(98)00054-1

Ibrahim, F. I., Omar, D., & Mohamad, N. H. N. (2015). Theoretical review on sustainable city indicators in Malaysia. *Procedia – Social and Behavioral Sciences, 202*, 322–329. http://doi.org/10.1016/j.sbspro.2015.08.236

Keirstead, J., & Leach, M. (2008). Bridging the gaps between theory and practice: A service niche approach to urban sustainability indicators. *Sustainable Development, 16*(5), 329–340. http://doi.org/10.1002/sd.349

Kennedy, C., Stewart, I. D., Ibrahim, N., Facchini, A., & Mele, R. (2014). Developing a multi-layered indicator set for urban metabolism studies in megacities. *Ecological Indicators, 47*, 7–15. http://doi.org/10.1016/j.ecolind.2014.07.039

Kılkış, Ş. (2016). Sustainable development of energy, water and environment systems index for Southeast European cities. *Journal of Cleaner Production, 130*, 222–234. http://doi.org/10.1016/j.jclepro.2015.07.121

Lazaroiu, G. C., & Roscia, M. (2012). Definition methodology for the smart cities model. *Energy, 47*(1), 326–332. http://doi.org/10.1016/j.energy.2012.09.028

Leal Filho, W., Platje, J., Gerstlberger, W., Ciegis, R., Kaaria, J., Klavins, M., & Kliucininkas, L. (2015). The role of governance in realising the transition towards sustainable societies. *Journal of Cleaner Production, 113*. http://doi.org/10.1016/j.jclepro.2015.11.060

Lee, Y. J., & Huang, C. M. (2007). Sustainability index for Taipei. *Environmental Impact Assessment Review, 27*(6), 505–521. http://doi.org/10.1016/j.eiar.2006.12.005

Li, F., Liu, X., Hu, D., Wang, R., Yang, W., Li, D., & Zhao, D. (2009). Measurement indicators and an evaluation approach for assessing urban sustainable development: A case study for China's Jining City. *Landscape and Urban Planning, 90*(3), 134–142. http://doi.org/10.1016/j.landurbplan.2008.10.022

Li, Y., & Qiu, L. (2015). A comparative study on the quality of China's eco-city: Suzhou vs Kitakyushu. *Habitat International, 50*, 57–64. http://doi.org/10.1016/j.habitatint.2015.08.005

Marans, R. W. (2015). Quality of urban life & environmental sustainability studies: Future linkage opportunities. *Habitat International, 45*(P1), 47–52. http://doi.org/10.1016/j.habitatint.2014.06.019

Marsal-Llacuna, M. L., Colomer-Llinàs, J., & Meléndez-Frigola, J. (2015). Lessons in urban monitoring taken from sustainable and livable cities to better address the Smart Cities initiative. *Technological Forecasting and Social Change, 90*, 611–622. http://doi.org/10.1016/j.techfore.2014.01.012

Mayer, A. L. (2008). Strengths and weaknesses of common sustainability indices for multidimensional systems. *Environment International, 34*(2), 277–291. http://doi.org/10.1016/j.envint.2007.09.004

McCool, S. F., & Stankey, G. H. (2004). Indicators of sustainability: Challenges and opportunities at the interface of science and policy. *Environmental Management, 33*(3), 294–305. http://doi.org/10.1007/s00267-003-0084-4

Meijering, J. V., Kern, K., & Tobi, H. (2014). Identifying the methodological characteristics of European green city rankings. *Ecological Indicators, 43*, 132–142. http://doi.org/10.1016/j.ecolind.2014.02.026

Michael, F. L., Noor, Z. Z., & Figueroa, M. J. (2014). Review of urban sustainability indicators assessment: Case study between Asian countries. *Habitat International, 44*, 491–500. http://doi.org/10.1016/j.habitatint.2014.09.006

Milman, A., & Short, A. (2008). Incorporating resilience into sustainability indicators: An example for the urban water sector. *Global Environmental Change, 18*(4), 758–767. http://doi.org/10.1016/j.gloenvcha.2008.08.002

Moher, D., Liberati, A., Tetzlaff, J., Altman, D., & The PRISMA Group (2009). Preferred reporting items for systematic reviews and meta-analyses: The PRISMA Statement. *PLoS Medicine*, *6*(7). http://doi.org/10.1371/journal.pmed.1000097

Moles, R., Foley, W., Morrissey, J., & O'Regan, B. (2008). Practical appraisal of sustainable development-methodologies for sustainability measurement at settlement level. *Environmental Impact Assessment Review*, *28*(2–3), 144–165. http://doi.org/10.1016/j.eiar.2007.06.003

Mori, K., & Christodoulou, A. (2012). Review of sustainability indices and indicators: Towards a new City Sustainability Index (CSI). *Environmental Impact Assessment Review*, *32*(1), 94–106. http://doi.org/10.1016/j.eiar.2011.06.001

Moussiopoulos, N., Achillas, C., Vlachokostas, C., Spyridi, D., & Nikolaou, K. (2010). Environmental, social and economic information management for the evaluation of sustainability in urban areas: A system of indicators for Thessaloniki, Greece. *Cities*, *27*(5), 377–384. http://doi.org/10.1016/j.cities.2010.06.001

Munier, N. (2011). Methodology to select a set of urban sustainability indicators to measure the state of the city, and performance assessment. *Ecological Indicators*, *11*(5), 1020–1026. http://doi.org/10.1016/j.ecolind.2011.01.006

Ness, B., Urbel-Piirsalu, E., Anderberg, S., & Olsson, L. (2007). Categorising tools for sustainability assessment. *Ecological Economics*, *60*(3), 498–508. http://doi.org/10.1016/j.ecolecon.2006.07.023

Pires, S. M., Fidélis, T., & Ramos, T. B. (2014). Measuring and comparing local sustainable development through common indicators: Constraints and achievements in practice. *Cities*, *39*, 1–9. http://doi.org/10.1016/j.cities.2014.02.003

Rametsteiner, E., Pülzl, H., Alkan-Olsson, J., & Frederiksen, P. (2011). Sustainability indicator development: Science or political negotiation? *Ecological Indicators*, *11*(1), 61–70. http://doi.org/10.1016/j.ecolind.2009.06.009

Rosales, N. (2011). Towards the modeling of sustainability into urban planning: Using indicators to build sustainable cities. *Procedia Engineering*, *21*, 641–647. http://doi.org/10.1016/j.proeng.2011.11.2060

Seifollahi, M., & Faryadi, S. (2011). Evaluating the quality of Tehran's urban environment based on sustainability indicators. *International Journal of Environmental Research*, *5*(2), 545–554.

Sharifi, A., & Yamagata, Y. (2014). Resilient urban planning: Major principles and criteria. *Energy Procedia*, *61*, 1491–1495. http://doi.org/10.1016/j.egypro.2014.12.154

Shen, L., & Guo, X. (2014). Spatial quantification and pattern analysis of urban sustainability based on a subjectively weighted indicator model: A case study in the city of Saskatoon, SK, Canada. *Applied Geography*, *53*, 117–127. http://doi.org/10.1016/j.apgeog.2014.06.001

Shen, L., & Zhou, J. (2014). Examining the effectiveness of indicators for guiding sustainable urbanization in China. *Habitat International*, *44*, 111–120. http://doi.org/10.1016/j.habitatint.2014.05.009

Shen, L.-Y., Jorge Ochoa, J., Shah, M. N., & Zhang, X. (2011). The application of urban sustainability indicators: A comparison between various practices. *Habitat International*, *35*(1), 17–29. http://doi.org/10.1016/j.habitatint.2010.03.006

Shmelev, S. E. (2017). Multidimensional sustainability assessment for megacities, in Shmelev S.E., ed. (2017). *Green Economy Reader: Lectures in Ecological Economics and Sustainability*, Springer, pp. 205–236.

Shmelev S. E., & Shmeleva, I.A. (2018). Sustainable development assessment: A multidimensional approach for smart and sustainable cities. *Sustainable Development*, *26*(6), 904–920.

Siddaway, A. (2014). What is a systematic literature review and how do I do one? https://pdfs.semanticscholar.org/2214/2c9cb17b4baab118767e497c93806d741461.pdf

Skea, J., & Nishioka, S. (2008). Policies and practices for a low-carbon society. *Climate Policy*, *8*(1), 5–16. http://doi.org/10.3763/cpol.2008.0487

Spaans, M., & Waterhout, B. (2016). Building up resilience in cities worldwide: Rotterdam as participant in the 100 Resilient Cities Programme. *Cities*, *61*, 109–116. http://doi.org/10.1016/j.cities.2016.05.011

Srivastava, M. (2016). Framework to assess city-scale sustainability. *Procedia Engineering*, *145*, 1440–1447. http://doi.org/10.1016/j.proeng.2016.04.181

Sun, X., Liu, X., Li, F., Tao, Y., & Song, Y. (2015). Comprehensive evaluation of different scale cities' sustainable development for economy, society, and ecological infrastructure in China. *Journal of Cleaner Production*, 1–9. http://doi.org/10.1016/j.jclepro.2015.09.002

Suzuki, H., Arish, D., Sebastian, M., & Nanae, Y. (2009). *Eco Cities, Ecological Cities as Economic Cities*. Washington, DC: The World Bank. http://siteresources.worldbank.org/INTEASTASIAPACIFIC/Resources/226262-1246459314652/Eco2Cities_FullReport_ConfEdition6-26-09_sm.pdf

Tan, S., Yang, J., Yan, J., Lee, C., Hashim, H., & Chen, B. (2016). A holistic low carbon city indicator framework for sustainable development. *Applied Energy*. http://doi.org/10.1016/j.apenergy.2016.03.041

Tanguay, G. A., Rajaonson, J., Lefebvre, J.-F., & Lanoie, P. (2010). Measuring the sustainability of cities: An analysis of the use of local indicators. *Ecological Indicators*, *10*(2), 407–418. http://doi.org/10.1016/j.ecolind.2009.07.013

Tran, L. (2016). An interactive method to select a set of sustainable urban development indicators. *Ecological Indicators*, *61*, 418–427. http://doi.org/10.1016/j.ecolind.2015.09.043

Turcu, C. (2012). Re-thinking sustainability indicators: Local perspectives of urban sustainability. *Journal of Environmental Planning and Management*, *56*(5), 695–719. http://doi.org/10.1080/09640568.2012.698984

United Nations Human Settlements Programme. (2009). *Planning Sustainable Cities: Policy Directions: Global Report on Human Settlements 2009*. London: Earthscan.

van Assche, J., Block, T., & Reynaert, H. (2010). Can community indicators live up to their expectations? The case of the Flemish City Monitor for Livable and Sustainable Urban Development. *Applied Research in Quality of Life*, *5*(4), 341–352. http://doi.org/10.1007/s11482-010-9121-7

Wacker, J. G. (1998). A definition of theory: Research guidelines for different theory-building research methods in operations management. *Journal of Operations Management*, *16*(4), 361–385. http://doi.org/10.1016/S0272-6963(98)00019-9

WHO (World Health Organisation). (2016). Urban population growth. www.who.int/gho/urban_health/situation_trends/urban_population_growth_text/en/

Wiek, A., & Binder, C. (2005). Solution spaces for decision-making: A sustainability assessment tool for city-regions. *Environmental Impact Assessment Review*, *25*(6), 589–608. http://doi.org/10.1016/j.eiar.2004.09.009

World Commission on Environment and Development. (1987). *Our Common Future*. New York.

Yigitcanlar, T., Dur, F., & Dizdaroglu, D. (2015). Towards prosperous sustainable cities: A multiscalar urban sustainability assessment approach. *Habitat International*, *45*, 36–46. http://doi.org/10.1016/j.habitatint.2014.06.033

Yuan, W., & James, P. (2002). Evolution of the Shanghai city region 1978–1998: An analysis of indicators. *Journal of Environmental Management*, *64*(3), 299–309. http://doi.org/10.1006/jema.2001.0526

Zhou, J., Shen, L., Song, X., & Zhang, X. (2015). Selection and modeling sustainable urbanization indicators: A responsibility-based method. *Ecological Indicators*, *56*, 87–95. http://doi.org/10.1016/j.ecolind.2015.03.024

Zoeteman, K., Mommaas, H., & Dagevos, J. (2016). Are larger cities more sustainable? Lessons from integrated sustainability monitoring in 403 Dutch municipalities. *Environmental Development*, *17*, 57–72. http://doi.org/10.1016/j.envdev.2015.08.003

4 Indicator-based multi-criteria urban sustainability assessment under varying policy priorities

Ellie Tonks and Stanislav E. Shmelev

1. Introduction

This chapter aims to operationalise the UN Sustainable Development Goals approach at the city level and offers policy advice on benchmarking urban sustainability using a compact and robust set of sustainability indicators. Based on an interdisciplinary approach to urban sustainability (Shmelev & Shmeleva, 2009, 2018) it seeks to assess and rank medium-sized European and North American cities on their sustainability performance using methodological principles developed in Shmelev (2017). Cities generate 80% of greenhouse gas emissions and claim 75% of global energy production (Lazaroiu & Roscia, 2012). Sustainable urban development is therefore a key political priority (Girardet, 2004; Girardet & Mendonca, 2009, Girardet, 2017), demanding increased consideration as the trend for urbanisation only continues to grow on a global scale (WHO, 2016). Sustainable urban development, takes precedence in European Union policy making, for example, as illustrated by the 7th Environmental Action Programme, entitled, *Sustainable Cities: 'Working together for Common Solutions'* (Priority Objective 8). The policy's aim is to increase the sustainability of EU cities by 2050, with a view to ensuring that all Europeans are 'living well, within the limits of the planet' (European Union, 2013).

According to the United Nations Human Settlement Program (UN-Habitat), a sustainable city is a city where urban planning addresses the environmental, demographic, economic, socio-spatial and institutional urban challenges (United Nations Human Settlements Programme, 2009). Echoing UN-Habitat, the European Commission considers a sustainable city to be one where urban development integrates physical urban renewal, education, economic development, social inclusion and environmental protection, whilst also fostering strong partnerships between citizens and civil society (European Commission, 2016). Following a literature review on the subject, Choon and colleagues (2011) define a sustainable city as a metropolis that fulfils both human and environmental well-being. Here, the sustainable city is seen as a complex hybrid of social, economic and ecological forms, producing and reproducing a sustained quality of life within a specific spatial context. This definition is in line with recommendations by Barbosa, Bragança and Mateus (2014) who stress the need to take into account the interrelationships and interdependences between environmental, social and economic actors when developing sustainable city models.

Despite the recent political momentum surrounding the sustainable city concept (Eurostat, 2013; Whitehead, 2003; World Commission on Environment and Development, 1987), analysis of sustainable urban development remains limited (Barbosa et al., 2014; Whitehead, 2003). Major gaps in the field of urban sustainability have been identified; however, this can be explained by the multidimensional nature of city systems in which innumerable objects and processes are temporally and spatially interacting. Such interconnectivity poses certain methodological difficulties for both study and management. An increasing number of cities are adopting sustainability as a goal with a view to changing current urban development; it thus becomes paramount to determine methods of assessment when measuring and reporting on metropolitan sustainability (Shmelev, 2017). It is necessary to identify areas in which a city is performing well, with regards to specific and wider sustainability goals (Hiremath et al., 2013). A systems approach, encompassing multidimensional urban interdependencies, is therefore required (Shmelev & Shmeleva, 2009).

The complex and multidimensional nature of sustainability lends itself to the use of indicators as a method of tracking and assessing a city's performance. Indicators have historically been used as tools for obtaining information on human health, economic welfare and weather; however, compared to socio-economic indicators, sustainable development and environmental indicators are a fairly new occurrence (Choon et al., 2011). The benefit of appropriately selected indicators is reducing the number of measurements needed to understand a system's performance (Guy & Kibert, 1998). Additionally, indicators are useful tools for decision makers, providing information on and giving insight into the progress of a system towards sustainable development, whilst also acting as communication tools, connecting scientists to authorities (Choon et al., 2011). The family of indicators are selected on the understanding that they provide the most comprehensive overview of the urban system, measure key characteristics of important social, environmental and economic subsystems, while keeping the number of indicators to a minimum (Guy & Kibert, 1998). Sustainability indicators have been used by academics, governmental institutions and non-governmental organisations alike in assessing the sustainability of numerous cities (Michael et al., 2014; Economist Intelligent Unit, 2009; Button, 2002; Shane & Graedel, 2000). This study – akin to that of Shane and Graedel (2000) – thus aims to develop a set of indicators that provide a snapshot of a city's sustainability status for both 'human' and 'environmental' well-being. These measures must be internationally applicable and must facilitate comparison between cities of differing sizes, geographies, histories and cultures.

The key challenge for researchers and practitioners, inherent to all sustainable development projects, is integrating the three pillars of a sustainable city into an effective framework for action and research (Michael et al., 2014). This chapter therefore seeks to address this challenge – 19 cities have been identified and a selection procedure developed in order to decide which of the cities is the most sustainable in terms of its economic, environmental and social performance. The procedure had to be able to deal with a number of urban indicators, to use the most recent data (subject to availability), and to take into account the fact that data was often inaccurate, given the differing temporal and spatial scales. In light of these factors, the proposed mechanism is the application of a multi-criteria decision aid (MCDA) tool. MCDAs have been employed before in assessing the environmental sustainability performance of cities (for a study on US and Canadian cities, see Egilmez et al., 2015). Shmelev (2017) performed three urban sustainability assessments of twelve large global cities using ELECTRE III, NAIADE and APIS MCDA tools.

This chapter describes the main features of the study that was undertaken to assess metropolitan sustainability across 19 cities, with the particular aim of defining and appraising urban sustainability criteria in order to produce a ranked order of cities under different sustainability priorities. It presents cities ranked against a set of identified sustainability criteria using a weak sustainability weighted-sum approach, before going on to produce a ranked order using a strong sustainability multi-criteria decision aid (specifically, ELECTRE III). Having then stated the 19 cities in question, the selected sustainability indicators are presented and the method of data collection described. The fourth section then outlines the two methods used to rank the cities under differing sustainability priorities. The results and sensitivity analysis are presented in the fifth section, followed by the discussion and conclusion.

2. Methodology

2.1 Alternatives

The cities selected for our assessment are presented in Table 4.1. The 19 cities have been selected on their comparability: most of them belong to the alpha and beta clusters of the Globalization and World Cities Research Network (GaWC, 2012) with many regarded as 'sustainability leaders'. Several cities in our set, namely Copenhagen and Stockholm have received a European Green Capital Award (City of Copenhagen, 2014; City of Stockholm, 2015).

2.2 Evaluation criteria

Seventeen sustainability criteria, denoted g_1, g_2, . . . , g_{17}, were selected (see Table 4.1). The selected indicators reflect the recently adopted UN Sustainable Development Goals (United Nations, 2015), the ISO 37120 standard 'Sustainable Development of Communities' (ISO, 2014), the Indicators for Sustainable Cities and the UNECE-ITU Smart Sustainable Cities Indicators (United Nations Economic Commission for Europe, 2017). The final set of 17 indicators is balanced from the point of view of giving equal prominence to economic, social and environmental dimensions and directly correspond to most Sustainable Development Goals, which can be seen in Table 4.1. It should be noted that in this chapter criteria are elements of the MCDA framework and indicators are elements of the sustainability city framework. This chapter gives them the same meaning; criteria and indicators are coupled together and used interchangeably to assess a cities performance using the MCDA tool. Each criterion represents the most accurate information about the status of the city system and, more specifically, the status of one of three elements; economic, environmental and social. Together, the criteria are intended to represent the varying viewpoints surrounding different policy priorities involved in the sustainable city concept:

- Criteria g_1–g_5 indicators of economic sustainability: $g_1(s)$ measures the size of a city's economy and, more precisely, represents the market value (in $million) of all final goods and services produced within metropolitan area s in a specified period of time; $g_2(s)$ is the amount of money that each individual within a household residing in city s has available for spending and saving after income taxes; $g_3(s)$ the share of the labour force of city s that is unemployed; $g_4(s)$ the number of patents per 1,000 inhabitants in city s; and, $g_5(s)$ rate of inflation.
- Criteria g_6–g_{12} are indicators of environmental sustainability: $g_6(s)$ represents the CO_2 emissions (in tonnes) from metropolitan area s per individual in city s; $g_7(s)$ is the share of renewable energy sources as a percentage of the overall energy mix of city s; $g_8(s)$ represents exposure to air pollutant $PM_{2.5}$ within metropolitan area s; $g_9(s)$ is the size (in m^2) of green space available per million inhabitants of city s; $g_{10}(s)$ represents water consumption (in m^3) of individuals living in city s; $g_{11}(s)$ represents the domestic municipal waste generation of individuals within metropolitan area s, and $g_{12}(s)$ is the percentage of waste generated within city s that is recycled.
- Criteria g_{12}–g_{17} are indicators of social sustainability: $g_{13}(s)$ represents the share of individuals within city s walking, cycling or taking public transport to work; $g_{14}(s)$ is the percentage of the population of city s with tertiary education; $g_{15}(s)$ is the extent to which the distribution of income of residents in city s deviates from a perfectly equal distribution (therefore used to represent equality within metropolitan area s); $g_{16}(s)$ represents the number of intentional homicides per 100,000 inhabitants of city s, and $g_{17}(s)$ represents life expectancy at birth of the inhabitants of metropolitan area s.

2.3 Data

The data for each criterion was collected through web-based searches according to the following steps. Initial consultation reviewed the metropolitan and regional statistics produced by the Organisation for Economic Co-operation and Development (OECD, 2016), and the region and city data produced by Eurostat. If neither database reported on a criterion, then another independent database was sought; for example, the Brookings 2014 Global Metro Monitor Map (Parilla et al., 2014) or the CDP (2015) Cities Electricity Mix Map. Finally, individual city reports and websites were used to source remaining missing criterion data. Table 4.1 presents the performance matrix of cities for each criterion.

Table 4.1 Performance matrix of cities (a1–a19)

	g1	g2	g3	g4	g5	g6	g7	g8	g9	g10	g11	g12	g13	g14	g15	g16	g17
Indicator	GDP	Income	Unemployment	Patents per 1000 inhabitants	Inflation	CO_2 emissions	Share of renewables	Exposure to air pollution	Green area per million people	Domestic water consumption	Volume of municipal waste	Share of waste recycled	Citizens walking, cycling or taking public transport to work	Tertiary educational attainment	Gini	Homicide rate	Life expectancy at birth
Unit	PPP, $million	US$ per capita	%	Number of patents per 1000 persons	%	Tonnes per capita	%	$PM_{2.5}$	m^2 per 1,000,000 person	m^3 per capita	kg per capita	%	%	%	Gini	Intentional homicides per 100,000 persons	Years
Sustainable Development Goal	8 Decent work and economic growth	1 No poverty	17 Partnership for the goals	9 Industry, innovation and infrastructure	2 Zero hunger	13 Climate action	7 Affordable and clean energy	15 Life on land	15 Life on land	6 Clean water and sanitation	12 Responsible consumption and production	14 Life below water	11 Sustainable Cities and Communities	4 Quality education	10 Reduced inequalities	16 Peace, justice and strong institutions	3 Good health and well-being
Cities, city class																	
Amsterdam, α	120671	18837	7.47	0.76	2.51	5.69	4.95	15.27	207.16	48.50	419.75	43.00	83.00	51.80	0.30	1.30	81.50
Barcelona, α-	171032	19447	19.98	0.47	1.41	3.30	2.17	11.12	2.29	38.90	498.30	9.00	69.40	47.00	0.30	1.00	83.20
Boston, α-	360110	46944	5.12	3.01	1.46	10.46	13.00	7.46	86.06	56.64	880.96	20.00	18.30	43.00	0.54	8.10	80.20
Copenhagen, β+	127001	18143	6.88	2.06	0.79	3.10	60.00	13.12	376.18	38.00	400.00	23.61	71.00	59.30	0.28	1.40	79.10
Edinburgh, γ+	32497	24767	4.65	0.06	2.55	6.50	34.60	8.28	204.42	48.60	454.22	30.70	55.10	58.20	0.30	2.40	80.50
Frankfurt, α	229985	25427	4.53	1.50	1.50	13.80	11.00	15.26	100.16	60.43	463.90	33.10	60.30	36.60	0.31	14.04	81.50
Madrid, α	262335	20925	19.02	0.32	1.41	1.83	40.75	10.86	24.23	47.82	354.32	9.88	54.00	54.00	0.32	0.70	84.30
Miami, α	262697	35631	6.28	0.42	1.46	12.87	2.30	6.10	140.90	188.81	1269.83	18.00	5.90	28.10	0.57	19.20	81.40

(continued)

Table 4.1 (continued)

Milan, α	312108	22642	8.19	0.52	1.22	9.80	33.40	28.36	22.40	83.10	480.00	42.90	54.36	25.90	0.30	0.60	83.50
Montreal, β+	155905	19706	8.20	0.54	0.94	9.60	96.74	7.17	712.34	99.30	572.68	38.00	28.80	42.70	0.29	1.30	81.70
Munich, α-	219943	30075	3.04	1.89	1.50	8.80	39.00	15.07	539.85	62.60	556.93	42.95	64.20	45.50	0.28	6.93	82.60
Rome, β+	163243	20818	11.43	0.24	1.22	3.46	33.40	10.46	224.78	75.00	242.25	19.50	53.35	31.60	0.33	0.80	82.80
Stockholm, β+	142953	23456	7.09	2.62	−0.04	1.44	70.00	6.50	114.66	73.00	597.00	31.00	93.00	58.00	0.30	1.00	82.80
Toronto, α	276313	22376	7.88	0.47	0.94	10.30	38.10	9.41	1297.80	157.23	353.60	52.00	28.00	57.80	0.31	1.20	80.60
Vancouver, β+	109805	23293	5.95	0.69	0.94	12.90	31.00	5.99	98.60	189.29	369.00	48.00	24.50	52.30	0.32	1.90	82.60
Vienna, α-	129515	23243	7.03	0.37	2.00	7.80	18.00	17.43	221.99	41.06	611.65	33.35	65.00	50.40	0.31	1.50	80.00
Warsaw, α-	141096	14136	7.19	0.11	1.03	7.43	16.01	16.76	1022.32	61.32	317.29	19.33	73.00	56.60	0.33	1.90	78.50
Washington, DC, α	442212	62002	5.02	0.65	1.46	12.69	0.00	12.13	973.99	206.56	344.73	26.00	17.90	46.80	0.54	15.90	77.90
Zurich, α-	109065	31154	4.19	1.51	−0.22	5.10	73.40	19.98	268.55	114.84	390.00	34.00	62.00	56.40	0.32	0.20	83.00
Thresholds																	
Goal	MAX	MAX	MIN	MAX	MIN	MIN	MAX	MIN	MAX	MIN	MIN	MAX	MAX	MAX	MIN	MIN	MAX
Indifference	20485.75	2393.30	0.85	0.15	0.14	0.62	4.84	1.12	64.78	8.43	51.38	2.15	4.36	1.67	0.01	0.95	0.32
Preference	40971.50	4786.60	1.69	0.30	0.28	1.24	9.67	2.24	129.55	16.86	102.76	4.30	8.71	3.34	0.03	1.90	0.64
Veto	410000	48000	17.00	7.00	3.00	13.00	97.00	23.00	1300.00	170.00	1030.00	50.00	90.00	40.00	0.30	20.00	7.00
Criteria weights (Indices of Importance) based on sustainability priority																	
Overall	0.067	0.067	0.067	0.067	0.067	0.048	0.048	0.048	0.048	0.048	0.048	0.048	0.056	0.056	0.056	0.056	0.056
Economic	0.100	0.100	0.100	0.100	0.100	0.036	0.036	0.036	0.036	0.036	0.036	0.036	0.050	0.050	0.050	0.050	0.050
Environmental	0.050	0.050	0.050	0.050	0.050	0.071	0.071	0.071	0.071	0.071	0.071	0.071	0.050	0.050	0.050	0.050	0.050
Social	0.050	0.050	0.050	0.050	0.050	0.036	0.036	0.036	0.036	0.036	0.036	0.036	0.100	0.100	0.100	0.100	0.100

The final criteria array, however, is an output bound by a number of limiting factors resulting from constraints on both data availability and the length of the data collection period. First, certain data values are temporally outdated and therefore, city comparisons are not necessarily made across the same time period. Second, data has been collected at the scale of the metropolitan area as far as reasonably possible; however, certain values represent larger spatial areas, for example, OECD regional data represents either NUTs levels 2 or 3, and data for North American cities is, for some criterion, at the state level. This chapter therefore emphasises the need for singular databases of regularly monitored metropolitan level sustainability indicators, more specifically information of both environmental and social sustainability are lacking for cities worldwide.

2.4 The four decision makers

To give meaning to a decision problem a particular actor D, the decision maker, must be introduced (Roy, 1991). This chapter incorporates four hypothetical decision makers whose preferences are used to inform the decision-aid problem: D_1 whose preferences favour overall sustainability; D_2 who favours economic sustainability over both environmental and social; D_3 who favours environmental sustainability over economic and social, and D_4 whose preferences favour social sustainability. The decision model provides a system with which it becomes possible to assess certain questions defined by the differing Ds (Roy, 1991). D_1 is not a pre-determined character but an ideal (closest akin to a median-voter), the general profiles of the other three decision makers, however, can be attributed to certain real-world examples: D_2 may be interpreted as a finance ministry, an economic or industrial think tank, the *Economist* magazine, or a conservative voter; D_3 could be taken as embodying the WWF, IUCN, Friends of the Earth, or the European Environment Agency, and, D_4 may be interpreted as Oxfam, left-wing politicians, unions, or social entrepreneurs.

This chapter, through the application of a weighted-sum evaluation method and a MCDA tool, therefore seeks to address four questions: (1) When D_1 favours overall metropolitan sustainability which city of A is the most satisfactory considering F? (2) When D_2 favours economic sustainability (g_1 and g_2) which city of A is the most satisfactory considering F? (3) When D_3 favours environmental sustainability (g_3, g_4, . . . g_{11}) which city of A is the most satisfactory considering F? (4) When D_4 favours social sustainability (g_{12}, g_{13}, . . . g_{17}) which city of A is the most satisfactory considering F?

3. Multi-criteria analysis

The previous sections have defined the set of 19 alternatives A and the family of 17 criteria F that were used to assess those alternatives. Here, the development and application of an elementary evaluation method and a multi-criteria decision aid (MCDA) are presented.

3.1 A weighted-sum approach

A preliminary weighted-sum approach, built under the assumption of perfect substitutability between sustainability dimensions, which did not incorporate a comprehensive model of preferences, was developed to answer the four questions. This approach resembles a simple cost-benefit analysis in its treatment of compensation between criteria and presents a partial case of the MCDA methods family. The data values for *F* were transformed to fall within a smaller common range [0.0, 1.0] to enable comparisons. A Min-Max normalisation procedure was undertaken, performing a linear transformation on the original data (Han et al., 2012). The goal function of the values was taken into account to allow for the two criteria goal options (either minimum or maximum values, for example, minimum values are most favourable for CO_2 emissions or water consumption). The relationships among the original data values are preserved when a Min-Max normalisation is applied (Han et al., 2012).

3.2 Weighting criteria

In order to incorporate the differing priorities of the four decision makers (D_1–D_4) weights were applied (see Table 4.1 for the weights under the preferences of *D*). The weights were set up in such a way that a prioritised dimension received a weight of 0.5 and the other dimensions – 0.25 with each criterion receiving a weight inversely related to the number of criteria in a dimension.

The normalised values for each criterion were then summed for each alternative (city) to gain a total score per city (on a scale of 0-1) according to which a complete pre-order of *A*, depending on *D*'s preferences, was formed. The weighted-sum approach was performed with MS Excel.

3.3 An outranking approach

Decision aiding can be defined as the activity of a person who, through the implementation of explicit but not necessarily fully formulated models, obtains elements of responses to the questions of stakeholders in a decision process. The generated elements help clarify the decision and usually work towards recommending, or favouring, an action that increases the consistency between the stakeholder's value system and objectives and the evolution of the decision process (Roy, 1996, 2005). Real-life decision-problems will, in general, involve several conflicting viewpoints (criteria) that need to be accounted for conjointly in order to attain an appropriate intervention (decision) (Govindan & Jepsen, 2016). The numerous viewpoints, represented by a family of criteria, are then used to evaluate potential intervention on a suitable qualitative or quantitative scale. In most circumstances, however, there is no widely accepted calculation to account for such heterogeneous scales when bringing in a common unit. Such a method is referred to as a weighted-sum approach and can, in many decision-making contexts, result in neglected aspects of realism, cause set-up equivalences between criteria and also represent features of a singular value system as objective. Through the implementation of a multi-criteria aid, however, such matters can be avoided in three ways: by defining a wide spectrum of viewpoints of involved actors to structure the decision-problem; by constructing a family of criteria that preserves the original meaning of each criterion; and finally, by facilitating debate on the role (weight, veto and threshold levels) that each criterion plays within the decision-aiding process (Roy, 2005).

The weighted-sum and multi-criteria approaches used in this chapter effectively represent weak and a strong sustainability assessment perspectives. It is widely accepted by ecological economists that decision makers need to adopt a strong sustainability position (assuming that human capital and natural capital are complementary, but not interchangeable) for the discussion and implementation of sustainable development policies (Pelenc & Ballet, 2015). A weighted-sum approach, alike to weak sustainability, allows for complementarity (compensation) between natural capital and manufactured, human and social capital in producing human well-being (Brand, 2009). On the other hand, a multi-criteria decision aid can account for those certain elements of capital that are critical due to their unique contribution to human well-being (such as ecosystem services provided by natural capital) by not allowing for compensation between the differing criteria. Consequently, this chapter now employs a strong sustainability multi-criteria tool by way of comparison to the initial weak sustainability weighted-sum method.

The problem of evaluating a finite set of alternatives according to a set of criteria, expressed in a quantitative or qualitative form, can be termed a discrete multi-criteria problem (Munda, 1995). MCDA methods are non-linear recursive four-step procedures: first, structuring the decision-problem; second, articulating and modelling preferences; third, aggregating alternative evaluations; and fourth, formulating recommendations (Guitouni & Martel, 1998). The numerous MCDA methods can be categorised into three groupings: (1) value measurement models; (2) goal, aspiration or reference level models, and (3) outranking models (Belton & Stewart, 2002).

In practice, a decision maker faces the primary decision-making situation to engage the stakeholders, understand the context, and then co-construct the decision-problem in order to select and justify the choice of MCDA method (Guitouni & Martel, 1998). This phase involves the determination and assessment of stakeholders, alternatives, consequences, important aspects (criteria), information quality, etc. (Guitouni & Martel, 1998). In other words, this step involves the structuring of the three fundamental concepts involved in MCDA: the problematique, the alternatives, and the criterion (Roy, 2005). In addition to the structuring of a decision-problem in aiding MCDA method selection, guidelines and specific questions have been developed to help the analyst select a MCDA method (see Guitouni & Martel, 1998; Figueira et al., 2013). Here, the guiding hierarchical questions formulated by Figueira and colleagues (2013) were referred to in the selection procedure.

3.3.1 Problematique description

Having already defined the alternatives and criteria it is now necessary to describe the problematique, which here refers to the way a decision-aid is envisaged and therefore determines its use (Roy, 1991). Four reference problematiques are currently used in practice – the description ($P.\delta$), choice ($P.\alpha$), sorting ($P.\beta$) and ranking ($P.\gamma$) problematics – though these four alone are not the only conceivable options (Roy, 2005) . This chapter seeks to address a $P.\gamma$ ranking problematic (Roy, 1991): to build a complete (or partial) pre-order as rich as possible on a subset A_0 among the cities of A which appear to be the most satisfactory considering F.

3.3.2 ELECTRE methods

The specific outranking method, given the decision-aid context, is primarily differentiated by the chosen problem statement (Roy, 1991): here defined by $P.\gamma$. Considering the problem statement and guiding questions (Figueira et al., 2013), three methods (ELECTRE II, III and IV) are in competition. ELECTRE III was selected as an appropriate decision-aid due to the following characteristics: the possibility of taking into account indifference and preference thresholds; the quantification of the relative importance of criteria; the use of a fuzzy binary outranking relation, and the production of a partial pre-order of alternatives (Roy, 1991). ELECTRE methods are relevant when facing decision situations with at least three criteria and if at least one of the four following conditions is verified (Figueira et al., 2005): (1) criteria performance is expressed in different units (for example, distance, price, colour, etc.) and the decision maker wishes to avoid defining a common scale; (2) compensation among criteria is not acceptable for the DM; (3) for at least one criterion, small differences are insignificant, although the sum of small differences is decisive (in this case, discrimination thresholds must be introduced) and (4) actions are evaluated on a scale presenting an order on a ordinal (or 'weak' interval) scale which are not suitable for the comparison of differences. The most recent studies focusing on ELECTRE family of methods include (Corrente et al., 2017; Figueira & Roy, 2002; Figueira et al., 2013; Roy & Słowiński, 2013; Roy et al., 2014). The main characteristic, and advantage, of the ELECTRE methods is the avoidance of any normalisation calculations, which distorts the original data, and of any compensation between criteria (Ishizaka & Nemery, 2013). The Multiple Criteria Aggregation Procedures of ELECTRE methods were conceived such that they do not allow for compensation of performances among criteria, in that, the degradation of performances on certain criteria cannot be compensated by performance improvements on other criteria (Figueira et al., 2013). ELECTRE methods have been criticised for the drawback of pressing the decision maker/researcher into defining a number of (difficult) technical parameters; this also means that they are not necessarily easy to fully comprehend (Figueira et al., 2005). Such difficulties however, in assigning essential numerical values, are in no way specific to multi-criteria aggregation methods of the ELECTRE category; the same limitations may be encountered in one form or another in all ways of modelling (Roy, 1991).

3.3.3 ELECTRE III

ELECTRE III is based on the construction of an outranking relation between two alternatives, compared by pairs (a', a). The model should allow the decision-maker to take into account hesitations between two of the three following cases (Roy, 1991):

$a'Ia$:	a' indifferent to a,
$a'Pa$:	a' strictly preferred to a,
aPa':	a strictly preferred to a'.

The credibility of the outranking relation, characterising each pair of alternatives, is assessed via a credibility index ($\in [0; 1]$). Let g_j be the criterion used for comparing the performance between the two alternatives. With a traditional criterion, knowing the sign of the difference

$$U = g_j(a') - g_j(a) \qquad (1)$$

is sufficient to determine which of the two is preferable (with a' and a only being judged equal if U=0). This 'true-criterion' model, however, does not address three phenomena – namely imprecision, indetermination and uncertainty – affecting the value of $g_j(a)$ of the jth criterion. The phenomena can, however, be studied by means of a 'pseudo-criterion' model, which incorporates threshold values. The performance of the two alternatives for each criterion j is evaluated using three thresholds: the indifference threshold (q_j) and the preference threshold (p_j), which make up the 'pseudo-criterion', and the veto threshold (v_j) whereby $q_j \leq p_j \leq v_j$ (B. Roy, Présent & Silhol, 1986). Two cases for a' are then possible: a greater $g_j(a')$ being preferred denotes a' a gain, or being inferior denotes an a' loss. Here, the first case is considered, supposing that $g_j(a') \geq g_j(a)$:

$a'Ia$ if the difference in their performance is less than q_j:

$$(a' I_j a) \Leftrightarrow g_j(a') - g_j(a) \leq q_j(g_j(a)) \qquad (2)$$

a' is weakly preferred to a if the difference in their performance lies between q_j and p_j:

$$(a' Q_j a) \Leftrightarrow q_j(g_j(a)) < g_j(a') - g_j(a) \leq p_j(g_j(a)) \qquad (3)$$

$a'Pa$ if the difference in their performance is greater than p_j:

$$(a' P a) \Leftrightarrow p_j(g_j(a)) < g_j(a') - g_j(a) \qquad (4)$$

Relations 4, 5 and 6 demonstrate how both the preference and indifference thresholds take the uncertainty of performance evaluations into account. The veto threshold (v_j) is then introduced (not necessarily for each criterion) to define the outranking relation S ($S \neq Sj$) that incorporates the whole family of criteria. v_j leads to refusing the outranking of *a* by *a'* when *a* performs sharply better than *a'* on g_j, even if *a'* outranks *a* according to all other criteria (Martin et al., 2007):

$$If\, g_j(a) - g_j(a') > v_j(g_j(a')) \Rightarrow non\,(aSb) \tag{5}$$

For validating a comprehensive outranking relation *S*, two indices are then calculated: concordance and discordance indices. The concordance index, *c(a'Sa)* (calculated from the difference in performance values q_j and p_j and the importance of a criteria, k_j) characterises the strength of the positive argument able to validate the assertation *a'Sa*. The discordance index, *d(a'Sa)* (calculated from p_j and v_j values, and the importance of a criteria,k_j) characterises the strength of the opposition to *a'Sa* (Roy, 1991). The two indices are integrated into the credibility index calculation, equalling 1 if, and only if, $a'S_ja$ equals 0 for all criteria when aS_ia' for all criteria, or when a veto effect occurs. Ascending and descending procedures are then employed to construct a pre-order ranking of the alternatives (i.e. not necessarily complete). Further details on technical aspects of the ELECTRE III methodology can be found in the literature (Figueira et al., 2005; Roy et al., 1986; Roy, 1991).

3.3.4 Determination of model parameters

Implementing ELECTRE III requires the determination of values for preference parameters. Here, the process for setting parameters in this study is described in detail.

3.3.4.1 INDIFFERENCE, PREFERENCE AND VETO THRESHOLDS

The indifference threshold defines the largest difference between the performances of *a* and *a'* on the *j*th criterion at which the alternatives remain indifferent for the decision maker. The preference threshold defines the largest difference between *a* and *a'* such that one is preferred over the other on consideration of the *j*th criterion (Ishizaka & Nemery, 2013).

The elicitation process of parameter values can be divided into two categories: direct and indirect elicitation techniques (Figueira et al., 2005). Developments concerning elicitation procedures have been recommended for ELECTRE Tri methods (see Dias & Mousseau, 2005; The & Mousseau, 2002). As the decision makers involved in this study are hypothetical, no elicitation techniques could be employed. Rather the values were set in accordance with $q_j \leq p_j$ (Roy et al., 1986), in the following manner. The indifference threshold, q_j, was set as 5% of the difference between the greatest and smallest values of criterion *j*, so that

$$q_j = \left(\max(j) - \min(j)\right) * 0.05 \quad (6)$$

This was repeated for all 17 criteria. A ratio of 2:1 (preference:indifference) was used to set the preference threshold. The indifference and preference thresholds are therefore defined as constant values.

The veto effect, as defined above, works on the all or nothing principle (Roy, 1991). The value assigned to the veto should be constant (except where particular conditions dictate otherwise) for each criterion, and can be determined with reference to the value of the preference threshold (Roy et al., 1986). This is because the discordance index registers above zero, for a given criterion, at the preference threshold and reaches its maximum (higher) value at the veto threshold. The higher veto threshold is often valued at three, five, or ten times the value of the preference threshold (Rogers & Bruen, 1998). Here, see Table 4.1: the veto threshold is set above the level of the maximum difference on each criterion because we have not found the dimensions on which the veto could be applied in this context.

Fixing threshold parameters is as subjective as the process of calculating error in the physical sense (Roy et al., 1986). Therefore, both the selection of the methodology behind threshold calculations and the final numerical values characterising parameters may result in some imprecision, contradiction, arbitrariness, and/or lack of consensus concerning parameter values (Rogers & Bruen, 1998; Roy, 1989). In order to verify that the arbitrariness of the final numerical thresholds does not significantly affect the final model outputs a robustness analysis of the indifference and preference thresholds was undertaken using extreme (10 and 20% plus and minus) values was conducted. A sensitivity analysis of the veto threshold was not applied, as it was set above the level of maximum difference on each criterion. The aim of the robustness analyses is to assess to what extent this arbitrariness influences the final results.

3.3.4.2 RELATIVE IMPORTANCE COEFFICIENTS

When validating a comprehensive outranking relation *S*, it is vital to account for the fact that the importance devoted to each criterion in the aggregation procedure is not necessary equal across criteria. It is therefore necessary to characterise 'the greater or lesser importance' given to each criterion of *F* (Roy, 1991). ELECTRE III does not employ weighting factors used in weighted totals methods, instead it employs relative importance coefficients, denoted k_j for criteria g_j. The relative importance coefficients are not part of a total compensation system: their role is to represent the degree of importance decision makers wish to assign to differing criteria whilst affecting only the definition of the concordance degree. Further details on technical aspects of the relative importance coefficients can be found in the relevant literature (see Roy et al., 1986). It must be emphasised that the researcher is confronted with similar difficulties when characterising the importance devoted to each criterion as to the determination of threshold values (Bottero et al., 2015; Roy, 1991).

Here, the preferences of four hypothetical decision makers, D_1–D_4, are modelled through the importance assigned to criteria. Values of k_j are used to reflect the sustainability preferences of the four decision makers (see Table 4.1).

3.3.4.3 DEGREE OF CREDIBILITY AND CUT-OFF LEVELS

Considering the concordance and discordance indices, the degrees of credibility indicate whether the outranking hypothesis is true. The degrees of credibility are then gathered into a credibility matrix that is used to produce a partial pre-ordered graph. For large numbers of alternatives, however, the graph can be highly complex (Giannoulis & Ishizaka, 2010). A cut-off level (λ_1) is therefore introduced for the values of credibility. The outranking values that are below λ_1 are then considered irrelevant (being assigned a 0) and the values corresponding to the relevant outranking relations are discovered (being assigned a value of 1) (Canals et al., 2011). With successive thresholds, the cut-off level λ_1 is progressively increased, it becomes much harder for alternative *a'* to be preferable to *a* (Giannoulis & Ishizaka, 2010). Here, three cut-off levels are employed to move the model from a state of weak ($\lambda_1 = 0.4$) to strong ($\lambda_3 = 0.8$), sustainability. The cut-off value of 0.6 shall also be tested.

To rank the alternatives, two indicators are then calculated: the Weakness (*W*) of *a* is the number of alternatives that outrank *a* and the Strength (*S*) of *a* corresponds to the number of alternatives that are outranked by *a*. The difference between S-W provides the final qualification (Canals et al., 2011).

The ELECTRE III computational results and experiments presented in this chapter were performed with an implementation of ELECTRE III in Diviz software (Bigaret & Meyer, 2015), a MCDM support software used in various fields (Preethi & Chandrasekar, 2015). The alternatives, criterions, performances of alternatives (performance table), and weights (relative importance coefficients) are defined in XML input files, and written in XMCDA language.

4. Results

4.1 Weighted-sum rankings

It can be seen that profiles of the ordered results obtained for D_1, D_3 and D_4 show convergence, although there are some exceptions (see Table 4.2). Within the top five highest scoring cities, the following four cities always appear: Stockholm, Zurich, Copenhagen and Munich. Stockholm ranks highest as the most sustainable. Across all four *D*s, Miami consistently scores lowest. A stark difference in order is produced under the weightings of D_2, economic priorities. For example, two American cities (Boston and Washington, DC) move from consistently ranking in the lowest half to ranking in the top eight (fifth and eighth respectively), and cities such as Amsterdam, Edinburgh and Warsaw move from consistent middle rankings (under D_1, D_3 and D_4) to ranking in the lowest seven.

The positioning of Boston and Washington, DC, under D_3 priorities, brings to light a limitation with the weighted-sum approach. Outperformed by 15 cities

Table 4.2 City rankings as predicted by a weighted-sum approach

Decision-maker								
Rank	*Overall SD* (D_1)		*Economic SD* (D_2)		*Environmental SD* (D_3)		*Social SD* (D_4)	
	City	*Score*	*City*	*Score*	*City*	*Score*	*City*	*Score*
1st	Stockholm	0.683	Stockholm	0.702	Stockholm	0.720	Stockholm	0.780
2nd	Zurich	0.623	Zurich	0.650	Zurich	0.646	Zurich	0.712
3rd	Copenhagen	0.595	Munich	0.609	Copenhagen	0.644	Copenhagen	0.671
4th	Munich	0.586	Copenhagen	0.599	Toronto	0.619	Munich	0.651
5th	Toronto	0.563	Boston	0.564	Munich	0.618	Madrid	0.630
6th	Montreal	0.534	Toronto	0.561	Montreal	0.600	Toronto	0.623
7th	Madrid	0.511	Montreal	0.521	Madrid	0.563	Amsterdam	0.616
8th	Amsterdam	0.504	Washington, DC	0.500	Amsterdam	0.555	Montreal	0.591
9th	Warsaw	0.492	Vancouver	0.496	Edinburgh	0.546	Edinburgh	0.582
10th	Vancouver	0.488	Milan	0.492	Warsaw	0.542	Barcelona	0.576
11th	Edinburgh	0.485	Frankfurt	0.490	Rome	0.538	Warsaw	0.574
12th	Boston	0.483	Madrid	0.482	Vancouver	0.515	Vancouver	0.571
13th	Rome	0.482	Warsaw	0.478	Vienna	0.504	Vienna	0.557
14th	Milan	0.473	Amsterdam	0.477	Barcelona	0.495	Rome	0.552
15th	Vienna	0.465	Rome	0.464	Milan	0.489	Milan	0.550
16th	Frankfurt	0.457	Vienna	0.453	Boston	0.481	Frankfurt	0.498
17th	Barcelona	0.453	Edinburgh	0.451	Frankfurt	0.472	Boston	0.460
18th	Washington, DC	0.424	Barcelona	0.424	Washington, DC	0.429	Washington, DC	0.377
19th	Miami	0.262	Miami	0.318	Miami	0.256	Miami	0.232

according to environmental sustainability and by 16 cities under social priorities, the weighted-sum model still allows for Boston to be ranked fifth under economic priorities. Boston's outstanding economic performances compensates for its poor performance on both environmental and social criterion; subsequently, due to compensation within the model Boston is ranked twelfth overall. The use of a MCDA that therefore does not allow compensation of performances among criteria, such as the ELECTRE methods (Ishizaka & Nemery, 2013), is therefore supported.

4.2 ELECTRE III rankings

Generally, the ranking results obtained for each *D* show convergence between the differing sustainability priorities, although there are some exceptions (see Table 4.3). Below the results of each sustainability priority are presented, as defined by the four questions posed by *D* (refer to section 2.4 'The four decision makers').

The influence of the cut-off level on each partial pre-order (see Table 4.3) should also be noted. For each *D*, as the cut-off level increases (from 0.4 to 0.8) it becomes harder for a city to be preferred to an alternate city. Additionally, the comparison between the two partial pre-orders graphically depicts the influence of the cut-off level increasing from 0.7 to 0.75 (see Figure 4.1). Consequently, under the highest cut-level (0.8), a number of outranking relations are eliminated (represented by a qualification of 0) as they are less confident. Cities with a qualification of 0 cannot be compared to the other alternatives, having no outranking relation to them (Canals et al., 2011). Here, a strong sustainability cut-off level is useful when the researcher wishes to define the highest ranking city; however, if the researcher wishes to produce a partial pre-order that includes a large number of the original cities, a lower cut-off level should be selected in order to preserve a greater number of outranking relations.

The partial pre-order in Figure 4.1 expands greatly on the information provided in Table 4.3, which ranks the cities by their ELECTRE III qualification, as it graphically outlines the relationships between all cities (arrows represent an out-ranking relationship). The dependencies seen in the partial pre-order are not captured in the weighted-sum approach. Three of the top performing cities – Munich, Stockholm and Zurich – are incomparable and not out-ranked by any other cities as none of the three out-performed the other two on the 17 indicators to such an extent that the threshold to an effect. Boston is in unique position, in that no city performs well enough to out-rank it but that it only performs well enough to out-rank one city (Miami). This is because Boston out-performs Miami in four out of five Economic indicators, four out of seven Environmental indicators, and three out of five Social with such confidence that at a cut-off level of 0.75, the out-ranking still holds. It should also be noted that Stockholm out-ranks the greatest number of cities, with 13 out-rankings, indicated by the number of arrows. Zurich, Copenhagen and Toronto all out-rank the second greatest number of cities, with all three outranking eight other cities.

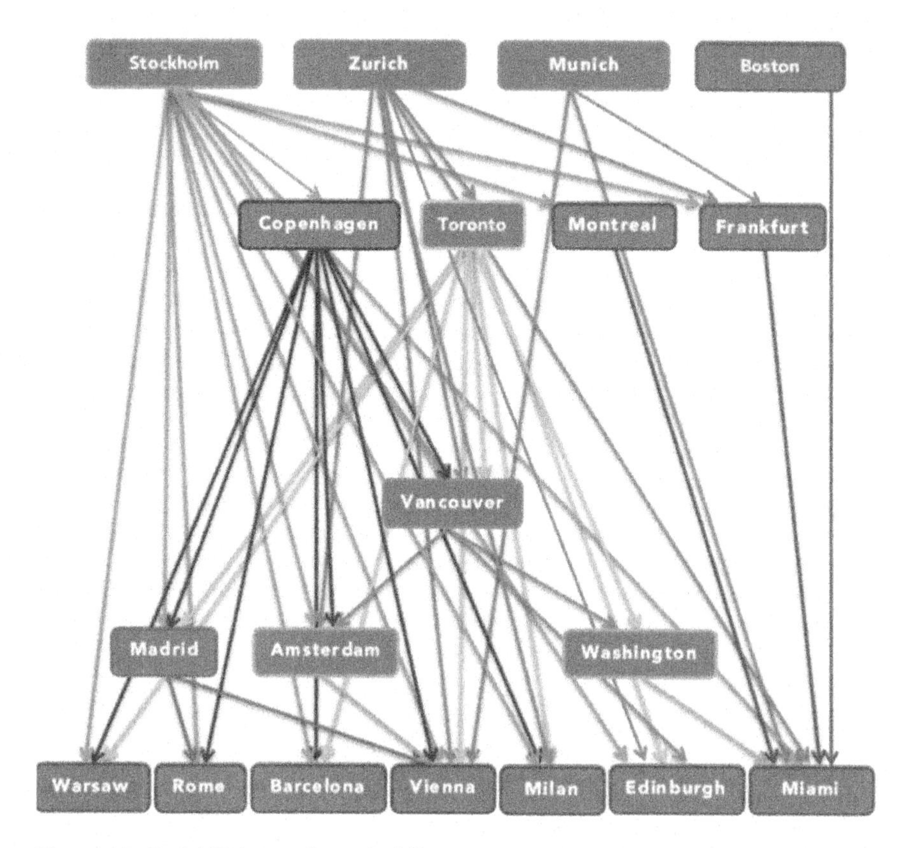

Figure 4.1 The MCDA overall sustainability assessment partial pre-order (cut-off level 0.75)

(a) *Overall sustainability partial pre-order.* With the common set of relative importance coefficients, under the priority of D_1, the application of ELECTRE III leads to the partial pre-order presented in Figure 4.1 (cut-off level 0.75). Under all cut-off levels, Stockholm ranks highest as the overall most sustainable city with Zurich, Munich and Copenhagen consistently ranking in the second, third and fourth positions (see Table 4.3). Similarly, across the three cut-off levels, Miami consistently ranks in the lowest position. In the MCDA partial pre-order, the top six ranked cities are the same as the top six in the weighted-sum rankings.

(b) *Economic sustainability partial pre-order.* Under the priority of D_2, Stockholm does not consistently rank first, rather Munich ranks first under the cut-off level of 0.6 (with Zurich and Stockholm ranking first under 0.4 and 0.6 cut-off levels respectively). Here the high performance of Munich in the economic indicators is evident as the city only ranks third and fourth in both the environmental and social partial pre-orders. Miami, Barcelona, Warsaw and Vienna order within the last four ranks under all cut-off levels.

(c) *Environmental sustainability partial pre-order.* Under the priority of D_3, Stockholm, Zurich and Copenhagen again rank in the top three, with Barcelona, Frankfurt, Miami, and Washington, DC ranking lowest across all cut-off levels.
(d) *Social sustainability partial pre-order.* Under the priority of D_4, similar to the weighted-sum approach, Stockholm, Zurich and Copenhagen consistently rank as highest. The cut-off level's influence is most stark on the social sustainability partial pre-orders with five incomparable cities under a 0.8 cut-off level.

Akin to the weighted-sum approach, the economic sustainability partial pre-order is the most different on comparison of all partial pre-orders. Under ELECTRE III (cut-off level of 0.6), however, the American cities do not rank as highly, sitting in the lower half of the rankings. As no compensation between criteria is allowed in the model, Boston orders eighth and Washington, DC ninth.

4.3 Robustness test

Here, the analysis of the 'robustness' of the conclusions is presented, through the examination of the effect of various – apparently arbitrary (Roy et al., 1986) – choices. Here, the term 'robust' refers to the capacity for withstanding imprecise approximations and/or areas of ignorance to prevent undesirable impressions, notably the degradation of the model output (Roy, 2010). The analysis of 'robustness' is therefore an analysis of the resistance or self-protection of the model's conclusions (city rankings) against fluctuations in model parameters. For each variable, the aim is to assess the model's overall stability, and justify in which cases the choice of numerical value is likely to significantly affect the output and subsequent recommendations of the model.

Influence of 'indifference' and 'preference' thresholds on the partial pre-order. The model appears to be relatively robust to fluctuations in both the indifference and preference threshold: neither threshold, when increased or decreased by 10 and 20 per cent, significantly affects the output (see Table 4.4). Consequently, changes in these thresholds do not alter the recommendations of the model, regarding which city is most sustainable.

5. Discussion

Across the spectrum of decision-maker priorities, Stockholm received the highest ranking both overall and under environmental and social policy priorities, and thus is the most sustainable city of the 19 cities assessed in this study. This result is supported by Stockholm's prior acknowledgements as a sustainable city. In 2009, Stockholm ranked second to Copenhagen in the European Green City Index. The index evaluates 16 quantitative and 14 qualitative indicators to assess both a city's current environmental performance as well as its intentions to become greener (Economist Intelligence Unit, 2009). Following the Green City Index, in 2010 Stockholm was the first city to be named 'European Green

Table 4.3 City rankings as predicted by ELECTRE III

	Decision-maker											
	Overall SD (D_1)			*Economic SD* (D_2)			*Environmental SD* (D_3)			*Social SD* (D_4)		
Cut-off level	*0.4*	*0.6*	*0.8*	*0.4*	*0.6*	*0.8*	*0.4*	*0.6*	*0.8*	*0.4*	*0.6*	*0.8*
Rank by qualification												
1st	Stockholm (14)	Stockholm (16)	Stockholm (10)	Zurich (15)	Munich (15)	Stockholm (10)	Stockholm (14)	Stockholm (16)	Stockholm (6)	Stockholm (15)	Stockholm (16)	Stockholm (11)
2nd	Munich (12) Zurich (12)	Copenhagen (15)	Zurich (5)	Munich (14) Stockholm (14)	Zurich (14)	Toronto (5) Zurich (5)	Zurich (12)	Copenhagen (14)	Zurich (5)	Zurich (12)	Copenhagen (14) Zurich (14)	Zurich (8)
3rd	Toronto (11)	Zurich (13)	Copenhagen (3) Munich (3)	Copenhagen (11) Toronto (11)	Stockholm (13)	Copenhagen (3)	Copenhagen (11) Toronto (11)	Zurich (12)	Copenhagen (4) Toronto (4)	Copenhagen (10) Munich (10) Toronto (10)	Munich (10) Toronto (10)	Copenhagen (3) Munich (3)
4th	Copenhagen (10)	Munich (12)	Toronto (2)	Montreal (6)	Toronto (11)	Munich (2)	Munich (10)	Munich (11) Toronto (11)	Munich (2)	Montreal (8)	Montreal (5)	Madrid (2) Toronto (2)
5th	Montreal (8)	Toronto (10)	Boston (1) Madrid (1)	Boston (2)	Copenhagen (9)	Boston (1) Milan (1) Washington, DC (1)	Montreal (9)	Montreal (7)	Boston (1) Madrid (1) Montreal (1)	Edinburgh (2)	Amsterdam (1)	Amsterdam (0) Boston (0) Milan (0) Montreal (0) Vancouver (0)
6th	Edinburgh (0)	Montreal (5)	Milan (0) Vancouver (0) Washington, DC (0)	Edinburgh (−1)	Montreal (5)	Madrid (0) Vancouver (0)	Edinburgh (1)	Edinburgh (3)	Amsterdam (0) Milan (0)	Amsterdam (−1) Madrid (−1) Warsaw (−1)	Edinburgh (0) Madrid (0)	Edinburgh (−2) Frankfurt (−2) Rome (−2)
7th	Madrid (−2) Rome (−2) Vancouver (−2)	Boston (−1)	Amsterdam (−1) Frankfurt (−1) Montreal (−1) Rome (−1)	Vancouver (−2)	Vancouver (2)	Edinburgh (−2) Frankfurt (−2) Montreal (−2) Rome (−2)	Madrid (−1) Rome (−1) Warsaw (−1)	Amsterdam (−1)	Edinburgh (−1) Frankfurt (−1) Rome (−1) Vancouver (−1) Washington, DC (−1)	Rome (−2) Vienna (−2)	Warsaw (−3)	Barcelona (−3) Warsaw (−3) Washington, DC (−3)

(continued)

Table 4.3 (continued)

Cut-off level	*Decision-maker*											
	Overall SD (D_1)			*Economic SD (D_2)*			*Environmental SD (D_3)*			*Social SD (D_4)*		
	0.4	*0.6*	*0.8*	*0.4*	*0.6*	*0.8*	*0.4*	*0.6*	*0.8*	*0.4*	*0.6*	*0.8*
8th	Amsterdam (−3) Boston (−3) Warsaw (−3)	Madrid (−2) Vancouver (−2)	Edinburgh (−2)	Amsterdam (−4) Frankfurt (−4)	Boston (1)	Warsaw (−3)	Amsterdam (−2)	Madrid (−2) Rome (−2)	Barcelona (−3) Warsaw (−3)	Vancouver (−3)	Vancouver (−4)	Vienna (−5)
9th	Vienna (−4)	Edinburgh (−3) Rome (−3)	Barcelona (−3) Warsaw (−3)	Madrid (−5) Rome (−5) Vienna (−5) Warsaw (−5)	Washington, DC (−3)	Amsterdam (−4) Barcelona (−4) Miami (−4)	Vancouver (−3)	Boston (−3)	Vienna (−5)	Boston (−6) Milan (−6)	Boston (−5) Rome (−5) Vienna (−5)	Miami (−9)
10th	Frankfurt (−6)	Amsterdam (−4)	Vienna (−6)	Milan (−8) Washington, DC (−8)	Edinburgh (−4) Frankfurt (−4) Madrid (−4)	Vienna (−5)	Boston (−5) Vienna (−5)	Warsaw (−4)	Miami (−8)	Barcelona (−7) Frankfurt (−7)	Barcelona (−6)	
11th	Barcelona (−8) Milan (−8)	Frankfurt (−6) Vienna (−6) Washington, DC (−6)	Miami (−7)	Barcelona (−10)	Amsterdam (−5) Milan (−5)		Frankfurt (−6)	Vancouver (−5)		Washington, DC (−14)	Milan (−7)	
12th	Washington, DC (−10)	Milan (−7) Warsaw (−7)		Miami (−16)	Rome (−7) Vienna (−7)		Barcelona (−8)	Frankfurt (−7) Vienna (−7) Washington, DC (−7)		Miami (−17)	Frankfurt (−8)	
13th	Miami (−16)	Barcelona (−8)			Barcelona (−8) Warsaw (−8)		Milan (−10) Washington, DC (−10)	Barcelona (−9)			Washington, DC (−10)	
14th		Miami (−16)			Miami (−15)		Miami (−16)	Milan (−11)			Miami (−17)	
15th								Miami (−16)				

Table 4.4 City qualifications with robustness testing (overall sustainability, cut-level 0.6)

Threshold	*Indifference*				*Preference*			
Rank	*+ 20%*	*+ 10%*	*– 10%*	*– 20%*	*+ 20%*	*+ 10%*	*– 10%*	*– 20%*
1st	Stockholm (16)	Stockholm (16)	Stockholm (16)	Copenhagen (16) Stockholm (16)	Stockholm (16)	Stockholm (16)	Stockholm (16)	Stockholm (16)
2nd	Copenhagen (14)	Copenhagen (14)	Copenhagen (15)	Zurich (13)	Zurich (14)	Zurich (14)	Copenhagen (14)	Copenhagen (15)
3rd	Zurich (13)	Zurich (13)	Zurich (13)		Copenhagen (13)	Copenhagen (13)	Zurich (13)	Zurich (13)
4th	Munich (12)	Munich (12)	Munich (12)	Munich (12)	Munich (12)	Munich (12)	Munich (12)	Munich (12)
5th	Toronto (11)	Toronto (11)	Toronto (10)	Toronto (10)	Toronto (11)	Toronto (11)	Toronto (11)	Toronto (11)
6th	Montreal (5)	Montreal (5)	Montreal (5)	Montreal (4)	Montreal (6)	Montreal (5)	Montreal (5)	Montreal (4)
7th	Boston (–1) Edinburgh (–1) Vancouver (–1)	Boston (–1) Vancouver (–1)	Boston (–1)	Boston (–1) Madrid (–1)	Boston (–1) Edinburgh (–1) Vancouver (–1)	Edinburgh (0)	Boston (–1)	Boston (–1) Madrid (–1)
8th	Madrid (–3)	Madrid (–2)	Madrid (–2) Vancouver (–2)	Vancouver (–2)	Madrid (–3)	Boston (–1)	Edinburgh (–2) Madrid (–2)	Edinburgh (–2) Vancouver (–2)
9th	Amsterdam (–4) Rome (–4)	Edinburgh (–3)	Edinburgh (–3) Rome (–3)	Edinburgh (–3) Rome (–3)	Amsterdam (–4)	Vancouver (–2)	Vancouver (–3)	Amsterdam (–4) Rome (–4)
10th	Frankfurt (–6) Vienna (–6) Washington, DC (–6)	Amsterdam (–4) Rome (–4)	Amsterdam (–5)	Amsterdam (–6) Frankfurt (–6) Vienna (–6)	Frankfurt (–5) Rome (–5)	Amsterdam (–3) Madrid (–3)	Rome (–4)	Frankfurt (–5)

(continued)

Table 4.4 (continued)

Threshold	Indifference				Preference			
Rank	+ 20%	+ 10%	– 10%	– 20%	+ 20%	+ 10%	– 10%	– 20%
				Warsaw (–6) Washington, DC (–6)				
11th	Milan (–7) Warsaw (–7)	Frankfurt (–6) Vienna (–6) Washington, DC (–6)	Frankfurt (–6) Vienna (–6) Warsaw (–6) Washington, DC (–6)	Milan (–7)	Warsaw (–6)	Rome (–5)	Amsterdam (–5) Frankfurt (–5)	Washington, DC (–6)
12th	Barcelona (–9)	Milan (–7) Warsaw (–7)	Milan (–7)	Barcelona (–8)	Barcelona (–7) Washington, DC (–7)	Frankfurt (–6) Washington, DC (–6)	Vienna (–6) Warsaw (–6) Washington, DC (–6)	Vienna (–7) Warsaw (–7)
13th	Miami (–16)	Barcelona (–8)	Barcelona (–8)	Miami (–16)	Milan (–8) Vienna (–8)	Barcelona (–7) Vienna (–7) Warsaw (–7)	Milan (–7)	Barcelona (–8) Milan (–8)
14th		Miami (–16)	Miami (–16)		Miami (–16)	Milan (–8)	Barcelona (–8)	Miami (–16)
15th						Miami (–16)	Miami (–16)	

Capital'. The Green Capital Award aims to recognise Europe's 'greenest' city based on its performance in relation to ten indicators (European Green Capital, 2010). Both the Index and Award assess future aims and policy intentions in their evaluations, it is therefore logical that this assessment five to six years later would begin to manifest – and thus rank highly – the implications of Stockholm's forward-thinking policies.

Further inspection of current sustainability efforts by the City of Stockholm highlights the importance placed both on the environment and sustainability by the Stockholm Stad (the municipality of the City of Stockholm). The City of Stockholm's Environmental Program for 2012–2015, for example, is based on the vision of Stockholm as a growing attractive city, where the needs of people and the environment complement each other. The Program focuses on six key priority areas, four of which coincide with indicators used in this study: efficient transport, goods and buildings free of dangerous substances, sustainable energy use, sustainable use of land and water, waste treatment with minimal environmental impact, and a healthy indoor environment (City of Stockholm, 2012). At the same time, Stockholm exhibits a very strong economic performance with Sweden ranked sixth in the 2016–2017 World Economic Forum's Global Competitiveness Index, attracting around 4.7% of GDP per annum in foreign direct investment (which exceeds the level of the US, Japan and Brazil) and investing around 3.7% of its GDP in research in development (compared to EU average of 1.8%). The city progressively aims to be fossil-fuel free by 2050 (Stockholm Stad, 2013) and is equally as focused on green urban mobility (City of Stockholm Traffic Administration, 2012). Stockholm and Sweden were able to decouple economic development from the growth in CO_2 emissions due to extensive use of hydropower and nuclear energy as well as creative use of environmental taxes (Sweden was one of the first countries to introduce a carbon tax in 1991). With regards to targets related to wider social and economic issues, in 2007 Stockholm City Council set a collective, long-term ambition of 'Vision 2030 – A World-Class Stockholm'. The vision focuses on three areas of development – 'versatile and full of experiences', 'innovation and growth' and 'citizens' (Stockholm City Council, 2007). The outcome of this study is therefore supported by the recognition of Stockholm in sustainability awards and by policies and targets, past and present, working towards sustainable development for both the environment and the people.

The results of this study highlight the non-compensatory nature of the ELECTRE methods, one of the their main characteristics and advantages for stronger sustainability policy making due to avoiding compensation between criteria (Ishizaka & Nemery, 2013). Following the above analysis of the same family of alternatives and criteria, it can be concluded that the weighted-sum approach and ELECTRE III MCDA method produce ranked orders with some similar characteristics (namely the highest ranking cities); however, one major limitation of the weighted-sum approach is that of compensation and thus it can be viewed as a weak sustainability decision-making method. ELECTRE methods have been conceived in a manner that does not allow for compensation of performances among criteria, with the existence of the veto threshold strengthening the non-compensatory effect (Figueira et al., 2013). In the context of the sustainable city – where the interrelationships and interdependences between environmental, social and economic actors must be considered holistically (Barbosa et al., 2014; Michael et al., 2014) – the possibility of offsetting a 'disadvantage' on some criteria by a sufficiently large 'advantage' on another criteria, such as a loss on social sustainability being counteracted by a gain on environmental, should be deemed unacceptable to the decision maker (Coveos, 1996; Figueira et al., 2005). Another feature of a non-compensatory procedure of particular relevance to a study of sustainable cities is that, within the context of complex decision processes, the decision maker is not forced to express trade-offs (for example, to favour life expectancy over unemployment) as the model only requires 'inter-attribute' information in terms of an importance relation and discordance set (Coveos, 1996). Consequently, the structuring and appraisal of the city system requires the use of non-compensatory aggregation procedures, such as the employment of the ELECTRE III MCDA in this study.

From a policy perspective, the dynamic relationships between economic, environmental and social urban dimensions make it difficult to track problems and to develop both short-term reactive plans and longer-term strategies (Button, 2002). Indicators are able to measure changes over time and provide critical real-world feedback – required by decision -makers, governments and organisations to enable more effective and directed choices to improve urban sustainability (Guy & Kibert, 1998). This study amply illustrates that a robust set of urban sustainability indicators based on SDGs and Smart and Sustainable approaches is a viable policy tool for sustainability assessment in keeping the balance between the availability of data and policy relevance. As it is accepted by ecological economists that decision makers need to adopt a strong sustainability position when discussing and implementing sustainable development policies (Pelenc & Ballet, 2015), this chapter proposes the application of a non-compensatory multi-criteria aid ELECTRE III as a strong sustainability-focused decision support tool.

6. Conclusion

This chapter presents a family of 17 urban sustainability indicators, based on UN Sustainable Development Goals, representing economic, social and environmental sustainability priorities, and sets out a methodology for the assessment of metropolitan sustainability. In doing so, it provides an answer to determine the level of success of 'sustainable cities' (Shane & Graedel, 2000) and demonstrates how cities compare with others across the globe. The multi-criteria aid allows a ranking of the various cities based on urban sustainability indicator evaluation. The chapter highlights Stockholm as the most successful sustainable city among the cities under consideration. This chapter supports the use of a multi-criteria decision aid in order to assess the complex and dynamic entity that is a city (Button, 2002). It is also concluded that total compensation among criteria is a major limitation of the weighted-sum approach, and therefore the non-compensatory qualities of the ELECTRE III method are a tremendous advantage for stronger sustainability analysis. The policy relevance of the results presented in this chapter stems from the fact that a tried and tested set of indicators we proposed, largely based on SDGs, could be used as an assessment framework by cities around the world for benchmarking against the cities presented here. Decision makers could learn from the policies, governance structures, and technological innovations that have led to Stockholm's success as opposed to directly replicating the pathways that led Stockholm to sustainability. The analysis presented in this chapter, therefore, not only assesses the sustainability of 19 global cities but also provides decision makers with a practical understanding of how to assess and benchmark their own city's sustainability.

References

Barbosa, J. A., Bragança, L., & Mateus, R. (2014). New approach addressing sustainability in urban areas using sustainable city models. *International Journal of Sustainable Building Technology and Urban Development*, *5*(4), 297–305. http://doi.org/10.1080/2093761X.2014.948528

Belton, V., & Stewart, T. J. (2002). *Multiple Criteria Decision Analysis: An Integrated Approach*. Kluwer Academic Publishers.

Bigaret, S., & Meyer, P. (2015). Decision Deck homepage. Retrieved March 2, 2016. www.decision-deck.org/diviz/

Bottero, M., Ferretti, V., Figueira, J. R., Greco, S., & Roy, B. (2015). Dealing with a multiple criteria environmental problem with interaction effects between criteria through an extension of the Electre III method. *European Journal of Operational Research*, *245*(3), 837–850. http://doi.org/10.1016/j.ejor.2015.04.005

Brand, F. (2009). Critical natural capital revisited: Ecological resilience and sustainable development. *Ecological Economics*, *68*, 605–612.

Button, K. (2002). City management and urban environmental indicators. *Ecological Economics*, *40*(2), 217–233. http://doi.org/10.1016/S0921-8009(01)00255-5

C40 Cities. (2016). The Power of C40 Cities. www.c40.org/cities/

Canals Ros, J., Supervisor, R., Mateu, V., Mar, L., Deim-rt-, I., Inform, E., . . . & This, V. (2011). *Introduction to Decision Deck-Diviz: Examples and User Guide*. www.diviz.org/_static/ReportDecisionDeck-DEIM-URV.pdf

CDP. (2015). 2015 Cities Electricity Mix Map. https://data.cdp.net/Cities/2015-Cities-Electricity-Mix-Map/kwjr-j78z/43

Choon, S.-W., Siwar, C., Pereira, J. J., Jemain, A. A., Hashim, H. S., & Hadi, A. S. (2011). A sustainable city index for Malaysia. *International Journal of Sustainable Development & World Ecology*, *18*(1), 28–35. http://doi.org/10.1080/13504509.2011.543012

City of Copenhagen. (2014). Copenhagen, European Green Capital 2014: A review. http://ec.europa.eu/environment/europeangreencapital/wp-content/uploads/2013/02/Copenhagen-Post-Assessment-Report-2014-EN.pdf

City of Stockholm. (2012). The Stockholm Environment Programme 2012–2015. *City of Stockholm*. https://international.stockholm.se/.../the-stockholm-environment-programme-2012-20...

City of Stockholm. (2015). The First European Green Capital. http://ec.europa.eu/environment/europeangreencapital/wp-content/uploads/2011/04/Stockholm-First-European-Green-Capital-.pdf

City of Stockholm Traffic Administration. (2012). Urban mobility strategy, 72.

Corrente S. et al. (2017). A robust ranking method extending ELECTRE III to hierarchy of interacting criteria, imprecise weights and stochastic analysis. *Omega*, http://dx.doi.org/10.1016/j.omega.2016.11.008i

Coveos, C. M. (1996). Some remarks on the notion of compensation in MCDAM. *Philosophical Investigations*, *19*(4), 308–317. http://doi.org/10.1111/j.1467-9205.1996.tb00422.x

Dias, L. C., & Mousseau, V. (2005). Inferring Electre's veto-related parameters from outranking examples. *European Journal of Operational Research*, *170*(1), 172–191. http://doi.org/10.1016/j.ejor.2004.07.044

Economist Intelligence Unit. (2009). *European Green City Index*. Munich. https://w3.siemens.no/home/no/no/presse/Documents/European_Green_City_Index.pdf

Egilmez, G., Gumus, S., & Kucukvar, M. (2015). Environmental sustainability benchmarking of the U.S. and Canada metropoles: An expert judgment-based multi-criteria decision making approach. *Cities*, *42*, 31–41. http://doi.org/10.1016/j.cities.2014.08.006

European Commission. (2016). The EU Emissions Trading System (EU ETS). http://ec.europa.eu/clima/policies/ets/index_en.htm

European Green Capital. (2010). *European Green Capital 2010*. Luxembourg.

European Union. (2013). Decision No 1386/2013/EU of the European Parliament and of the Council of 20 November 2013 on a General Union Environment Action Programme to 2020 'Living well, within the limits of our planet.' *Official Journal of the European Union*, (1600), 171–200. http://doi.org/10.2779/57220

Eurostat. (2013). *Sustainable Development in the European Union*. Luxembourg. http://bookshop.europa.eu/en/sustainable-development-in-the-european-union-pbKS0213237/

Figueira, J. R., & Roy, B. (2002). Determining the weights of criteria in the ELECTRE type methods with a revised Simos' procedure. *European Journal of Operational Research*, *139*, 317–326.

Figueira, J., Mousseau, V., & Roy, B. (2005). Electre Methods. In C. A. Bana E Costa, J.-M. Corte, & J.-C. Vansnick (eds), *Multiple Criteria Decision Analysis: State of the Art Surveys* (pp. 133–162). New York: Springer Science & Business Media, Inc. http://doi.org/Doi 10.1007/0-387-23081-5_4

Figueira, J. R., Greco, S., Roy, B., & Slowinski, R. (2013). An overview of Electre methods and their recent extensions. *Journal of MultiCriteria Decision Analysis*, *20*, 61–85. http://doi.org/10.1002/mcda.1482

GaWC (Globalization and World Cities Research Network). (2012). The world according to GaWC. www.lboro.ac.uk/gawc/world2012t.html

Giannoulis, C., & Ishizaka, A. (2010). A Web-based decision support system with Electre III for a personalised ranking of British universities. *Decision Support Systems*, *48*(3), 488–497. http://doi.org/10.1016/j.dss.2009.06.008

Girardet, H. (2004). *Cities People Planet: Liveable Cities for a Sustainable World*, John Wiley and Sons.

Girardet, H. (2017). Regenerative cities, in Shmelev, S. E., ed. (2017) *Green Economy Reader: Lectures in Ecological Economics and Sustainability*, Springer.

Girardet, H., & Mendonca, M. (2009). *A Renewable World: Energy, Ecology, Equality – A Report for the World Future Council*, Green Books.

Govindan, K., & Jepsen, M. B. (2016). Electre: A comprehensive literature review on methodologies and applications. *European Journal of Operational Research*, *250*(1), 1–29. http://doi.org/10.1016/j.ejor.2015.07.019

Guitouni, A., & Martel, J.-M. (1998). Tentative guidelines to help choosing an appropriate MCDA method. *European Journal of Operational Research*, *109*(2), 501–521. http://doi.org/10.1016/S0377-2217(98)00073-3

Guy, G. B., & Kibert, C. (1998). Developing indicators of sustainability: US experience. *Building Research & Information*, *26*(1), 39–45. http://doi.org/10.1080/096132198370092

Han, J., Kamber, M., & Pei, J. (2012). Data preprocessing. In *Data Mining: Concepts and Techniques* (3rd ed., pp. 83–123). Morgan Kaufmann.

Hiremath, R. B., Balachandra, P., Kumar, B., Bansode, S. S., & Murali, J. (2013). Indicator-based urban sustainability: A review. *Energy for Sustainable Development*, *17*(6), 555–563. http://doi.org/10.1016/j.esd.2013.08.004

Ishizaka, A., & Nemery, P. (2013). Electre, in *Multi-criteria Decision Analysis Methods and Software*. John Wiley & Sons, Ltd. http://doi.org/10.1002/9781118644898.ch7

ISO (International Organization for Standardization). (2014). ISO 37120: 2014 Sustainable development of communities: Indicators for city services and quality of life. www.iso.org/obp/ui/#iso:std:iso:37120:ed-1:v1:en

Lazaroiu, G. C., & Roscia, M. (2012). Definition methodology for the smart cities model. *Energy*, *47*(1), 326–332. http://doi.org/10.1016/j.energy.2012.09.028

Martin, C., Ruperd, Y., & Legret, M. (2007). Urban stormwater drainage management: The development of a multicriteria decision aid approach for best management practices. *European Journal of Operational Research*, *181*(1), 338–349. http://doi.org/10.1016/j.ejor.2006.06.019

Miami-Dade County. (2010). *Climate Change Action Plan*. www.miamidade.gov/greenprint/pdf/climate_action_plan.pdf

Miami-Dade Water & Sewer Department. (2015). *Water Conservation Plan 2015 Annual Report*. Miami-Dade County. www.miamidade.gov/waterconservation/library/reports/2015-water-conservation-plan.pdf

Michael, F. L., Noor, Z. Z., & Figueroa, M. J. (2014). Review of urban sustainability indicators assessment: Case study between Asian countries. *Habitat International*, *44*, 491–500. http://doi.org/10.1016/j.habitatint.2014.09.006

Munda, G. (1995). *Multicriteria Evaluation in a Fuzzy Environment*. Physica-Verlag HD. http://doi.org/10.1007/978-3-642-49997-5

OECD (Organisation for Economic Co-operation and Development). (2016). OECD.Stat. http://stats.oecd.org/Index.aspx?datasetcode=REG_DEMO_TL2#

Parilla, J., Trujillo, J. L., Berube, A., & Ran, T. (2014). 2014 Global Metro Monitor Map. www.brookings.edu/research/reports2/2015/01/22-global-metro-monitor

Pelenc, J., & Ballet, J. (2015). Weak sustainability versus strong sustainability. *Brief for GSDR 2015*. https://sustainabledevelopment.un.org/content/documents/6569122-Pelenc-Weak%20Sustainability%20versus%20Strong%20Sustainability.pdf

Preethi, G. A., & Chandrasekar, C. (2015). Seamless Handoff using Electre III and Promethee Methods. *International Journal of Computer Applications*, *126*(13), 32–38.

Rogers, M., & Bruen, M. (1998). Choosing realistic values of indifference, preference and veto thresholds for use with environmental criteria within Electre. *European Journal of Operational Research*, *107*(3), 542–551. http://doi.org/10.1016/S0377-2217(97)00175-6

Roy, B. (1989). Main sources of inaccurate determination, uncertainty and imprecision in decision models. *Mathematical and Computer Modelling*, *12*(10–11), 1245–1254. http://doi.org/10.1016/0895-7177(89)90366-X

Roy, B. (1991). The outranking aproach and the foundations of Electre methods. *Theory and Decision*, *31*, 49–73. http://doi.org/10.1007/BF00134132

Roy, B. (1996). Problematics as guides in decision aiding. In *Multicriteria Methodology for Decision Aiding: Nonconvex Optimization and Its Applications*, vol 12 (pp. 57–74). Springer. http://doi.org/10.1007/978-1-4757-2500-1_6

Roy, B. (2005). Paradigms and challenges, in *Multiple Criteria Decision Analysis: State of the Art Surveys* (Vol. 1, pp. 3–24). http://doi.org/10.1007/0-387-23081-5_1

Roy, B. (2010). Robustness in operational research and decision aiding: A multi-faceted issue. *European Journal of Operational Research, 200*(3), 629–638. http://doi.org/10.1016/j.ejor.2008.12.036

Roy, B., & Słowiński, R. (2013). Questions guiding the choice of a multicriteria decision aiding method. *EURO Journal on Decision Processes, 1*, 69–97. http://doi.org/10.1007/s40070-013-0004-7

Roy, B., Présent, M., & Silhol, D. (1986). A programming method for determining which Paris metro stations should be renovated. *European Journal of Operational Research, 24*(2), 318–334. http://doi.org/10.1016/0377-2217(86)90054-8

Roy, B., Figueira, J. R., & Almeida-Dias, J. (2014). Discriminating thresholds as a tool to cope with imperfect knowledge in multiple criteria decision aiding: Theoretical results and practical issues. *Omega* , *43*, 9–20. http://dx.doi.org/10.1016/j.omega.2013.05.003

Shane, A. M., & Graedel, T. E. (2000). Urban environmental sustainability metrics: A provisional set. *Journal of Environmental Planning and Management, 43*(5), 643–663. http://doi.org/10.1080/713676586

Shmelev, S. E. (2017). Multidimensional sustainability assessment for megacities, in Shmelev S. E., ed., *Green Economy Reader: Lectures in Ecological Economics and Sustainability*, Springer. http://doi.org/10.1017/CBO9781107415324.004

Shmelev, S. E., & Shmeleva, I. A. (2009). Sustainable cities: Problems of integrated interdisciplinary research. *International Journal of Sustainable Development, 12*(1), 4. http://doi.org/10.1504/IJSD.2009.027526

Shmelev, S. E., & Shmeleva, I. A. (2018) Sustainable development assessment: A multidimensional approach for smart and sustainable cities. *Sustainable Development, 26*(6), 904–920.

Stockholm City Council. (2007). *A Guide to the Future*. Stockholm. www.jstor.org/stable/458906?origin=crossref

Stockholm Stad. (2013). Roadmap for a fossil fuel-free Stockholm 2050. www.stockholm.se/.../Roadmap%20for%20a%20fossil%20fuel-free%20Stockholm%2...

The, A. N., & Mousseau, V. (2002). Using assignment examples to infer category limits for the Electre TRI method. *Journal of Multi-Criteria Decision Analysis, 11*(1), 29–43. http://doi.org/10.1002/mcda.314

United Nations. (2015). *Transforming Our World: The 2030 Agenda for Sustainable Development*. https://sustainabledevelopment.un.org/content/documents/21252030%20Agenda%20for%20Sustainable%20Development%20web.pdf

United Nations Economic Commission for Europe. (2017). *Collection Methodology for Key Performance Indicators for Smart Sustainable Cities*, Geneva. www.unece.org/fileadmin/DAM/hlm/documents/Publications/U4SSC-CollectionMethodologyforKPIfoSSC-2017.pdf

United Nations Human Settlements Programme. (2009). *Planning Sustainble Cities: Policy Directions: Global Report on Human Settlements 2009*. Earthscan.

Whitehead, M. (2003). (Re)Analysing the sustainable city: Nature, urbanisation and the regulation of socio-environmental relations in the UK. *Urban Studies, 40*(7), 1183–1206. http://doi.org/10.1080/0042098032000084550

WHO (World Health Organisation). (2016). Urban population growth. www.who.int/gho/urban_health/situation_trends/urban_population_growth_text/en/

World Commission on Environment and Development. (1987). *Our Common Future*. New York.

5 Multidimensional sustainability assessment for the cities of the Global South

The PROMETHEE approach

Bei Zhang and Stanislav E. Shmelev

1. Introduction

1.1 Sustainable development

The modern concept of sustainable development was introduced in the *Our Common Future* report, released by the United Nations Commission on Environment and Development (WCED, 1987, p.41). Daly (1999) distinguished the concepts of 'development' and 'growth', referring to the former as improvement and to the latter as expansion. He described sustainable development as a process of social improvement in which economic growth does not exceed the environmental carrying capacity.

Inevitably, there is a range of impacts, such as environmental and societal, as a result of economic development. In this respect, the concept of sustainable development accounts for both short- and long-term effects in the achievement of economic, environmental and social goals (Shmelev & Shmeleva, 2009; Shmelev, 2017c; Shmelev & Shmeleva, 2018). Therefore, sustainable development is a multidimensional issue, which considers environmental, economic and social impacts simultaneously.

Urban sustainability has garnered significant attention since the Rio Summit of 1992 (Shmelev, 2017c). Since the conference took place, the potential of international cooperation in achieving sustainability has become evident and cities are thought to have a key role in addressing environmental problems. A sustainable city harmoniously integrates economic, social and environmental aspects in its development. It is a complex and dynamic system, which connects various objects and processes and is integral to day-to-day life.

Research on sustainable urban development issues has been conducted across several areas: energy, transportation, material flows and urban metabolism, landscape, green space, economic activity and planning, environmental design and architecture, environmental psychology, public participation and globalization impacts (Shmelev & Shmeleva, 2009). Scott (2012) asserted that third-wave urbanization can lead to a range of urban dilemmas and social issues such as environmental pollution, land use problems and transportation problems. Hall (2013) summarized the five challenges that cities now face: (1) how to provide enough jobs and promote economic growth; (2) how to meet demand for housing; (3) how to solve travel needs efficiently through land use planning and transportation; (4) how to limit anthropogenic environmental impacts and resource consumption and (5) how liaisons between public and private agencies in city development should take place. The author illustrates several cases of how urban policies and planning interventions, based on the real-world understanding of the problems of daily life, lead to urban success. Based on the empirical findings on the corporate hierarchy and networks, as well as the flow of information of polycentric regions in Europe, Hall and Pain (2006) also presented an analysis on polycentric metropolis which can be viewed as a blueprint for urban development.

According to United Nations statistics, developing countries are expected to face extraordinary urbanization in the coming years (UNDESA/PD, 2012). The United Nations Conference on Housing and Sustainable Urban Development (Habitat III) was held in Quito in 2016. A new Agenda, which set global standards for sustainable urban development was adopted at this conference. The Agenda aims to achieve no poverty and zero hunger, reduce the social and gender inequalities, promote sustainable economic growth, improve health and protect the environment. As part of its list of actions, the Agenda acknowledges that developing countries are facing unique and emerging challenges in their urban development. Therefore, it is necessary to recognize the importance of urban sustainability in developing countries. While much of the research in this field has focused primarily on the success of developed counties in urban sustainability, this chapter aims to shed light on the reasons behind both the success and failure of Global South cities in sustainability. Seventeen cities are selected for comparison with respect to the sustainability indicators. The cities are: Istanbul, Mumbai, Delhi, Shanghai, Beijing, Shenzhen, Almaty, Johannesburg, Kampala, Nairobi, Mexico City, Bogota, Buenos Aires, Lima, São Paulo, Rio de Janeiro and Quito. Ranking these cities by their sustainability allows the most and least sustainable cities to be identified. Since Global North cities and Global South cities use different strategies and policies to achieve sustainability, the experiences of sustainable cities in the Global South area provide a more suitable basis from which other Global South cities can learn.

1.2 Urban sustainability analysis methods

Several well-developed methods have been used in urban sustainability analysis. Material flows analysis (MFA), quantifying the stocks and flows of metals in a certain timescale in a well-defined system (Müller et al., 2014), has been applied to solve city-scale problems. For example, Rosado et al. (2014) accounted for the material stocks and flows of the Lisbon Metropolitan Area and analyzed the correlation between material flows and economic activities in this area. Optimization is a mathematical method to identify the inputs for which a function achieves its minimum or maximum value (Banos et al., 2011). This method can help decision makers choose the most suitable alternative to meet their goals. Chang et al. (2012) applied this method to choose the most appropriate municipal solid waste management strategy in Taichung City, considering both economic factors and environmental impacts. Multi-criteria decision aid (MCDA) methods, which could help decision makers find the best compromise alternative between conflicting goals, are widely applied in urban sustainability analysis. Munda (2005) provided the first application in comparing cities by sustainability assessment through MCDA methods. Shmelev (2017c) applied three MCDA methods (ELECTRE III, NAIADE and APIS) to assess the urban sustainability of twelve large cities in the world.

A city can be viewed as a system of interacting dimensions, such as energy, transport, etc. The sustainability of cities reflects how the city operates in economic, environmental and social dimensions simultaneously (Shmelev, 2017c). Moreover, in the case of sustainable development decision making, there are usually conflicting economic, environmental and societal objectives involved (De Montis et al., 2005). MCDA methods can evaluate alternatives based on conflicting criteria, which allows them to be widely used to solve sustainability issues. Therefore, due to the multidimensional nature of urban sustainability, multi-criteria decision aid (MCDA) methods are applied in this chapter to compare and assess the sustainability of Global South cities.

1.3 Multi-criteria decision aid (MCDA)

It is important to define two concepts in a multi-criteria analysis: alternative and criterion. In a multi-criteria problem, alternative represents an option, a set of actions or an object to be compared to find the best compromise solution to this problem (Shmelev, 2017b). It has different characteristics that could be evaluated by a set of criteria. In this study, different cities are compared to obtain the ranking of the cities in terms of their sustainability. Therefore, the alternatives will be cities in this study. Moreover, criteria are defined as values, attributes or objectives that can measure the performance of different alternatives. In this study, the criterion represents the sustainability indicators such as GDP per capita, $PM_{2.5}$ annual mean concentration or life expectancy at birth.

Solving a discrete multi-criteria problem can be viewed as a process of assessing the possible alternatives based on a set of quantitative or qualitative criteria (Munda, 1995). It assists individuals or groups in decision making, which can assign both quantitative and qualitative values to its criteria, allowing multiple criteria in incommensurable units (Hyde et al., 2003). The MCDA process generally has four steps: (1) structuring the problem, (2) building the analysis model, (3) implementing the analysis, and (4) summarizing the results and making recommendations (Guitouni & Martel, 1998; Belton & Stewart, 2002). In the first step, the stakeholders, goals, alternatives, values and constraints need to be identified. In the second step, model building and preferences should be specified, followed by defining the criteria and eliciting the values. The third step involves implementing the analysis which includes alternative evaluations, aggregation and sensitivity analysis. In the final step, a recommendation is made based on the results of analysis.

1.4 Selection of MCDA methods

Various multi-criteria decision analysis methods have been developed to support decision makers, most of which belong to one of the following categories: continuous methods (e.g. the optimization methods), the inputs of which are described by numerical criteria, and discrete methods (e.g. MAUT, AHP, the ELECTRE methods and the PROMETHEE methods), the inputs of which are described by discrete criteria (Dombi & Zsiros, 2005; Guitouni & Martel, 1998).

Having a clear objective and identifying the relevant problem can help the decision maker choose the appropriate method. To identify the objective, the decision maker needs to be clear about what kinds of problems he/she needs to solve. Roy (1996) proposed four kinds of problems that a decision maker could encounter:

1 The choice problem, in which the decision makers aim to choose the appropriate alternatives among several viable options;
2 The sorting problem, in which the decision makers need to assign the alternatives to their corresponding categories according to certain criteria;
3 The ranking problem, in which the decision makers put alternatives into ordered equivalence classes, and
4 The description problem, in which the decision makers evaluate the alternatives based on the existing criteria.

The decision maker should also be clear about the types of results one would expect from an MCDA method before it is selected. Roy and Słowiński (2013) summarized five types of results: (1) a numerical value, (2) the set of alternatives is ranked, (3) a subset of alternatives to be selected as a final choice, (4) the alternatives to be assigned to corresponding categories, and (5) a subset of alternatives with some outstanding features to be selected as a base for a subsequent decision aiding process.

Rowley et al. (2012) also proposed a detailed workflow for sustainability analysts to choose the most appropriate methods according to their needs. Based on the workflow and the nature of this research, there are four factors which must be addressed in the selection of the MCDA method: (1) whether the decision maker is an individual or group, (2) the objectives of the decision process, (3) whether the veto threshold approach is sufficient, and (4) whether the aggregation logic is appropriate.

With respect to Factor 1, this research does not include a participatory process or several decision makers and, as such, there is no need to determine how to represent different preferences; a method which can account for an individual decision maker's preference is sufficient for this research.

The objective of this research is to identify which city ranks first in sustainability among the Global South cities. In terms of Factor 2, this research is therefore concerned with the ranking problem; or, Problem 3 as discussed above (Roy, 1996). As such, the method used should have the capacity to make comparisons between alternatives according to various criteria. The objective also explains that the type of results that the decision maker can expect is a ranked set of alternatives; or, Type 2 as discussed (Roy & Słowiński, 2013). For the purpose of having a ranked result, Roy and Słowiński (2013) recommended the following method: ELECTRE III, IV (Figueira et al., 2005, 2013), PROMETHEE I and II (Brans & Mareschal, 2005) and Robust Ordinal Regression methods (Greco et al., 2010), etc.

A veto threshold in this research is the minimum sustainability level, which is accepted. However, this is difficult to determine and can only be utilized to identify cities, which are not sustainable. For Factor 3, since it cannot rank cities in terms of sustainability, the veto threshold approach is not used in this research.

Various levels of sustainability, weak or strong, determine the appropriateness of aggregation logic. Martinez-Alier et al. (1998) noted the difference in sustainability in terms of compensation among different dimensions or criteria: more compensation is accepted in weak sustainability, whereas less compensation is accepted in strong sustainability. The author indicated that the incommensurability of conflicting values is the 'cornerstone' in ecological economics. Ayres (2007) also supports the view of strong sustainability as there are important limits with this concept. For Factor 4, with respect to the threshold in sustainability, the strong sustainability is considered in this research; however, methods based on the weak sustainability will also be applied for comparison.

There are different methods with diverse sustainability (weak/strong) in their methodologies. Polatidis et al. (2006) summarized the relationships between MCDA methods and weak or strong sustainability based on the degree of compensation allowed in the methods (see Figure 5.1).

Cost-benefit analysis (CBA) allows full compensation between its criteria; in other words, the high-performance value in one criterion can fully counteract the low-performance value on some other criterion. For example, the high GDP could offset the low air quality from pollution of cities in this methodological assumption. Therefore, this method operationalizes the strong sustainability.

The simple additive weighted method, or the weighted-sum method (WSM), allows trade-offs between different criteria so it also operationalizes the strong sustainability. However, the distinction from cost-benefit analysis is that the trade-offs in this method depend on the weights assigned to the criteria.

Multi-attribute utility theory (MAUT), based on the performance aggregation, permits trade-offs between alternatives and their effects on objectives. The decision maker's preference can be considered by defining the utility functions and weights for each attribute (Von Neumann & Morgenstern, 1945; Keeney, 1972; Cinelli et al., 2014).

The preference ranking organization method of enrichment evaluation (PROMETHEE) methods do not allow full compensation between criteria (Brans et al., 1986). By using indifference and preference thresholds, the criteria could only be partially compensated so this method could operationalize the strong sustainability concept.

The *Elimination Et Coix Traduisant la Realite* (ELECTRE) methods are outranking approaches as they are based on the pair-wise comparisons of the alternatives behavior (Roy et al., 1975; Roy & Hugonnard, 1982; Roy, 1968,

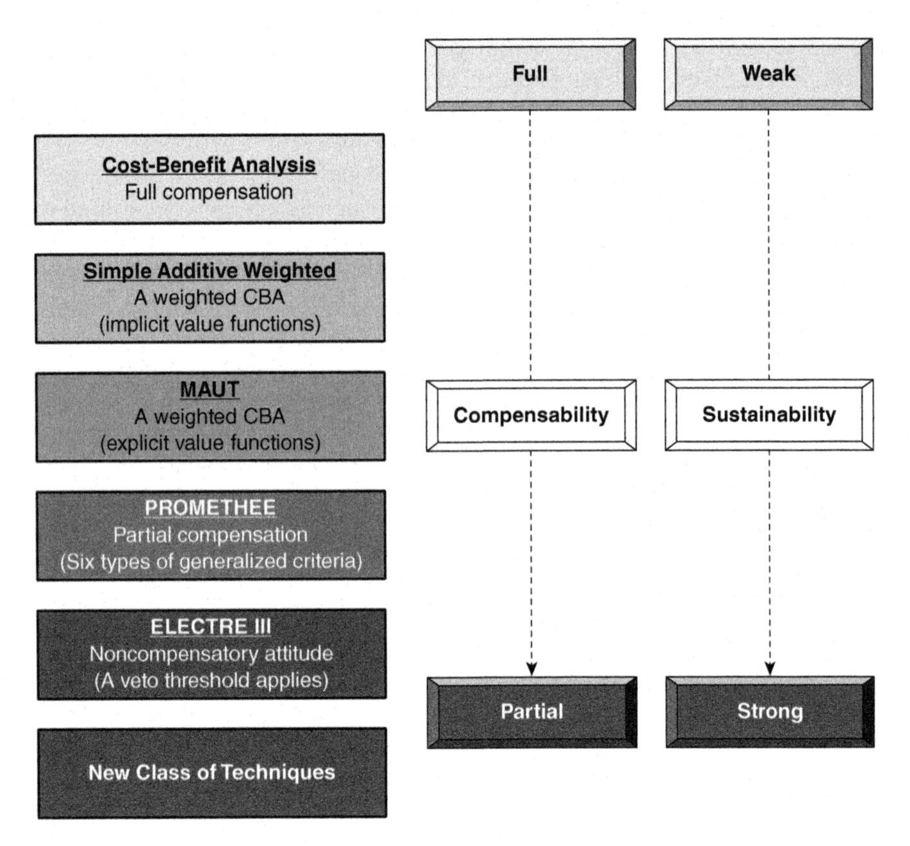

Figure 5.1 MCDA methods facing compensability and sustainability

Source: Polatidis et al. (2006)

1973, 1977a, 1977b, 1978; Figueira et al., 2005). The concordance and discordance indexes are applied to identify the outranking relations between two alternatives. It is also a strong sustainability-based approach which prevents the compensation.

This research is mainly based on the PROMETHEE methods. These methods have a transparent computational process so it is easily understood and widely applied by the researchers and decision makers. (Georgopoulou et al., 1998; Ozelkan & Duckstein, 1996). Three PROMETHEE methods are used in this study: PROMETHEE II is used to give a complete ranking of the cities; PROMETHEE I is used to provide information about the incomparability between cities, and PROMETHEE-GAIA is used to offer visualized results obtained by the PROMETHEE method (Mareschal & Brans, 1988). The weighted-sum method (WSM) will also be applied to compare the results from strong sustainability and weak sustainability methods.

1.5 Chapter roadmap

This chapter is composed of five sections. Section 1 is this introduction. A detailed literature review focusing on the PROMETHEE methods is presented in Section 2. Section 3 accounts for factors such as the selection of criteria, reasoning for the weights, which parameters are changed in sensitivity analysis, and the methodologies of the PROMETHEE methods as well as the weighted-sum method. Section 4 discusses the results of selected criteria, the ranking obtained by the weighted-sum method and different PROMETHEE methods. The results of sensitivity analysis are also outlined. Section 5 consists of three parts: the first discusses the difference of results obtained by different MCDA methods, while the second discusses the reasons behind the success of Global South cities in sustainability. In the third part, the policies of most sustainable Global South cities, as well as the sustainable Global North city (Singapore), are also compared to identify similarities or differences in success.

2. Literature review of the PROMETHEE methods

2.1 Introduction of PROMETHEE

PROMETHEE (Preference Ranking Organization Method for Enrichment Evaluation) is an outranking method which compares alternatives in pairs on each criterion (Brans et al., 1986). The numerical value of the preference functions is used to express the preference which represents the decision-maker's view. This value ranges from 0 to 1. When the value of the preference function is 0, the decision maker is indifferent between the two alternatives on that criterion. When the value is 1, one alternative is strictly preferred to the other. Each criterion is assigned with its corresponding weight which represent its relative importance. The alternatives are then ranked by the comparison of their positive rank flow and negative rank flow, or the net outranking flow. For a detailed explanation of the computational process, see Section 3.6.

This method is suitable for solving problems where a finite number of alternatives need to be ranked based on various, even conflicting criteria (Albadvi et al., 2007). Moreover, it can avoid small differences between some criteria having a significant influence on the ranking of the alternatives, by involving the indifference threshold (Sun & Han, 2010).

The PROMETHEE I and PROMETHEE II methods were proposed by J. P. Brans. They were presented at the conference *L'ingénièrie de la decision* organized at the Université Laval, Québec, Canada in 1982 for the first time. In the same year, G. Davignon used this method in several practical applications in the field of health care (Greco, Ehrgott & Figueira, 2005). Several years later, Brans, together with B. Mareschal, proposed PROMETHEE III and PROMETHEE IV. In 1988, they developed GAIA, a visual interactive module, which present a graphic interpretation of the PROMETHEE method. In 1992 and 1994, the same authors further developed two modifications,

PROMETHEE V and PROMETHEE VI. PROMETHEE methods have been widely used in numerous fields such as environmental management, investments, health care, chemistry and tourism. (Greco, Ehrgott & Figueira, 2005).

Each PROMETHEE method has its own purpose. PROMETHEE I focuses on partial ranking, while PROMETHEE II deals with the full ranking results. PROMETHEE III associates an interval to each alternative to emphasize the relation of indifference between alternatives in the rankings. PROMETHEE IV is used in the continuous case. PROMETHEE-GAIA provides a graphical representation, PROMETHEE V is the method which includes segmentation constraints and PROMETHEE VI has the function of representing the human brain. Figueira et al. (2004) put forth two extensions based on the PROMETHEE method: the PROMETHEE TRI which solves the sorting problem and the PROMETHEE CLUSTER which is used for nominal classification.

In this research, PROMETHEE II is used to provide a complete ranking of cities, PROMETHEE I illustrates possible incomparability between cities, and PROMETHEE-GAIA provides a visual depiction of the results obtained by PROMETHEE on a two-dimensional plane (Hyde et al., 2003).

2.2 Comparison between PROMETHEE methods and other MCDA methods

Brans and Vincke (1985) argued that the ELECTRE methods (Roy et al. 1975; Roy & Hugonnard, 1982; Roy 1968, 1973, 1977a, 1977b, 1978) are complicated since they need many parameters. Some of the parameters (e.g. concordance discrepancies and discrimination thresholds) do not have an economic value, but a technical representation, so decision makers find it difficult to understand their own impacts on the results. The PROMETHEE I & II, which only need two parameters to be fixed maximally, have economic meanings on these parameters (e.g. preference threshold and indifference threshold); therefore, these methods can be rather easily understood by the decision maker. Al-Shemmeri et al. (1997) also indicated that management of PROMETHEE methods is simple relative to that of ELECTRE III.

Brans et al. (1986) compared the stability of PROMETHEE and ELECTRE III by testing the two methods on 21 problems. Random deviations were made on the initial thresholds. The results show that the stability of PROMETHEE is better than that of ELECTRE III for generalized criteria with linear preference and indifference area. This is mainly because λ-qualification has the 'discrete' character in ELECTRE III.

Gilliams et al. (2005) applied PROMETHEE II, ELECTRE III and AHP for the selection of afforestation strategies and found that PROMETHEE II is slightly preferable to the other two methods in terms of user friendliness, simplicity and implementation.

Macharis et al. (2004) compared the PROMETHEE methods and AHP; it was found that AHP is better in terms of structuring the problem and determination of the weights, while PROMETHEE methods need fewer inputs and could provide a better analysis of the problem. The authors also proposed improving PROMETHEE by combining a tree-like structure which is similar to AHP.

Salminen et al. (1998) made comparisons between the ELECTRE III, PROMETHEE I, II, and SMART decision aids in four applications to evaluate the applicability of the methods in aiding environmental decision making. It was found that there was not much difference between SMART and PROMETHEE, while ELECTRE III was recommended due to its extra functionality.

2.3 Limitations of the PROMETHEE methods and extensions of PROMETHEE

A criterion performance value represents the performance of an alternative according to a certain criterion. These values are generally given by experts and decision makers, which are based on expert knowledge and mathematical models (Hyde et al., 2004; Kheireldin & Fahmy, 2001). This may lead to the imprecision of the parameters (Mousseau et al., 2003). In coping with this potential imprecision, PROMETHEE introduced generalized criterion functions to model the decision-maker's evaluation between two alternatives. However, the decision makers have encountered difficulty in choosing the appropriate generalized criterion functions and imprecision would be generated in determining preference thresholds and indifference thresholds used in the functions (Salminen et al., 1998). The criteria weight represents the importance of a certain criterion compared with other criteria. In addition to the criteria performance values, criteria weights are another source of imprecision which cannot be adequately addressed by the generalized criterion functions. When multiple decision makers are involved in a multi-criteria decision, consensus would be difficult to achieve due to decision makers having different evaluations of each criterion (Hyde et al., 2004). These two kinds of imprecision would affect the reliability and robustness of the final results.

Although researchers such as Ringuest (1997) and Triantaphyllou and Sanchez (1997) introduced methods to analyze the sensitivity of chosen alternatives in a MCDA problem, there are still drawbacks, such as: most methods mainly focus on the uncertainty of criteria weighting, but criteria performance values could also influence the result; in these methods, some of the parameters are changed while others remain fixed to check the difference of ranking in various situations, which cannot identify the combined effects of various parameter changes simultaneously. Hyde et al. (2004) proposed a reliability-based approach to test the robustness of the ranking of alternatives as a solution to the uncertainty problem. This approach used probability distributions as the input values and conducted the sensitivity analysis by Monte Carlo simulation. A significance analysis was also used to analyze the contribution of each input parameter, such as criteria weights, in the value of an alternative by Spearman rank correlation coefficient. This approach can ensure the optimal performance of the final decision under various scenarios.

Several studies improve PROMETHEE by combining it with other techniques. Dağdeviren (2008) proposed an integrated approach which combines analytic hierarchy process (AHP) and the PROMETHEE together to solve the problem of selecting milling machines. After analyzing the problem and determining weights by AHP, the PROMETHEE method is used to provide the ranking of machines and perform sensitivity analysis. Other studies combining PROMETHEE with AHP include those by Babic and Plazibat (1998); Wang et al. (2006); Bilsel et al. (2006); Wang and Yang (2007). Goletsis et al. (2003) proposed a hybrid of ELECTRE III and PROMETHEE methods as a group-ranking methodology; a similar study which made use of a PROMETHEE and ELECTRE combination is that by van Huylenbroeck (1995).

2.4 Applications of the PROMETHEE methods

The PROMETHEE methods have been widely used and successfully applied in numerous areas. Based on 217 scholarly papers, Behzadian et al. (2010) categorized the applications of PROMETHEE methods into several categories: environment management, hydrology and water management, chemistry, energy management and business and financial management, among others.

In the field of environment management and sustainability, many researchers use PROMETHEE methods to find the best compromise alternative for decision making. Vego et al. (2008) applied PROMETHEE and GAIA in waste management planning in Dalmatia, Croatia based on ecological, economic, social and functional criteria. Diakoulaki and Karangelis (2007) applied PROMETHEE methods to find the best compromise alternative scenarios among four mutually exclusive scenarios for the Greek power generation sector. Beynon and Wells (2008) applied PROMETHEE with sensitivity analysis in ranking the motor vehicles based on components of their emissions. The authors also present a visual illustration of how the changes of criteria values impacts the results.

There have been a few studies on the use of PROMETHEE to rank and compare cities/areas in terms of sustainability is limited. Vaillancourt and Waaub (2004) applied PROMETHEE to rank regions and countries to allocate equitable international greenhouse gas (GHG) emission entitlements based on eight equity principles. Moreover, Diakoulaki et al. (2007) applied the PROMETHEE II method to identify priority countries and investment opportunities for the exploitation of the Kyoto Protocol's Clean Development Mechanism (CDM). However, the body of research in this area is otherwise limited.

3. Methodology and data collection

3.1 The analysis process

Based on the workflow for sustainability analysts using MCDA proposed by Rowley et al. (2012), this research follows these steps:

Step 1: Selecting the criteria by comparing three sustainable development indices

Step 2: Collecting the data of 17 cities according to the selected criteria

Step 3: Determining the weights of each criterion

Step 4: Applying the weighted sum method

Step 5: Applying the PROMETHEE methods

Step 6: Sensitivity analysis

3.2 Selecting criteria

Bouyssou (1990) and Rowley et al. (2012) put forth characteristics of the criteria in the MCDA method. Based on their findings, this research follows five attributes in selecting criteria:

- Exhaustive: the criteria contain sufficient information which can distinguish alternatives; the criteria should also provide the essential foundation for decision making.
- Minimal: redundant or unnecessary criteria should be excluded.
- Monotonic and cumulative: partial preference should be consistent with the full preference; comparing alternatives by several criteria should be consistent with comparing them by a single criterion.
- Independent: each criterion should have its unique value; the criteria which have similar functions will be excluded to avoid double-counting.
- Available: since the data are collected from online open sources, it is important to select the criteria for which all alternatives have the corresponding data.

Bouyssou (1990) also acknowledged that it could be impossible to find a set of criteria to fulfill all these properties simultaneously; it could be especially difficult to find the criteria which are both exhaustive and independent.

In this research, the criteria are used to assess the sustainability of these cities; the alternatives are the global south cities. Shmelev and Rodríguez-Labajos (2009) indicated that it is important to evaluate the alternatives based on a unified standard. Based on the questions that need to be addressed in selecting their criteria, the following questions to be considered in this research are:

1. Which dimensions should be considered in evaluating the sustainability of cities?
2. Which criteria are most suitable for representing the sustainability of global south cities?
3. Should each dimension use the same number of criteria? How many criteria should be used to represent each dimension?

Since the indicators of sustainable development are used to evaluate the sustainability, these indicators can be used as the criteria for assessing the sustainability of global south cities in the MCDA process.

Three sustainable development indices are compared in this research:

- Sustainable Development Goals Index (Sdgindex, 2016);
- The UNECE–ITU Smart Sustainable Cities Indicators (UNECE, 2015);
- ISO 37120:2014 Sustainable development of communities – Indicators for city services and quality of life (ISO, 2014).

A comparison of these indices could then be used for selecting the criteria; subsequent analysis also provides a comprehensive pattern of assessing the urban sustainability (see Section 4.1).

3.3 Data collection

The data are all collected from online open sources. Since the data for each city changes each year, 2014 is selected to be the standard year for data collection. However, due to availability issues, data from other years close to 2014 are also used where necessary. In the cases when the data cannot be found directly, it is calculated based on other data; for example, the municipal waste per capita generation of Mumbai city is calculated by the total volume of municipal waste generation of Mumbai city divided by the population of Mumbai city. There are no international standards for the collection of statistical data, so certain data might not be surveyed or counted in some cities. When data for a city are not available from appropriate sources, it is represented by national data; for example, the share of renewable energy sources of Nairobi is replaced by the share of renewable energy sources of Kenya. Despite these methods' potential uncertainties or imprecisions in the analysis, the indifference threshold (explained in Section 3.6.1) can mitigate these to an extent. Small deviations in performance values which do not exceed the indifference threshold can be omitted so the imprecisions of data would not influence the overall ranking too much.

3.4 Assigning weights to criteria

Assigning weights to criteria can reflect its relative importance or, in other words, reflect its evaluation by the decision maker. If the criteria are not equally important in assessing the alternatives, three options can be used to distinguish the importance of the criteria: a veto threshold, a hierarchical structure, and weighting (Benoit & Rousseaux, 2003).

A veto threshold refers to the minimum performance value for the criteria. If an alternative performance value cannot surpass this threshold in each criterion, this alternative will not be selected as one of the feasible options (Rowley et al., 2012). This threshold can be used to exclude the non-sustainable cities, but not to rank the cities by sustainability. This veto threshold can also be defined by the decision maker. For example, the cities in Japan have high performance values in sustainability criteria but since there is a nuclear disaster in Fukushima, the cities near Fukushima could not be selected as the most sustainable city. However, the veto threshold is hard to be defined objectively in this research so it is not used.

Another option to assign weights to criteria is using a hierarchical structure. It refers to the assessment of alternatives in the order of importance of the criteria. (Rowley et al., 2012) For example, GDP per capita is the most important criterion, followed by air quality, followed by life expectancy, and so on. First, the cities which have low GDP per capita would be excluded, then those with low air quality, then those with low life expectancy, and so on. This process continues until the most sustainable city is found, as used in the choice problem. However, since this research aims to rank the cities, this method would not be used.

Weighting refers to assigning a numerical value to each criterion. There are two types of weights in MCDA method: substitution rates and importance coefficients. Substitution rates are used in compensatory aggregation methods as they refer to capacity for trade-offs between the criteria. Importance coefficients are used in non-compensatory aggregation methods as they refer to the relative importance of criteria (Roy, 1996; Rowley et al., 2012). Therefore, the weights in the weighted-sum analysis represent the substitution rates while the weights used in the PROMETHEE method represent the importance coefficients.

There are two basic ways of determining weights: direct weighing and trade-off between criteria. Experiments show that decision makers find it difficult to provide confident answers to trade-off questions; as such, the results do not fully represent the decision makers' accurate evaluation of the criteria (Choo et al., 1999; Hämäläinen & Salo, 1997). Madlener and Stagl (2005) assumed equal weights for each criterion in their study, where each dimension represents the value of different stakeholders, each of whom have equal importance (Munda, 1995). However, a shortcoming of this assumption is that the number of criteria involved will change each criterion's importance and, therefore, significantly influence the outcome of the analysis. To mitigate this drawback, this research classifies the criteria into three dimensions: environment, society and economy; moreover, it assigns equal weights to these dimensions. In each dimension, the criteria will be assigned with equal weights. For instance, if there are three dimensions and five criteria in the environmental dimension, the weight of each criterion in the environmental dimension will be: 1/3/5.

Assigning equal weights to the three groups is based on the assumption that three dimensions (economy, environment and society) are equally important in sustainability. New criteria can influence the weights of other criteria in its group; nevertheless, each group holds the same importance (the importance of each dimension will not change). In sensitivity analysis, the weights of each dimension will be changed to observe the impact of weights on the ranking of cities.

3.5 The weighted-sum method

The weighted-sum method is arguably the most widely used MCDA method (Yoon & Hwang, 1995). The process adds the value of each indicator or criterion based on its corresponding weight. However, before this is done, normalization needs to be performed since indicators are in different units (Mitropoulos & Prevedouros, 2016). Normalization is used to calculate the utility value on a scale of 0 to 1 for each indicator; the values would be dimensionless after normalization (Maxim, 2014).

The value of 1 represents this highest sustainability performance, while, on the contrary, a value of 0 means the lowest sustainability performance. Some indicators have a direct correlation with utility in that they have positive impacts on sustainability (e.g. GDP per capita and life expectancy at birth) while other indicators are inversely correlated with utility in that they have negative impacts on sustainability (e.g. unemployment rate and Gini coefficient). Therefore, there are two corresponding normalization equations (Krajnc & Glavič 2005):

$$N_{ij}^{+} = \frac{I_{ij}^{+} - I_{min,j}^{+}}{I_{max,j}^{+} - I_{min,j}^{+}} \tag{3.1}$$

$$N_{ij}^{-} = \frac{I_{max,j}^{-} - I_{i,j}^{-}}{I_{min,j}^{-} - I_{max,j}^{-}} \tag{3.2}$$

where:

N_{ij}^{+} is the normalized indicator with has positive impacts on sustainability achieved by *i* th city, with respect to the criteria (indicator) *j*;

I_{ij}^{+} is the criteria performance value belonging to the *i* th city evaluated by the criteria (indicator) *j*;

$I_{max,j}^{+}$ is the highest criteria performance value under the criteria (indicator) *j*;

$I_{min,j}^{+}$ is the lowest criteria performance value under the criteria (indicator) *j*.

The normalized value represents the performance of each city relative to the most sustainable city. After the normalization process, the weighted-sum method was used to aggregate normalized values of indicators to get an overall sustainability index for each city. This method is based on the additive utility assumption; moreover, the indicators are favorably independent to avoid double counting. The overall sustainability index S_i for each city is calculated by equation 3.3 (Mitropoulos & Prevedouros, 2016):

$$S_i = \sum_{j=1}^{n} w_j N_{ij} \qquad i = 1,\ldots,m \tag{3.3}$$

where:

w_j is the assigned weight for each criteria (indicator) *j*, as discussed;

N_{ij} is the normalized performance value of criteria (indicator) *j* for *i* th city.

3.6 The PROMETHEE method

3.6.1 Preference modeling

$$\text{Max}\left\{f_1(a), f_2(a)\ldots, f_i(a) \mid a \in K\right\} \tag{3.4}$$

Equation 3.4 illustrates a multi-criteria problem in which k is a finite set of alternatives. For each alternative $a \in k$, $f_i(a)$ is the evaluation of this alternative. The results of comparison between two alternatives $a, b \in k$ can be expressed by preference. Then, the preference function P, which shows the difference between the evaluation of two alternatives (a and b) based on a particular criterion, can be created. This preference degree is expressed by a real number from 0 to 1 (Brans & Mareschal, 2005). When the value is 0, there is no preference between alternatives, whereas a value of 1 indicates that one alternative is strictly preferred to another. Therefore:

$$P_i(a,b) = F_i\left[f_i(a) - f_i(b)\right] a, b \in K \tag{3.5}$$

where:

$$d_i(a,b) = f_i(a) - f_i(b);$$

$P_i(a,b)$ ranges from 0 to 1.

When the criterion is to be maximized, the function shows the preference of a over b for observed deviations between $f_i(a)$ and $f_i(b)$. The shape of this preference function can be seen in Figure 5.2. When the deviations are negative, the value of preferences function will be 0 (Brans & Mareschal, 2005).

When the criterion is to be minimized, the function will be changed to:

$$P_i(a,b) = F_i\left[-d_i(a,b)\right] \tag{3.6}$$

The pair $\{f_i(\cdot), P_i(a,b)\}$ is referred to as the generalized criterion. A particular preference function should be defined for each criterion. Brans et al. (1986) proposed six types of preference functions: usual function, U-shape function, V-shape function, level function, linear function and Gaussian function. Each type of preference function corresponds to a type of generalized criterion.

The parameters used in some of these functions need to be defined. There are three kinds of parameters: the indifference threshold q, the strict preference

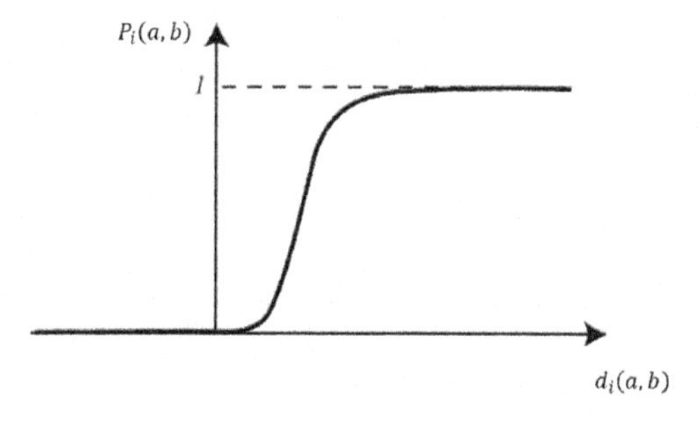

Figure 5.2 Preference function

threshold p and the intermediate value s between indifference threshold and strict preference threshold.

The strict preference threshold represents the minimum deviation which is sufficient to yield a preference of one alternative over another. This threshold can be equal to zero (when in the case of usual criteria) (Roy et al., 2014). The indifference threshold is the largest deviation between two alternatives which can be neglected on a certain criterion. This threshold can be equal to zero but cannot exceed the preference threshold (Roy et al., 2014). The preference threshold also represents the degree of compensation among different criteria. When the preference threshold is larger, more compensation is allowed.

The six types of preference functions proposed by Brans et al. (1986) are described as follows:

(1) Usual function (usual criterion)

$$P(d) = \begin{cases} 0 & d \le 0 \\ 1 & d > 0 \end{cases} \tag{3.7}$$

In this case, when $f_i(a)$ is larger than $f_i(b)$, the decision maker strictly prefers alternative a to alternative b. No parameter needs to be determined in this function.

(2) U-shape function (Quasi-criterion)

$$P(d) = \begin{cases} 0 & d \le q \\ 1 & d > q \end{cases} \tag{3.8}$$

In this case, when the deviation of the evaluations between two alternatives does not exceed the indifference threshold, the decision maker is indifferent between these two alternatives; otherwise, one alternative will be strictly preferred to the other. The indifference threshold q needs to be determined in this function.

(3) V-shape function (Criterion with linear preference)

$$P(d) = \begin{cases} 0 & d \le 0 \\ \dfrac{d}{p} & 0 < d < p \\ 1 & d \ge p \end{cases} \tag{3.9}$$

In this case, when d is smaller than or equal to 0, the decision maker is indifferent between these two alternatives; when d is greater than 0 but smaller than p, the preference increases linearly with d. When d is greater than or equal to p, the decision maker strictly prefers alternative a to alternative b. The preference threshold p needs to be determined in this function.

(4) Level function (Level criterion)

$$P(d)=\begin{cases}0 & d \le q \\ \frac{1}{2} & q < d \le p \\ 1 & d > p\end{cases} \tag{3.10}$$

In this case, if d lies between the indifference threshold and preference threshold, the value of preference function will be equal to 0.5, which is a weak preference. Both thresholds need to be determined in this case.

(5) Linear function (Criterion with linear preference and indifference area)

$$P(d)=\begin{cases}0 & d \le q \\ \frac{d-q}{p-q} & q < d \le p \\ 1 & d > p\end{cases} \tag{3.11}$$

In this case, the preference increases linearly from indifference to strict preference. Both thresholds need to be determined.

(6) Gaussian function (Gaussian criterion)

$$P(d)=\begin{cases}0 & d \le 0 \\ 1-e^{-\frac{d^2}{2s^2}} & d > 0\end{cases} \tag{3.12}$$

In this case, the preference increases for all deviations. The result is stable since it has no discontinuities. The parameter s needs to be determined in this function.

The linear function is chosen in this research. For further explanation, see Section 3.6.2.

Generalized criterion and preference functions are summarized in Table 5.1.

3.6.2 Choice of preference function and parameters

In this research, the criteria are the indicators which evaluate the sustainability of cities. The performance values of the cities are statistical data which are all quantitative. The indifference threshold is necessary since the small deviation between two alternatives on a certain criterion does not indicate that one is strictly better than another in terms of sustainability.

For example, we assume that a city with 10 mg/m^3 annual PM_{10} concentration is as sustainable as another city with 10.5 mg/m^3 annual PM_{10} concentration in terms of air quality. This threshold can also cope with the imperfect data obtained from a survey which unavoidably carries an imprecision margin (Roy et al., 2014). The preference threshold is also important in another respect: when the

Table 5.1 Types of generalized criterion and preference function

Generalized criterion	*Preference function*		*Parameters*
Usual criterion	Usual function	$P(d)=\begin{cases}0 & d\le 0\\ 1 & d>0\end{cases}$	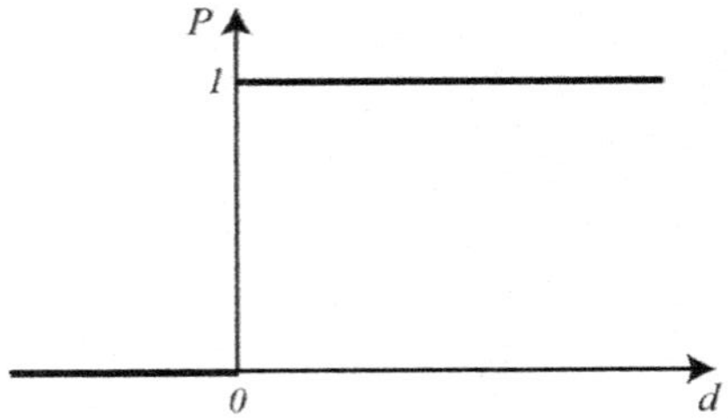 /
Quasi-criterion	U-shape function	$P(d)=\begin{cases}0 & d\le q\\ 1 & d>q\end{cases}$	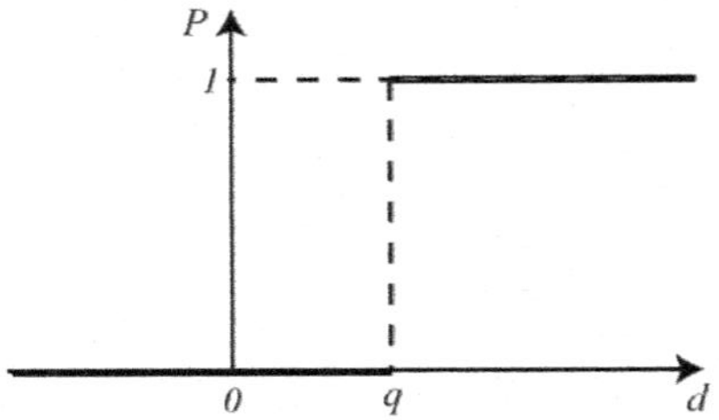 q
Criterion with linear preference	V-shape function	$P(d)=\begin{cases}0 & d\le 0\\ \frac{d}{p} & 0<d<p\\ 1 & d\ge p\end{cases}$	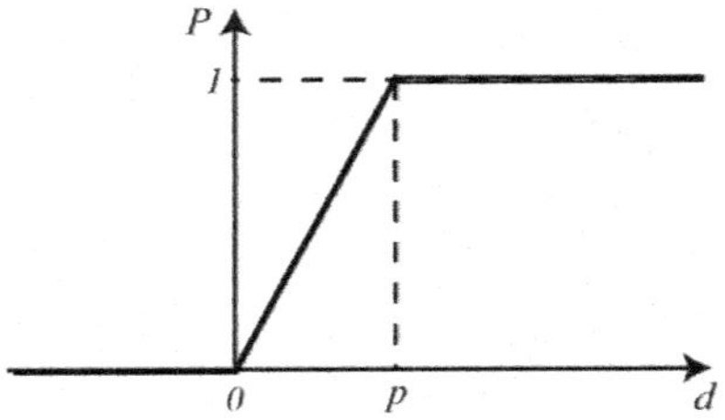 p

Level criterion	level function	$P(d)=\begin{cases}0 & d \le q\\ \frac{1}{2} & q < d \le p\\ 1 & d > p\end{cases}$		p, q
Criterion with linear preference and indifference area	linear function	$P(d)=\begin{cases}0 & d \le q\\ \frac{d-q}{p-q} & q < d \le p\\ 1 & d > p\end{cases}$		p, q
Gaussian criterion	Gaussian function	$P(d)=\begin{cases}0 & d \le 0\\ 1-e^{-\frac{d^2}{2s^2}} & d > 0\end{cases}$	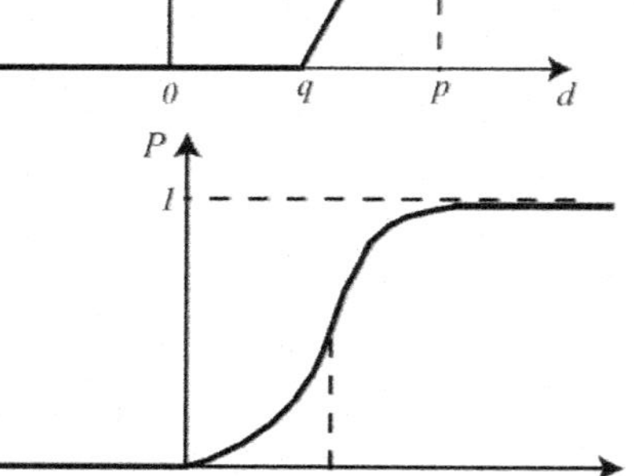	s

Source: Brans & Mareschal, 2005

deviation between two cities on a certain criterion becomes large enough, it is sufficient to indicate that one city is strictly more sustainable than another. For example, we assume the city with 10 mg/m^3 annual PM_{10} concentration is more sustainable than another city with 20 mg/m^3 annual PM_{10} concentration in terms of air quality. The functions with indifference threshold and preference threshold are either level function and linear function. Since the level function is suitable for qualitative data and linear function is suitable for quantitative data, the linear function is chosen for all criteria in this research.

In this research, the indifference threshold will be 5% of the maximum difference between pairs of alternatives on each criterion and the preference threshold will be 25% of the maximum difference between pairs of alternatives on each criterion. In sensitivity analysis, the thresholds will be changed to test the robustness of results.

3.6.3 Aggregated preference indices

After determining the preference function $f_i(\cdot)$ and weights w_i ($i = 1,2...,n$) for each criterion, the computation of PROMETHEE can be applied. For alternative $a,b \in K$, let:

$$\begin{cases} \pi(a,b) = \sum_{i=1}^{n} P_i(a,b) w_i \\ \pi(b,a) = \sum_{i=1}^{n} P_i(b,a) w_i \end{cases} \tag{3.13}$$

In this function, n is the number of alternatives. $\pi(a,b)$ represents how alternative a is preferred to alternative b based on all criteria, and vice versa. If $\pi(a,b)$~0, alternative a is weakly preferred to b as a whole. If $\pi(a,b)$~1, alternative a is strongly preferred to b as a whole.

3.6.4 Outranking flows

Each alternative needs to be compared with (n–1) other alternatives in K. There are two outranking flows:

$$\varnothing^{+}(a) = \frac{1}{n-1} \sum_{x \in K} \pi(a,x) \tag{3.14}$$

$$\varnothing^{-}(a) = \frac{1}{n-1} \sum_{x \in K} \pi(x,a) \tag{3.15}$$

The leaving ranking flow (or positive ranking flow) $\varnothing^{+}(a)$ measures how alternative a is outranking other alternatives. The entering ranking flow (or negative ranking flow) $\varnothing^{-}(a)$ measures how other alternatives are outranking alternative a.

3.6.5 The PROMETHEE I

The PROMETHEE I can provide a partial ranking of alternatives. Three kinds of relationships between alternatives can be seen in The PROMETHEE I: *aPb* (*a* is preferred to alternative *b*), *aIb* (*a* is indifferent to alternative *b*) and *aRb* (*a* is incomparable with alternative *b*).

In PROMETHEE I:

$$\begin{cases} aPb \quad iff \begin{cases} \varnothing^+(a) > \varnothing^+(b) \text{ and } \varnothing^-(a) < \varnothing^-(b), or \\ \varnothing^+(a) = \varnothing^+(b) \text{ and } \varnothing^-(a) < \varnothing^-(b), or \\ \varnothing^+(a) > \varnothing^+(b) \text{ and } \varnothing^-(a) = \varnothing^-(b); \end{cases} \\ aI \quad biff \quad \varnothing^+(a) = \varnothing^+(b) \text{ and } \varnothing^-(a) = \varnothing^-(b) \\ aRb \quad iff \begin{cases} \varnothing^+(a) > \varnothing^+(b) \text{ and } \varnothing^-(a) > \varnothing^-(b), or \\ \varnothing^+(a) < \varnothing^+(b) \text{ and } \varnothing^-(a) < \varnothing^-(b) \end{cases} \end{cases} \tag{3.16}$$

When the leaving ranking flow of alternative *a* is greater than *b*, or the entering ranking flow of alternative *a* is smaller than *b*, alternative *a* outranks *b*.

When the two alternatives have the same leaving ranking flow and entering ranking flow, there is indifference between the two alternatives.

When the leaving ranking flow of alternative *a* is greater than *b* and the entering ranking flow of alternative *a* is smaller than *b*, *or* the leaving ranking flow of alternative *a* is smaller than *b* and the entering ranking flow of alternative *a* is greater than *b*, the two alternatives are incomparable. Therefore, a partial ranking could be given by PROMETHEE I.

3.6.6 The PROMETHEE II

The PROMETHEE II provides a complete ranking of alternatives. The net outranking flow is introduced in the PROMETHEE II:

$$\varnothing(a) = \varnothing^+(a) - \varnothing^-(b) \tag{3.17}$$

When the net outranking flow of alternative *a* is greater than the net outranking flow of alternative *b*, alternative *a* outranks *b*. When net outranking flows of the two alternatives are equal, the two alternatives are ranked the same. Therefore:

$$\begin{cases} aPb \text{ iff } \varnothing(a) > \varnothing(b) \\ aIb \text{ iff } \varnothing(a) = \varnothing(b) \end{cases} \tag{3.18}$$

In the PROMETHEE II, all alternatives are comparable.

3.6.7 The GAIA Plane

The GAIA Plane can graphically display the alternatives according to their contributions to the criteria (Brans & Mareschal, 1994). Principal components analysis is applied in this method to limit the number of dimensions while minimizing the loss of information after projection (Brans & Mareschal, 2005). When all weights are assigned to a single criterion $f_i(\cdot)$, the single criterion net flow $\varnothing_i(a)$ is obtained. It shows how an alternative is preferred to all other alternatives (when $\varnothing_i(a)>0$) or how all other alternatives are preferred to this alternative (when $\varnothing_i(a)<0$). Therefore:

$$\varnothing_i(a) = \frac{1}{n-1} \sum_{x \in K} P_i(a,x) - P_i(x,a) \tag{3.19}$$

A two-dimensional plot is generated and the relative position of the alternatives will be shown on this graph. This tool can also help the decision maker to understand the conflicts among criteria and to show the impacts of weights on the final ranking results (Mareschal & Brans, 1988).

3.7 Sensitivity analysis

There are various sources of uncertainty in the application of PROMETHEE, including the criteria performance values and criteria weights. It is standard practice to apply sensitivity analysis at the final stage to test the robustness of the results.

In this study, the weights of each group/dimension (economy, environment and society) will be changed to see if the ranking of cities are sensitive to a certain dimension. The indifference and preference threshold values will also be altered to see if the final ranking is sensitive to the selection of threshold values.

There are six scenarios in the sensitivity analysis. Scenario 1 is the basic scenario. In this scenario, the three dimensions (economy, environment and society) are assigned equal weights with 5% of the maximum difference of the performance values between alternatives on each criterion as the indifference threshold q, and 25% of the maximum difference of the performance values between alternatives on each criterion as the preference threshold p.

Scenario 2 (ENV), Scenario 3 (ECON) and Scenario 4 (SOC) each emphasize a different dimension. For example, Scenario 3 (ECON) emphasizes the economic dimension so the weight for this dimension is 50% and the weight for the other dimensions are both 25%. The indifference threshold q and the preference threshold p of these scenarios are the same as the thresholds for Scenario 1. Scenario 5 (5%/10%) and Scenario 6 (5%/50%) change the preference threshold p compared with Scenario 1 (Basic). The preference threshold p represents the minimum difference that one alternative is strictly preferred to another on that criterion. As such, increasing the preference threshold will allow more compensation among the criteria. Therefore, in the ranking of cities, 10% represents a situation with relatively less compensation allowed, whereas 50% represents a situation with relatively more compensation allowed.

Table 5.2 The weighting and thresholds for different scenarios

Scenario	*Weighting*			*Thresholds*	
	Economy	*Environment*	*Society*	*Indifference (q)*	*Preference (p)*
1 (Basic)	1/3	1/3	1/3	5%	25%
2 (ENV)	1/2	1/4	1/4	5%	25%
3 (ECON)	1/4	1/2	1/4	5%	25%
4 (SOC)	1/4	1/4	1/2	5%	25%
5 (5%/10%)	1/3	1/3	1/3	5%	10%
6 (5%/50%)	1/3	1/3	1/3	5%	50%

In the sensitivity analysis for the weighted sum method, only the weighting will be changed since there is no indifference or preference threshold in this method. Therefore, Scenarios 1, 2, 3 and 4 will be applied in weighted-sum analysis without thresholds. All scenarios will be applied in the PROMETHEE analysis.

The criteria weights and thresholds for different scenarios are illustrated in Table 5.2.

4 Results

4.1 Comparison of indices and the selection of criteria

4.1.1 Introduction of the indices

The Sustainable Development Goals (SDG) Index is based on 17 Sustainable Development Goals with 169 targets proposed by the United Nations. The 17 Sustainable Development Goals are: 1) No poverty; 2) Zero hunger; 3) Good health and well-being; 4) Quality education; 5) Gender equality; 6) Clean water and sanitation; 7) Affordable and clean energy; 8) Decent work and economic growth; 9) Industry, innovation and infrastructure; 10) Reduced inequalities; 11) Sustainable cities and communities; 12) Responsible consumption and production; 13) Climate action; 14) Life below water; 15) Life on land; 16) Peace, justice and strong institutions and 17) Partnerships for the goals (United Nations, 2015). These goals are part of the 2030 Agenda for Sustainable Development. The SDG Index could help countries recognize their SDG starting point for 2015 (SDGIndex, 2016). The UNECE–ITU Smart Sustainable Cities Indicators are developed by the International Telecommunication Union (ITU) and UNECE (United Nations Economic Commission for Europe) (UNECE, 2015). They are the first indicators to evaluate both the decisions and sustainability of cities. ISO 37120:2014 (Sustainable development of communities – Indicators for city services and quality of life) is published by international Organization for Standardization (ISO, 2014). This set of indicators provides a uniform standard to track and monitor progress on city performance. It is developed for a holistic and integrated approach to sustainable development (ISO, 2014).

4.1.2 Comparison of the indicator categories

The SDG Index has 63 indicators which are grouped into 17 categories. The UNECE–ITU Smart Sustainable Cities Indicators have 47 core indicators and 54 supporting indicators. The indicators are also grouped into 17 categories. ISO 37120:2014 has 47 core indicators and 29 additional indicators. Its indicators are grouped into 19 categories. The indicator categories can be seen in Figure 5.3.

The indicator categories can be grouped into three dimensions: economic and technological dimension, environmental dimension, and social and cultural dimension. The white cells belong to the economic and technological dimension, the light grey cells belong to the environmental dimension and the dark grey cells belong to the social and cultural dimension. Figure 5.3 also shows that both SDG index and ISO 37120:2014 emphasize the social and cultural dimension, while the UNECE–ITU Smart Sustainable Cities Indicators have equal indicator categories for all dimensions.

The indicator categories that are frequently used in these indices are identified. Although the names of indicator categories may not be identical, those categories with similar keywords will be acknowledged together. For example, the categories of **Education** in the UNECE–ITU Smart Sustainable Cities Indicators and **Quality Education** in the SDG index are considered to be a single category in Table 5.3.

Table 5.3 shows the indicator categories that are used by two indices, or all three indices.

Table 5.3 shows that five categories are used in all three indices, which illustrates the importance of these topics in sustainability. However, the individual categories with same keywords may differ in the contents of their indicators. For example, the category of *Safety* in ISO 37120:2014 emphasizes public security; as such, it includes indicators such as 'Number of police officers per 100 000 population' and 'Number of homicides per 100 000 population'. On the other hand, the category of *Safety* in the UNECE–ITU Smart Sustainable Cities Indicators emphasizes disaster and emergency as well as information security; as such, it includes indicators such as 'Disaster mitigation plans' and 'Information security and privacy protection'. Moreover, individual categories with different names in different indices may include similar indicators. For example, the indicator '$PM_{2.5}$ concentration' is included in the category of *Environment* in ISO 37120:2014, but the similar indicator '$PM_{2.5}$ in urban areas' is grouped into the category of *Sustainable Cities and Communities* in the SDG index. Therefore, it is also necessary to acknowledge the indicators in all three indices.

4.1.3 Comparison of the indictors

Since indicators can be described or measured in different ways, there is no fully identical indicator among the indices. However, there are some similar

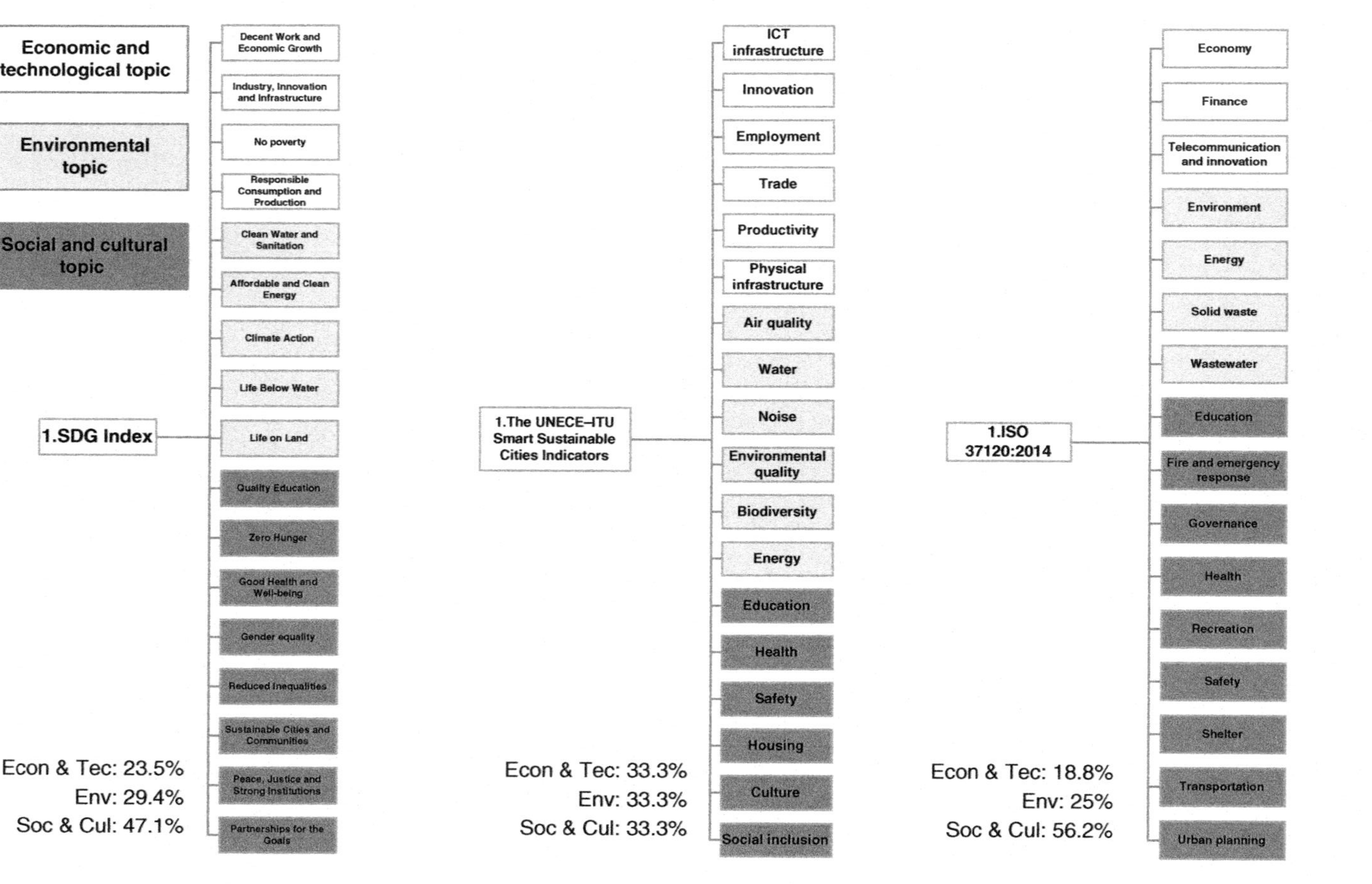

Figure 5.3 The indicator categories

Table 5.3 Indicator categories used by two indices or all three indices

Category used in all three indices	*Category used by two indices*
Innovation	Infrastructure
Water	Employment
Energy	Environment
Health	Safety
Education	Economy
	Governance (strong institutions)

indicators shared by the indices. The shared indicators among all three indices are identified in Table 5.4. Each row shows the shared indicators which belong to different indices. For example, the city's unemployment rate from ISO 37120:2014 is similar to the unemployment rate (%) from the SDG index. Since there is no indicator from the UNECE–ITU Smart Cities Indicators similar to unemployment rate, the column in that row will be empty.

Table 5.4 Shared indicators among three indices

ISO 37120:2014	*SDG Index*	*The UNECE–ITU Smart Cities Indicators*
City's unemployment rate	Unemployment rate (%)	
Percentage of city population living in poverty	Poverty headcount ratio at \$1.90 a day (%)	
Fine particulate matter ($PM_{2.5}$) concentration	$PM_{2.5}$ in urban areas ($\mu g/m^3$)	
Greenhouse gas emissions measured in tonnes per capita		GHG emissions
Average life expectancy		Life expectancy
Number of in-patient hospital beds per 100,000 population		In-patient hospital beds
Under age 5 mortality per 1,000 live births	Under 5 mortality (per 1,000 live births)	
Number of homicides per 100,000 population	Homicides (per 100,000)	
Total collected municipal solid waste per capita	Municipal solid waste (kg/person/year)	
Transportation fatalities per 100,000 population	Traffic deaths (per 100,000)	
Green area (hectares) per 100,000 population		Green areas and public spaces
Percentage of population with access to improved sanitation	Access to improved sanitation (%)	
	R&D expenditures (% GDP)	R&D expenditure
	Gini index (0–100)	Gini coefficient

By summarizing Tables 5.3 and 5.4, the key topics and indicators in sustainability can be identified. The topics and indicators that are shared by two or more indices are: Innovation (R&D expenditures), Water (percentage of population with access to improved sanitation), Energy (greenhouse gas emissions measured in tones per capita), Health (number of in-patient hospital beds per 100,000 population, under age 5 mortality per 1,000 live births, average life expectancy), Education, Employment (city's unemployment rate), Environment (fine particulate matter ($PM_{2.5}$) concentration, total collected municipal solid waste per capita, green area (hectares) per 100,000 population), Safety (number of homicides per 100,000 population), Economy (percentage of city population living in poverty), Governance (Gini coefficient), Transportation (Transportation fatalities per 100,000 population) and Infrastructure. These topics and criteria show the key considerations in assessing urban sustainability.

Based on this finding, as well as on other requirements in selecting criteria (see Section 3.2), 14 sustainability indicators are chosen as criteria in this study. These criteria are grouped into three dimensions: economic, environmental and social.

4.1.4 Selection of criteria

In the economic dimension, the chosen criteria are: GDP per capita (US$ per head, constant PPP), inflation (%) and unemployment rate (%). This selection is based on the key topic *Employment and Economy* which identified by comparing three indices. Unemployment rate (%) is directly used as the criterion for measuring employment. Since the definition of poverty varies by region and the data collected differs across cities, GDP per capita and inflation are used to evaluate the poverty status of a city.

In the environmental dimension, the chosen criteria are: CO_2 emissions (tonnes per capita), share of renewable energy sources (%), $PM_{2.5}$ annual mean concentration (mg/m^3), PM_{10} annual mean concentration (mg/m^3), green area (square meters per person), domestic water consumption (m^3 per capita per year) and volume of municipal waste (kilograms per capita). This selection is based on the key topic *Water, Energy and Environment*. Compared with using fossil fuels, using renewable energy sources has relatively negligible environmental impacts; the share of renewable energy sources can reflect the sustainability of the city in its energy source decisions. However, CO_2 emissions is used as a replacement of the criterion, as most cities tend to collect this data. PM_{10} annual mean concentration is also chosen as the criterion because it has a different source than that of $PM_{2.5}$: PM_{10} comes from industry processes such as crushing or grinding operations or dust on roads, while $PM_{2.5}$ comes from the emissions of vehicles. Therefore, PM_{10} annual mean concentration is also used to reflect the air condition. The data for percentage of population with access to improved sanitation is not often collected by every city. This criterion is replaced by domestic water consumption to represent water in this dimension.

In the social dimension, the chosen criteria are population with tertiary educational attainment (%), Gini coefficient, number of homicides per 100,000 population, life expectancy at birth (years) and underground stations per 100,000 population. Population with tertiary educational attainment represents the topic *Education*. Gini coefficient represents the topic *Governance* as it shows how governments aim to mitigate inequalities in society. The number of homicides per 100,000 population and life expectancy at birth represent the topics *Safety* and *Health*, respectively. Underground stations per 100,000 population represents the topic *Traffic*, as public transport services play a key role in urban sustainability.

The topics *Innovation* and *Infrastructure* are not represented by any criterion in this study due to the lack of availability of data for the indicators of these two topics in Global South cities.

Therefore, the criteria selected for assessing the urban sustainability are GDP per capita (US$ per head, constant PPP), inflation (%) and unemployment rate (%), CO_2 emissions (tonnes per capita), share of renewable energy sources (%), $PM_{2.5}$ annual mean concentration (mg/m^3), PM_{10} annual mean concentration (mg/m^3), green area (square meters per person), domestic water consumption (m^3 per capita per year), volume of municipal waste (kilograms per capita), population with tertiary educational attainment (%), Gini coefficient, number of homicides per 100,000 population, life expectancy at birth (years) and underground stations per 100,000 population. This is also based on the research of Shmelev (2017a) in the multidimensional sustainability assessment of the world's large cities and also closely linked to the 17 Sustainable Development Goals proposed by the United Nations (United Nations, 2015).

4.2 Performance of Global South cities in sustainability indicators

The values of sustainability indicators of the Global South cities are listed in Figure 5.3. Seventeen cities are included in this research: seven Asian cities, three African cities, one North American city and six South American cities in the dataset.

Several findings are evident from Figure 5.3. In the economic dimension, only Shenzhen and Almaty have more than 30,000 US$ per person in 2014 among the Global South cities in the dataset ($35,298 and $33,698 per person correspondingly). This is already close to the GDP per capita of some Global North cities such as Barcelona and Adelaide (around $36,000 per person). Four cities have less than $10,000 per person; Nairobi has the lowest performance in GDP with only $3,824 per person. In terms of unemployment, Nairobi has significantly poor performance in this indicator (48%). The city with second highest unemployment rate is Johannesburg (25%). Both are African cities. The unemployment rates of the rest of the Global South cities are all around 10% or lower. The inflation ranges from 0.3% to 10.6%. Buenos Aires has the highest inflation rate among these cities.

In the environmental dimension, Shenzhen and Almaty are the only two cities which have more than 10 tonnes of CO_2 emissions per capita. This is partly due to their high economic performances. Since renewable energy has relatively low carbon emissions compared with fossil fuels, the share of renewable energy sources is another factor which influences carbon emission. The cities with more than 80% renewable energy sources have significantly low carbon emissions (lower than 2 tonnes per capita). These cities are Bogota (84.5%), Kampala (91.1%) and São Paulo (100%). It also can be seen that Asian cities have a relatively lower share of renewable energy sources than South American cities. The dataset also shows that the Asian cities have relatively worse air quality than South American cities. Delhi has the highest $PM_{2.5}$ annual mean concentration (122.00 mg/m^3) which is almost 10 times higher than that of Buenos Aires (14.00 mg/m^3), which has the best air quality among the Global South cities in the dataset. In terms of green space, Quito is significantly higher than other Global South cities, with 1,491 square meters per person. However, São Paulo, Lima, Buenos Aires, Bogota, Mumbai and Istanbul each have less than 10 square meters' green area per person. Buenos Aires and Almaty have relatively higher domestic water consumption compared with other Global South cities with 244.30 and 226.50 m^3 per capita per year, respectively. Mumbai is the city with the lowest municipal waste generation, generating only 84.70 kg per capita per year.

In the social dimension, Beijing and Shanghai have more than 30% of tertiary educational attainment while Johannesburg only has 7.9%. The Gini coefficient of most Global South cities in the dataset ranges from 0.4 to 0.5, which already exceeds the international warning level (0.4). Almaty has the lowest Gini coefficient (0.27) while the Gini coefficient of Delhi has reached 0.6. In terms of homicide rate, Johannesburg is the only city which has more than 20 homicides per 100,000 people. Most Global South cities in this dataset have around 75 years' life expectancy. Shanghai has the longest life expectancy (82 years), while Mumbai has the shortest life expectancy (56.8 years). There are five cities which have no underground stations as of 2014: Johannesburg, Kampala, Nairobi, Bogota and Quito. All African cities in this dataset have no underground stations as of 2014.

4.3 Results of weighted-sum analysis (Scenario 1)

The results of weighted-sum analysis (Scenario 1) are shown in Table 5.5.

In this table, un-highlighted cells denote Asian cities, cells highlighted in light grey denote South American cities, cells highlighted in dark grey denote African cities, and the cell highlighted in black denotes the North American city. The cities are ranked based on their aggregated performance values of sustainability indicators, which is the sustainability index, from largest to the smallest. This is based on the equal weighting on each dimension.

According to the ranking of cities, the cities from China (Shanghai, Shenzhen and Beijing) have high performance in sustainability and rank among the top Global South cities. Shanghai and Shenzhen have similar levels of sustainability based on values they achieved in the weighted-sum analysis; Beijing ranks relatively lower. The cities from South America also rank relative high among the

Table 5.5 Ranking based on WSM (Scenario 1)

City	*Sustainability index*
Shenzhen	0.709797
Shanghai	0.708439
Beijing	0.671265
São Paulo	0.604176
Lima	0.601931
Quito	0.601873
Bogota	0.591113
Mexico City	0.576679
Istanbul	0.555553
Almaty	0.540419
Buenos Aires	0.527386
Rio de Janeiro	0.502782
Mumbai	0.493196
Kampala	0.488941
Delhi	0.470344
Nairobi	0.401961
Johannesburg	0.331292

Global South cities. However, the cities from Africa have relatively poor performance in sustainability.

4.4 Sensitivity analysis of the weighted-sum method (Scenarios 2, 3, 4)

Table 5.6 shows the ranking of Scenarios 2 (ENV), 3 (ECON) and 4 (SOC) based on the weighted-sum method.

Scenarios 2, 3 and 4 emphasize the environmental, economic and social dimensions, respectively. In each case, the emphasized dimension is assigned with 50% of the total weight. Table 5.6 shows that Shanghai and Shenzhen are ranked among the top Global South cities in all scenarios. This means the two cities have similar levels of sustainability in each dimension. Specifically, Shanghai is slightly more sustainable than Shenzhen in terms of environment and society, while Shenzhen has higher performance in the economic dimension.

It is also important to acknowledge that Quito ranks first in the environmental dimension due to the city's high performance in most environmental criteria, especially in terms of the criterion of green space. With the exceptions of share of renewable energy sources and volume of municipal waste (around 0.6), Quito's normalized performance values of environmental indicators are approximately 0.8 or higher. Table 5.6 also shows that the African cities tend to have relative low ranking in all scenarios due to relatively poor performance in all dimensions.

Table 5.6 Ranking based on WSM (Scenarios 2, 3 and 4)

Scenario 2 (ENV)		*Scenario 3 (ECON)*		*Scenario 4 (SOC)*	
City	*Sustainability index*	*City*	*Sustainability index*	*City*	*Sustainability index*
Quito	0.655707728	Shenzhen	0.7726236	Shanghai	0.746831
Shanghai	0.641137586	Shanghai	0.73734856	Shenzhen	0.718756
Shenzhen	0.638011309	Beijing	0.72227438	Beijing	0.685878
Bogota	0.624217168	Lima	0.62848649	Lima	0.580452
São Paulo	0.623333642	Bogota	0.61749755	São Paulo	0.574057
Beijing	0.605642094	São Paulo	0.61513635	Almaty	0.561504
Lima	0.596853412	Mexico City	0.60787243	Mexico City	0.556657
Mexico City	0.565508893	Almaty	0.59781929	Buenos Aires	0.554007
Istanbul	0.553423383	Quito	0.59622435	Istanbul	0.55376
Rio de Janeiro	0.518363041	Istanbul	0.55947711	Quito	0.553688
Kampala	0.516256709	Rio de Janeiro	0.52444822	Bogota	0.531624
Mumbai	0.515526424	Buenos Aires	0.51840053	Mumbai	0.471128
Buenos Aires	0.509750047	Kampala	0.49293596	Delhi	0.469701
Nairobi	0.481794729	Mumbai	0.49293523	Rio de Janeiro	0.465535
Almaty	0.46193197	Delhi	0.48407206	Kampala	0.45763
Delhi	0.457259718	Johannesburg	0.37837895	Nairobi	0.375394
Johannesburg	0.359770343	Nairobi	0.34869266	Johannesburg	0.255727

4.5 Results from the PROMETHEE method

4.5.1 PROMETHEE II complete ranking (Scenario 1)

The complete ranking of the Global South cities based on their sustainability (see Table 5.7 and Figure 5.5).

This ranking is based on Scenario 1 in which each dimension is weighted equally. In terms of the difference in performance values between alternatives, a 5% maximum difference is taken as the indifference threshold q, while a 25% maximum difference is taken as the preference threshold p.

Figure 5.4 presents a clear visualization of the level of sustainability of these cities. Table 5.7 shows the complete ranking of the cities with their regional attributes (Asia, Africa, South America . . ., etc.). According to the figure and table, Shenzhen is the most sustainable city among these Global South cities. Shanghai is slightly lower than Shenzhen in sustainability but it is still much higher than all other cities. Beijing ranks third in the dataset. Four South American cities (Lima, Bogota, São Paulo and Quito) and one North American city (Mexico City) have similar levels of sustainability. The cities mentioned here achieve positive value in their net outranking flow; in other words, they outrank the cities from the dataset. However, the rest of the Global South cites (Istanbul, Almaty, Buenos Aires, Kampala, Mumbai, Delhi, Rio de Janeiro, Nairobi, Johannesburg) have a negative value in their net outranking flow since they are relatively outranked on these sustainability indicators.

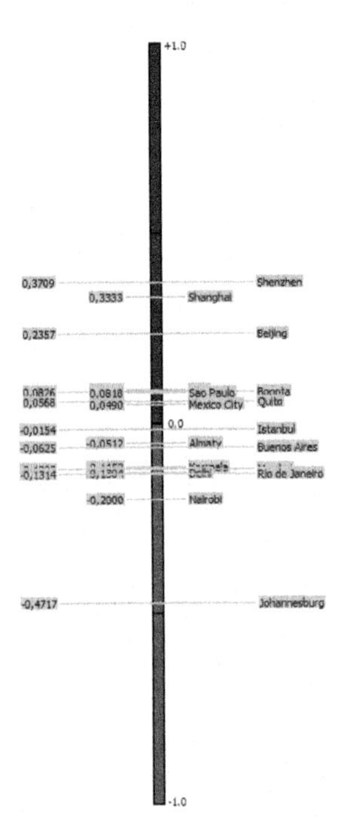

Figure 5.4 PROMETHEE II complete ranking (Scenario 1)

Table 5.7 Complete ranking based on PROMETHEE II (Scenario 1)

PROMETHEE II ranking	*Net outranking value*
Shenzhen	0.3709
Shanghai	0.3333
Beijing	0.2357
Lima	0.0881
Bogota	0.0826
São Paulo	0.0818
Quito	0.0568
Mexico City	0.0490
Istanbul	−0.0154
Almaty	−0.0512
Buenos Aires	−0.0625
Kampala	−0.1152
Mumbai	−0.1205
Delhi	−0.1304
Rio de Janeiro	−0.1314
Nairobi	−0.2000
Johannesburg	−0.4717

Among these cities are Kampala, Mumbai, Delhi and Rio de Janeiro, which have similar levels of sustainability. Johannesburg, with the lowest net outranking flow, is much less sustainable than other Global South cities.

4.5.2 Sensitivity analysis of PROMETHEE II complete ranking (Scenario 2,3,4)

In this section, the weighting of each dimension will be changed for different scenarios, while the thresholds remain the same as those applied in Scenario 1.

Table 5.8 outlines the complete ranking in different scenarios. Shenzhen and Shanghai rank top in all scenarios. Beijing ranks third in Scenario 1 (Basic), Scenario 3 (ECON) and Scenario 4 (SOC) while it only ranks sixth in Scenario 2 (ENV), the environmental dimension. Beijing has relatively poor performance values on environmental criteria compared with the criteria in other dimensions. Johannesburg ranks at the bottom in all scenarios which indicate that it is less sustainable than other cities in all dimensions. According to Scenario 2, where the environmental dimension is assigned more weight, the difference among the net outranking value of these Global South cities (except Johannesburg) is small since their positions are close, seen in Figure 5.5. These cities are considered to have similar levels of sustainability

(a)

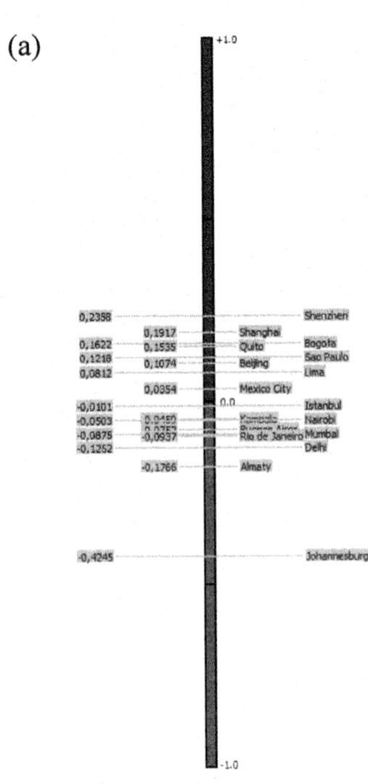

+1.0
0,2398
Shenzhen
0,1917
Shanghai
0,1622
Bogota
0,1535
Quito
0,1218
Sao Paulo
0,1074
Beijing
0,0812
Lima
0,0354
Mexico City
0.0
-0,0101
Istanbul
-0,0503
Nairobi
-0,0875
Mumbai
-0,0937
Rio de Janeiro
-0,1252
Delhi
-0,1766
Almaty
-0,4245
Johannesburg
-1.0

(b)

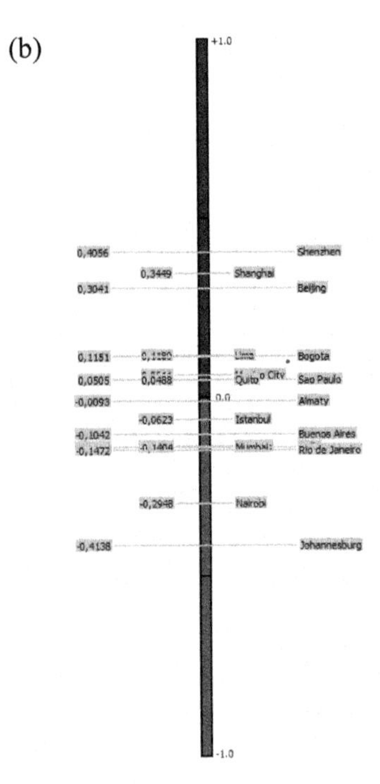

+1.0
0,4056
Shenzhen
0,3449
Shanghai
0,3041
Beijing
0,1151
Lima
Bogota
0,0505
0,0488
Quito
Sao Paulo
0.0
-0,0093
Almaty
-0,0623
Istanbul
-0,1042
Buenos Aires
-0,1472
Rio de Janeiro
-0,2948
Nairobi
-0,4138
Johannesburg
-1.0

(c)

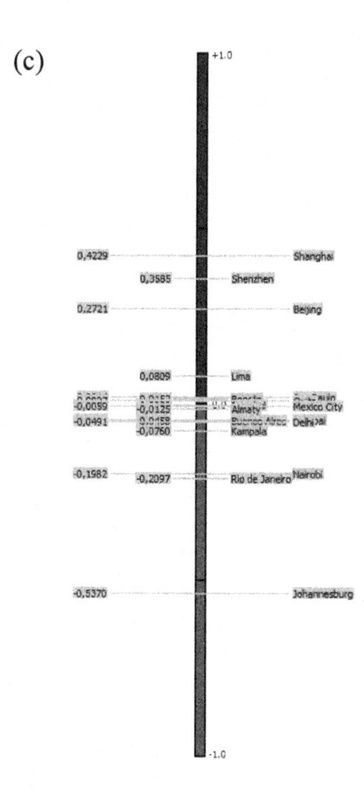

Figure 5.5 PROMETHEE II complete ranking. Scenario 2 (a), Scenario 3 (b), Scenario 4 (c)

Table 5.8 Complete ranking based on PROMETHEE II (Scenarios 2, 3 and 4)

Scenario 2 (ENV)	*Scenario 3 (ECON)*	*Scenario 4 (SOC)*
Shenzhen	Shenzhen	Shanghai
Shanghai	Shanghai	Shenzhen
Bogota	Beijing	Beijing
Quito	Lima	Lima
São Paulo	Bogota	São Paulo
Beijing	Mexico City	Bogota
Lima	São Paulo	Quito
Mexico City	Quito	Istanbul
Istanbul	Almaty	Mexico City
Kampala	Istanbul	Almaty
Nairobi	Buenos Aires	Mumbai
Buenos Aires	Kampala	Buenos Aires
Mumbai	Delhi	Delhi
Rio de Janeiro	Mumbai	Kampala
Delhi	Rio de Janeiro	Nairobi
Almaty	Nairobi	Rio de Janeiro
Johannesburg	Johannesburg	Johannesburg

in the environmental dimension in PROMETHEE II. In Scenario 4, where the social dimension is emphasized, there is a vast disparity between the net outranking values of the most and least sustainable cities, while the positions of eleven cities (Lima, São Paulo, Bogota, Quito, Istanbul, Mexico City, Almaty, Mumbai, Buenos Aires, Delhi, Kampala) are quite close to one another other. In this scenario, there is an exceptional difference in sustainability between the most and least sustainable cities in the social dimension, in which eleven other cities have similar levels of sustainability.

4.5.3 Sensitivity analysis of PROMETHEE II complete ranking (Scenarios 5 and 6)

In Scenario 5 (5%/10%) and Scenario 6 (5%/50%), each dimension is equally weighted, which is the same as Scenario 1. The indifference threshold is still 5% of the maximum difference of the performance values between alternatives but the preference threshold is changed to 10% in Scenario 5 and 50% in Scenario 6. The complete ranking of Scenario 5 and Scenario 6 is shown in Table 5.9 and the visualized ranking result is shown in Figure 5.6.

Table 5.9 Complete ranking based on PROMETHEE II (Scenarios 5 and 6)

Scenario 5 (5%/10%)	*Scenario 6 (5%/50%)*
Shenzhen	Shenzhen
Shanghai	Shanghai
Beijing	Beijing
Quito	Lima
Lima	São Paulo
São Paulo	Bogota
Bogota	Quito
Mexico City	Mexico City
Mumbai	Istanbul
Kampala	Almaty
Nairobi	Buenos Aires
Rio de Janeiro	Mumbai
Istanbul	Rio de Janeiro
Buenos Aires	Kampala
Delhi	Delhi
Almaty	Nairobi
Johannesburg	Johannesburg

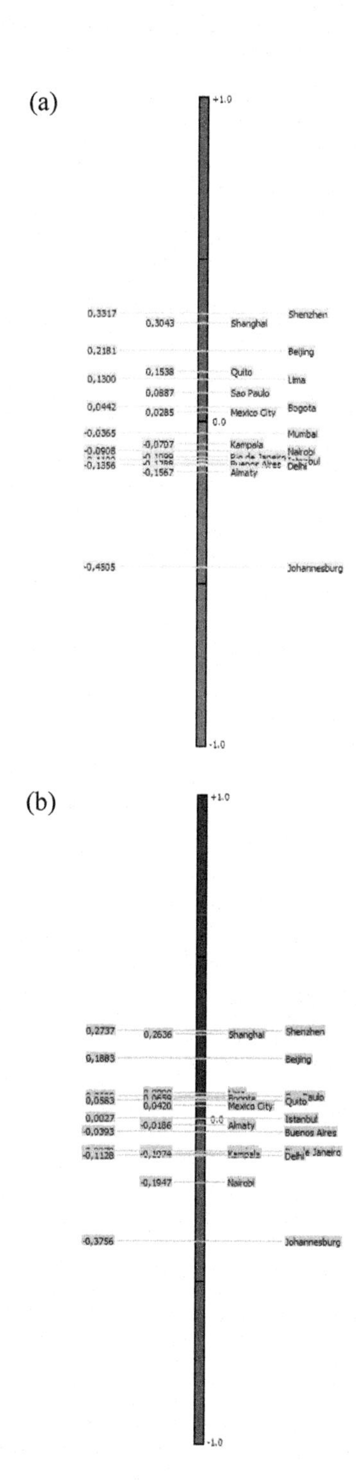

Figure 5.6 PROMETHEE II complete ranking. Scenario 5 (a), Scenario 6 (b)

Scenarios 5 and 6 use different preference thresholds, but the ranking of the top three cities (Shenzhen, Shanghai and Beijing) remains unchanged. When more compensation is allowed, these cities do not change their relative sustainability, which means they are not particularly better or worse than other cities on a certain criterion. Also worth consideration is that the ranking of Kampala and Nairobi dropped from 10th and 11th to 14th and 16th respectively when more compensation is allowed. This is because some indicators of these two cities have particularly poor performance value. For example, Kampala has poor performance in the indicators such as GDP per capita, $PM_{2.5}$ annual mean concentration, life expectancy at birth and underground stations per 100,000 population, while Nairobi has poor performance in the indicators such as GDP per capita, unemployment rate, green area and underground stations per 100,000 population. In the opposite case, Almaty significantly improves its ranking from 16th to 10th in Scenario 6 as it has particularly high performance value in indicators such as GDP per capita, unemployment rate and Gini coefficient which can compensate for its 'weak' indicators.

4.5.4 PROMETHEE I partial ranking (Scenario 1)

Incomparability is considered in PROMETHEE I. The partial ranking of the cities in Scenario 1 is shown in Figure 5.7. The PROMETHEE Network, based on PROMETHEE I and as shown in Figure 5.8, can outline the preferences among the Global South cities.

There are eight cities (Shenzhen, Shanghai, Beijing, Lima, Bogota, São Paulo, Quito and Mexico City) which achieve positive values in net outranking flow in the complete ranking based on PROMETHEE II (Scenario 1). These cities are considered more sustainable than other Global South cities in the dataset. To obtain a clear visualization of the incomparability among these cities, the eight cities are selected and ranked by PROMETHEE I, isolated from the other cities in the dataset. The results are shown in Figure 5.9.

In Figure 5.7, the left column corresponds to the value of leaving (or positive) ranking flow and the right column corresponds to the value of entering (or negative) ranking flow. The value of leaving (or positive) ranking flow is represented by Phi+ and the value of entering (or negative) ranking flow is represented by Phi–. Only when city A with both higher value in leaving ranking flow and lower value in entering ranking flow, when compared to city B, is city A considered to be more sustainable than city B.

According to Figures 5.7 and 5.8, Shenzhen has the highest value in leaving ranking flow and the lowest value in entering ranking flow, so it is considered the most sustainable city. Shanghai ranks second, followed by Beijing at third. However, the four South American cities (Lima, Bogota, São Paulo, Quito), despite being considered less sustainable than Beijing, are incomparable with each other in PROMETHEE I. From Figure 5.10, it can be inferred that Quito has a higher value in leaving ranking flow (0.22) than São Paulo (0.20) but its value in entering ranking flow (0.33) is also higher than São Paulo's (0.31). It is

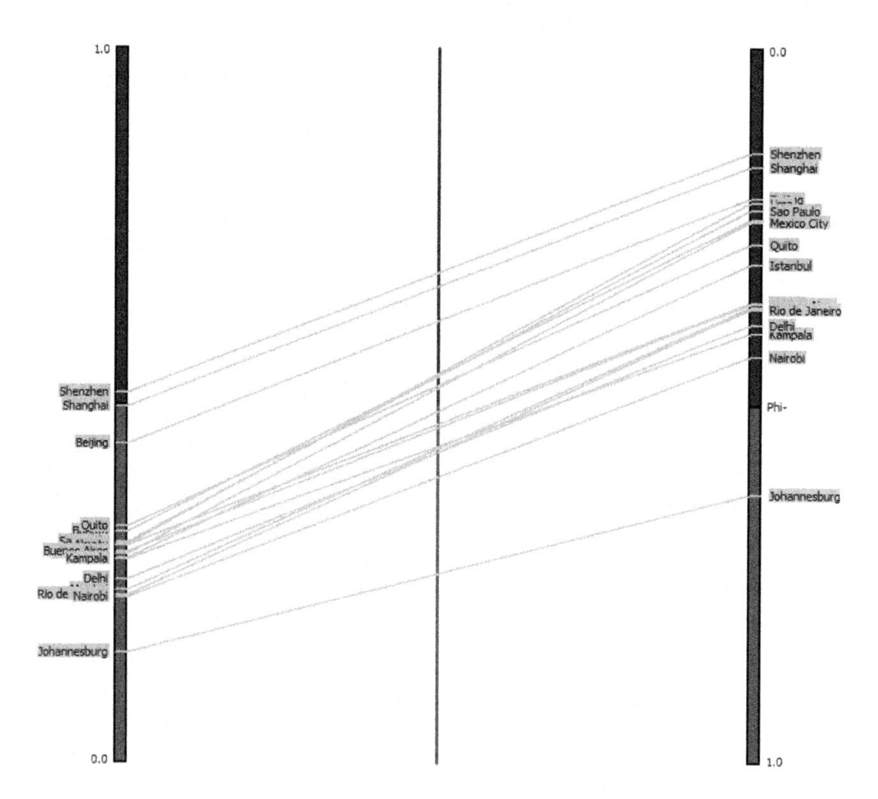

Figure 5.7 PROMETHEE I partial ranking (Scenario 1)

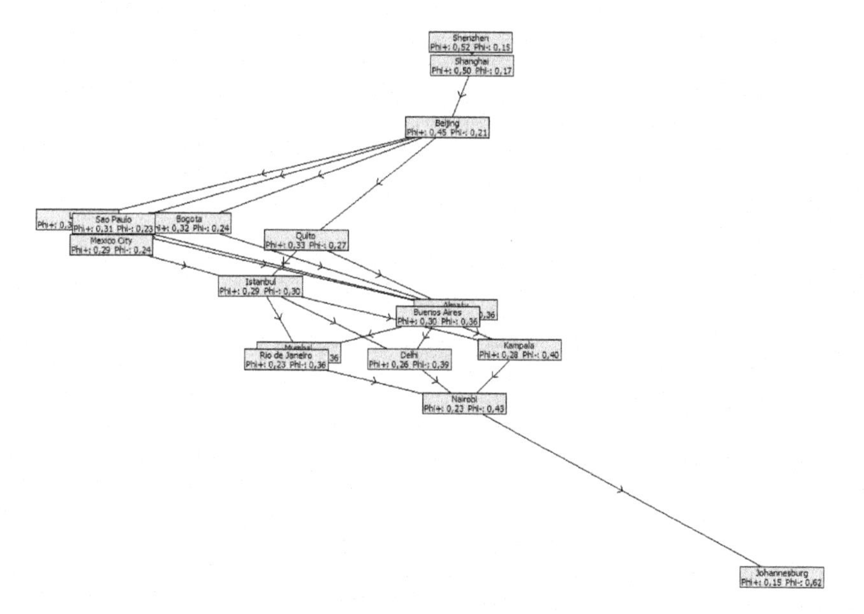

Figure 5.8 PROMETHEE network (Scenario 1, all cities)

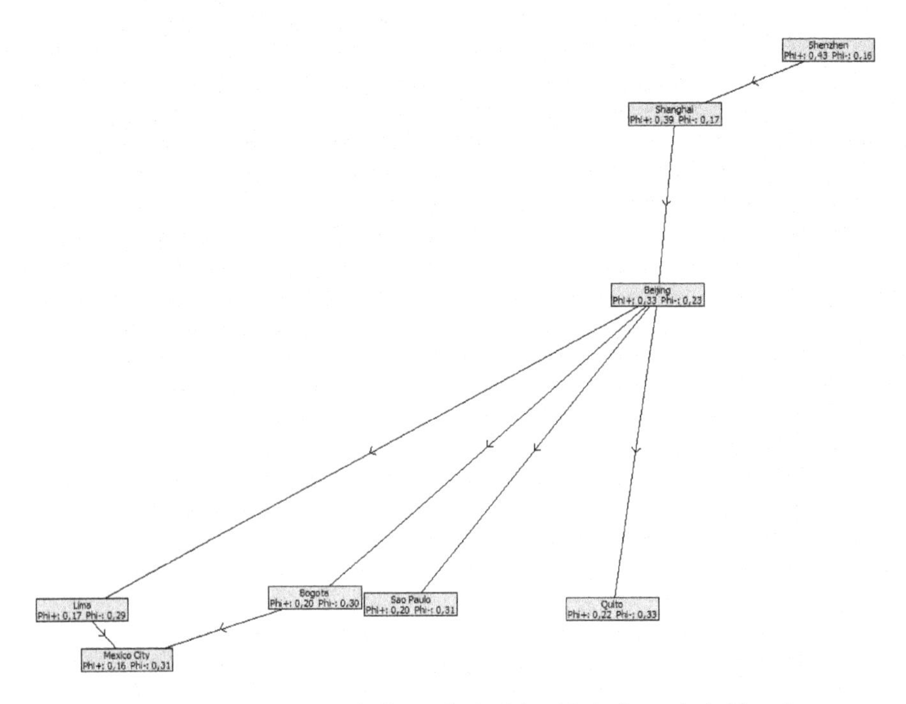

Figure 5.9 PROMETHEE network (Scenario 1, eight cities). Scenario 1, Shenzhen (Phi+:0.426 Phi−:0.161), Shanghai (Phi+:0.386 Phi−:0.170)

clear that Quito dominates most cities in the sustainability indicators like green area, unemployment rate and $PM_{2.5}$ annual mean concentration, but has considerably poor performance in GDP per capita, population with tertiary educational attainment and underground stations per 100,000 population among all Global South cities. São Paulo has high performance value in criteria such as share of renewable energy sources, CO_2 emissions and life expectancy at birth, but it achieves low scores for green area and volume of municipal waste. Therefore, Quito is incomparable with Lima. Mexico City, with lower value in leaving ranking flow and higher value in entering ranking flow, is less sustainable than Lima and Bogota. When all cities are ranked in PROMETHEE I (Figure 5.8), Nairobi and Johannesburg are less sustainable than all the other Global South cities in this research, regardless of incomparability.

4.5.5 Sensitivity analysis of PROMETHEE I partial ranking (Scenarios 5 and 6)

Since changes to the weights are used to check cities' sensitivity to dimensions as discussed in Section 4.5.2, Scenarios 2 (ENV), 3 (ECON) and 4 (SOC) are not applied in PROMETHEE I partial ranking. Rather, Scenarios 5 (5%/10%)

and 6 (5%/50%) are applied in PROMETHEE I partial ranking to observe the impacts of changes to the preference threshold on the incomparability among Global South cities. Only the eight cities (Shenzhen, Shanghai, Beijing, Lima, Bogota, São Paulo, Quito and Mexico City) which achieve positive values in net outranking flow in the complete ranking are ranked in Scenarios 5 and 6; the reason for this process is that the analysis focuses on testing whether Shenzhen is still the most sustainable city when incomparability and different preference thresholds are considered. The results of PROMETHEE I partial ranking based on Scenario 5 are presented in Figure 5.10 and the results of PROMETHEE I partial ranking based on Scenario 6 are presented in Figure 5.11.

Compared with the partial ranking based on Scenario 1 (basic, 5%/25%), changes to the preference threshold leads to incomparability between Shenzhen and Shanghai. Explanations of both a decrease and increase of the preference threshold are discussed here.

Figure 5.10 shows that Shanghai is incomparable with Shenzhen and Lima is incomparable with São Paulo in Scenario 5 when the PROMETHEE I is applied. However, Shanghai and Shenzhen are still the top two cities among the seven cities. Figure 5.11 shows that Shanghai and Shenzhen, which are the top two cities in terms of sustainability, are still incomparable with each other. It’s also interesting to see that the four South American cities are incomparable with each other in Scenario 6.

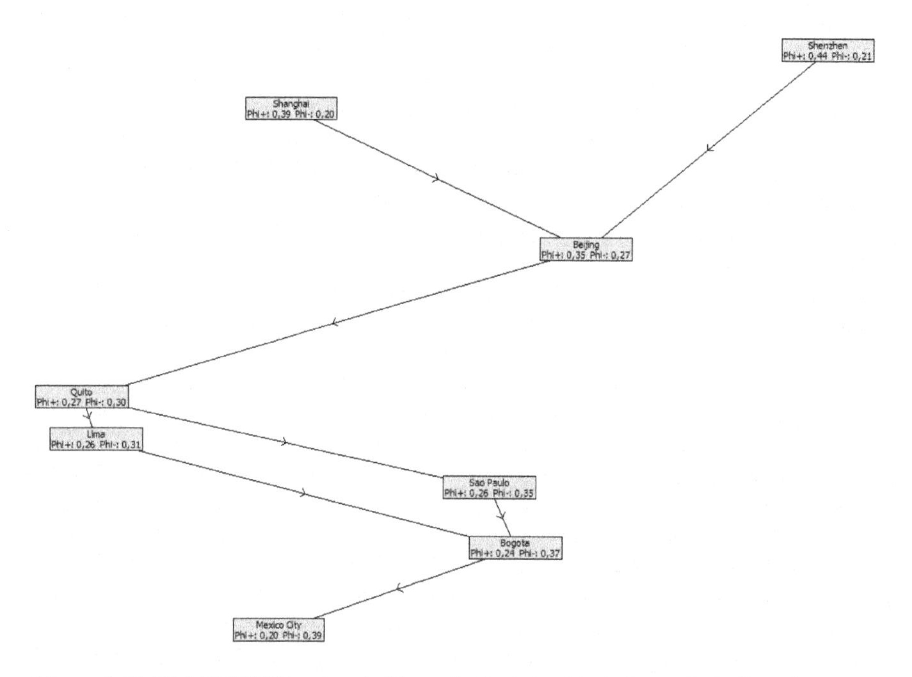

Figure 5.10 PROMETHEE network (Scenario 5, eight cities). Scenario 5, Shenzhen (Phi+:0.437 Phi−:0.210), Shanghai (Phi+:0.392 Phi−:0.203)

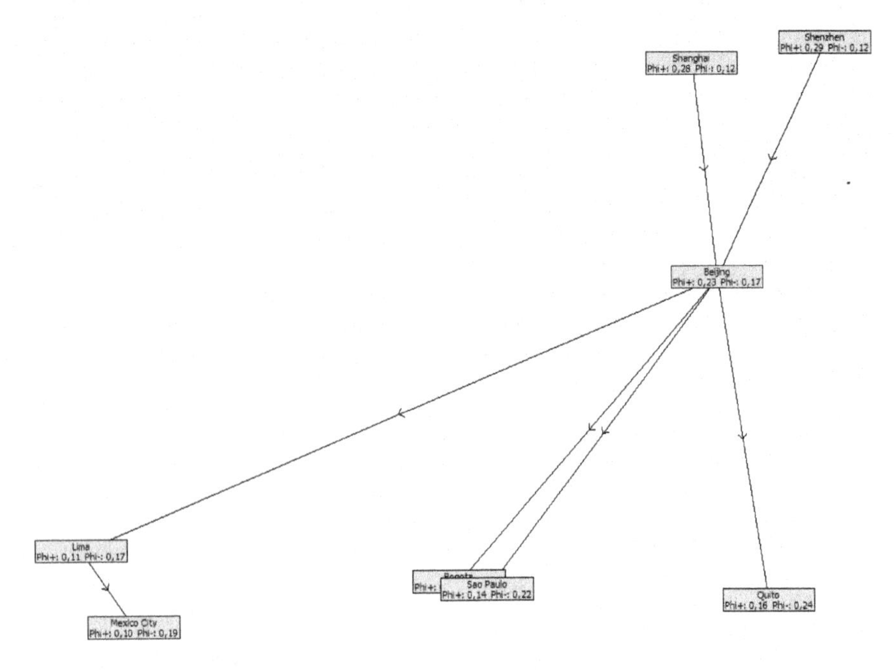

Figure 5.11 PROMETHEE network (Scenario 6, eight cities). Scenario 6, Shenzhen (Phi+:0.292 Phi−:0.124), Shanghai (Phi+:0.276 Phi−: 0.117)

4.5.6 Results of PROMETHEE-GAIA (Scenario 1, all cities)

By transferring a multidimensional problem into a two-dimensional problem, PROMETHEE-GAIA provides the visualization of results obtained by the PROMETHEE method. The results obtained by PROMETHEE-GAIA in Scenario 1 (all cities) is shown in Figure 5.12.

Principal components analysis is used in PROMETHEE-GAIA to reduce the dimensions of the problem. Three principal components are computed in the principal components analysis: U, V and W. U contains the maximum possible quantity of information. V provides the maximum additional information orthogonal to U. W provides the maximum additional information orthogonal to both U and V. Figure 5.12 shows that the U–V plane contains about 58% of total information, the U–W plane contains 45% of total information and the W–V plane only contains 38% of total information.

The left plane in Figure 5.12 shows the results of the U–V plane. The positions of the alternative cities show the relationships between them; the cities which have similar profiles are close to each other. Therefore, Kampala, Mumbai and Delhi are similar based on their evaluations from the sustainability indicators. Johannesburg, Quito, Bogota and Nairobi are also similar, but the remainder of cities are quite different from each other.

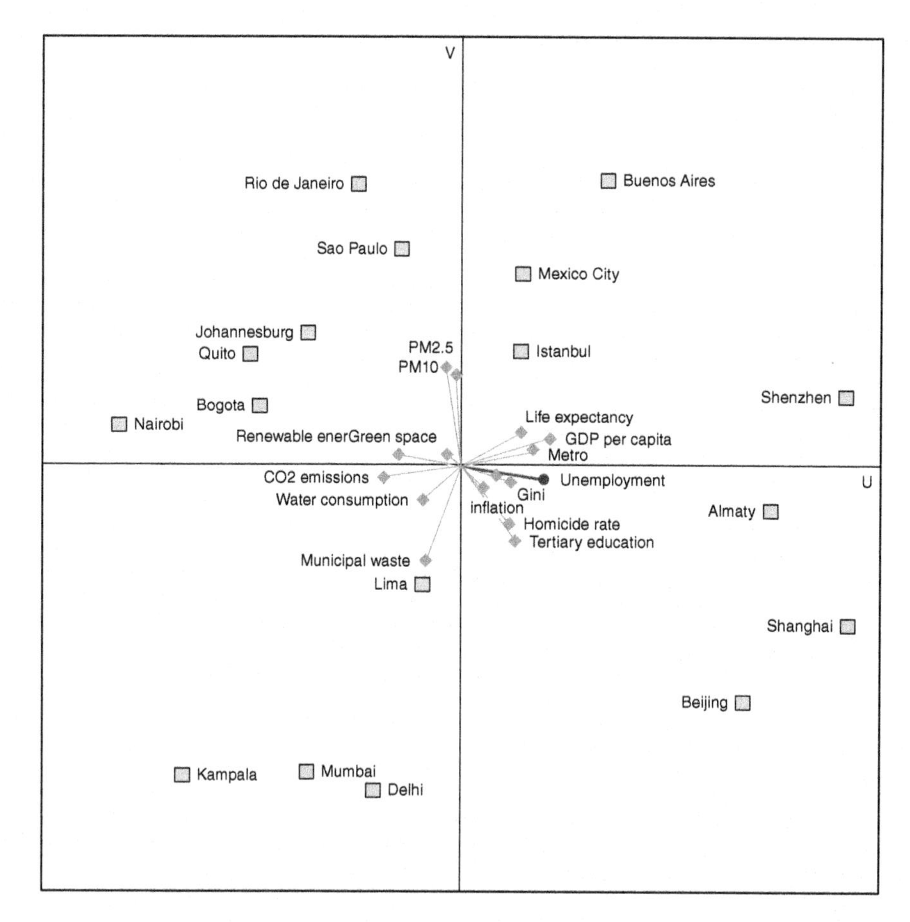

Figure 5.12 PROMETHEE-GAIA (Scenario 1, all cities)

In this U–V plane, only 57.7% of information is included. This is mainly because there are too many criteria in this analysis and the alternatives cities are quite different from each other. When the information level is close to or above 70%, the results of the GAIA plane are considered to be robust. Therefore, the eight cities (Shenzhen, Shanghai, Beijing, Lima, Bogota, São Paulo, Quito and Mexico City) are selected and applied in PROMETHEE-GAIA. The results are shown in Section 4.5.7.

4.5.7 Results of PROMETHEE-GAIA (Scenario 1, eight cities)

The results of applying eight cities in PROMETHEE-GAIA are shown in Figure 5.13.

Figure 5.13 shows this U–V plane gathers 75% of the total information which is more reliable than the initial proportion of 57.7%. According to Figure 5.13, Mexico City and São Paulo are similar based on the sustainability indicators, while Bogota and Quito are also close to each other. The remainder of cities are

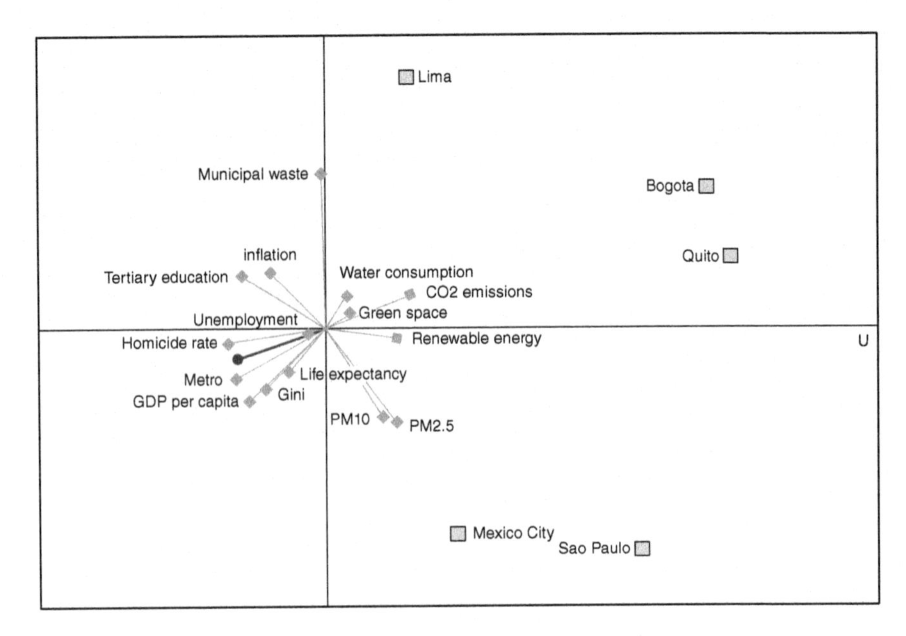

Figure 5.13 PROMETHEE-GAIA (Scenario 1, eight cities)

somewhat different from each other which means each city tends to have its own feature in sustainability.

The criteria axes in the U–V plane show how well the criteria are related to each other. The axes' proximity to each other indicates the similarity of preferences in the criteria: similar criteria have axes in the same direction, whereas conflicting criteria have axes in opposite directions. Figure 5.13 shows that the $PM_{2.5}$ concentration and PM_{10} concentration are expressing similar preferences, while these criteria are conflicted with inflation, and population with tertiary educational attainment. The share of renewable energy is also conflicted with inflation and population with tertiary educational attainment, but is expressing different preferences with both $PM_{2.5}$ concentration and PM_{10} concentration.

Unemployment rate, Gini coefficient, number of homicides per 100,000 population, life expectancy at birth, underground stations per 100,000 population and GDP per capita are all expressing similar preferences but these criteria are conflicted with green area, domestic water consumption and CO_2 emissions. The social criteria are also conflicted with the criterion volume of municipal waste; however, this criterion does not express similar preferences with other environmental criteria.

Knowledge of the conflicting criteria can help decision makers to implement trade-offs based on their objectives. In the case when only these eight Global South cities are considered, the cities with better air quality and a large share of renewables as their energy sources would tend to have poor performance in education and Gini coefficient. Similarly, the cities which have high performance in social dimension and GDP per capita would tend to have high carbon emissions, low green space and high water consumption and waste generation. These conflicts are analyzed based on the characteristics of cities in Section 5.2.

The length of the criteria axes is also of importance: when the axis is longer, its corresponding criterion is more discriminative. Figure 5.13 shows that the volume of municipal waste is certainly a discriminative criterion while the green area and unemployment rate cannot effectively differentiate the alternatives to this extent. The highest and lowest unemployment rates of these cities are 48% and 1.3% respectively, but most cities' unemployment rates are around or below 10%. This is the case for the green area criterion. The performance values of green area ranges from 1 to 1,494 square meters per person but the green area of most cities is less than 60 square meters per person. However, the performance values on the municipal waste criterion of the cities are generally well-distributed from 84kg to 584kg per capita per year. Therefore, the volume of municipal waste is a discriminative criterion.

The decision axis is the dark-coloured axis in the U–V plane in Figure 5.13. The direction of this axis indicates whether the criteria are in accordance with the PROMETHEE rankings. The criteria which have the axes in similar orientation with the decision axis are unemployment rate, Gini coefficient, number of homicides per 100,000 population, life expectancy at birth, underground stations per 100,000 population and GDP per capita. This means the cities with high rankings in terms of sustainability among these eight Global South cities achieve high scores in these social indicators and GDP per capita. Conversely, the axes of the criteria green area, domestic water consumption and CO_2 emissions have opposite directions with the decision axis. This indicates that the cities with high rankings have relatively poor performance in these environmental criteria.

5 Discussion and conclusion

5.1 Which is the most sustainable Global South city?

5.1.1 Comparing results of strong and weak sustainability methods

Based on equal weighting on each dimension, the ranking obtained from the strong sustainability method (weighted-sum method) and weak sustainability method (PROMETHEE II) are summarized in Table 5.10.

Scenarios 5 and 6 use equal weighting but different thresholds compared with Scenario 1. Therefore, the ranking based on these two scenarios are also compared in Table 5.10. From the table, it is clear that Shenzhen ranks first, Shanghai ranks second and Beijing ranks third in all equal weighting scenarios

Table 5.10 Ranking based on equal weighting on each dimension by different methods

WSM Scenario 1	*PROMETHEE II Scenario 1 (5%/25%)*	*PROMETHEE II Scenario 5 (5%/10%)*	*PROMETHEE II Scenario 6 (5%/50%)*
Shenzhen	Shenzhen	Shenzhen	Shenzhen
Shanghai	Shanghai	Shanghai	Shanghai
Beijing	Beijing	Beijing	Beijing
São Paulo	Lima	Quito	Lima
Lima	Bogota	Lima	São Paulo
Quito	São Paulo	São Paulo	Bogota
Bogota	Quito	Bogota	Quito
Mexico City	Mexico City	Mexico City	Mexico City
Istanbul	Istanbul	Mumbai	Istanbul
Almaty	Almaty	Kampala	Almaty
Buenos Aires	Buenos Aires	Nairobi	Buenos Aires
Rio de Janeiro	Kampala	Rio de Janeiro	Mumbai
Mumbai	Mumbai	Istanbul	Rio de Janeiro
Kampala	Delhi	Buenos Aires	Kampala
Delhi	Rio de Janeiro	Delhi	Delhi
Nairobi	Nairobi	Almaty	Nairobi
Johannesburg	Johannesburg	Johannesburg	Johannesburg

based on PROMETHEE II and WSM. This consistency illustrates the robustness of the results. The relative rankings of South American cities São Paulo, Lima, Quito and Bogota change, although their overall positions do not; all four nevertheless rank high among these Global South cities in Table 5.10, just after Beijing. These four cities can be viewed as having the same level of sustainability. The African cities of Kampala, Nairobi and Johannesburg all rank relatively low in the dataset. Johannesburg ranks last in all cases, indicating that it is something of an outlier compared with other Global South cities in terms of sustainability. Nairobi ranks second last in all but one case; however, in Scenario 5, analyzed by PROMETHEE II, Nairobi is ranked relatively higher. In this scenario, the lowest preference threshold is applied among all scenarios which means least compensation is allowed in this scenario. Therefore, indicators like GDP per capita, unemployment rate, life expectancy at birth and underground stations per 100,000 population which achieve very poor performance values would not be compensated too much by other indicators. This makes Nairobi achieve a higher net outranking value in PROMETHEE II compared with other scenarios.

The smallest preference threshold allows the least compensation among criteria. Scenario 1 (5%/25%) allows more compensation than Scenario 5 (5%/10%)

and Scenario 6 (5%/50%) allows the highest compensation among these scenarios in PROMETHEE II. The weighted sum allows compensation depending on the weights assigned to the criteria. Therefore, when different degrees of compensation are considered, Shenzhen and Shanghai remain the first- and the second-ranked among the Global South cities. This shows the results are reliable in both strong sustainability and weak sustainability.

Table 5.11 shows a comparison of the ranking obtained by different methods when a certain dimension is assigned more weighted.

Table 5.11 shows that Shenzhen and Shanghai achieve a high level of sustainability in all scenarios by different methods. Specifically, Shenzhen ranks higher than Shanghai in Scenario 3 which emphasizes economic criteria and Shanghai ranks higher than Shenzhen in Scenario 4 which emphasizes social criteria. This means Shenzhen has higher sustainability in terms of the economic dimension while Shanghai has higher sustainability in terms of the social dimension. It must also be acknowledged that Quito ranks the first in Scenario 2 based on the weighted-sum method, but ranks fourth based on PROMETHEE I. This is due to the fact that high performance in criteria such as green area and $PM_{2.5}$ annual mean concentration cannot compensate other criteria sufficiently when less compensation is allowed.

By comparing the results obtained from the WSM and PROMETHEE II in different scenarios, Shenzhen and Shanghai lead the way in terms of sustainability. Considering the results gained by equal weighting scenarios, Shenzhen ranks higher than Shanghai in both strong and weak sustainability methods. However, this does not necessarily conclude that Shenzhen is definitively more sustainable than Shanghai; the incomparability between the cities is discussed here.

5.1.2 Comparing results of PROMETHEE I and PROMETHEE II

Since this chapter aims to find whether Shenzhen is the most sustainable city among the Global South cities, only the ranking of the eight cities which achieve positive value in outranking flow are discussed here. In Scenario 1, both PROMETHEE I and PROMETHEE II rank Shenzhen higher than Shanghai. However, compared with Scenario 1 (basic, 5%/25%), changing the preference threshold leads Shenzhen and Shanghai to become incomparable.

Decreasing the preference threshold makes one city 'easier' to be strictly preferred to another one (in this situation, it is 'easier' for the value of preference functions on each criterion to equal 1). This would increase both the values of the leaving ranking flow and entering ranking flow. In Scenario 5 (5%/10%), the leaving ranking flow of Shenzhen rises from 0.426 to 0.437 and its entering ranking flow rises from 0.161 to 0.210. At the same time, the leaving ranking flow of Shanghai rises from 0.386 to 0.392 and its entering ranking flow rises from 0.170 to 0.203. Although the ranking flows of both cities increase, the increment of the entering ranking flow of Shenzhen is smaller than that of Shanghai. Therefore,

Table 5.11 Ranking based on unequal weighting on each dimension by different methods

Weighted-sum method			*PROMETHEE II*		
Scenario 2 (ENV)	*Scenario 3 (ECON)*	*Scenario 4 (SOC)*	*Scenario 2 (ENV)*	*Scenario 3 (ECON)*	*Scenario 4 (SOC)*
Quito	Shenzhen	Shanghai	Shenzhen	Shenzhen	Shanghai
Shanghai	Shanghai	Shenzhen	Shanghai	Shanghai	Shenzhen
Shenzhen	Beijing	Beijing	Bogota	Beijing	Beijing
Bogota	Lima	Lima	Quito	Lima	Lima
São Paulo	Bogota	São Paulo	São Paulo	Bogota	São Paulo
Beijing	São Paulo	Almaty	Beijing	Mexico City	Bogota
Lima	Mexico City	Mexico City	Lima	São Paulo	Quito
Mexico City	Almaty	Buenos Aires	Mexico City	Quito	Istanbul
Istanbul	Quito	Istanbul	Istanbul	Almaty	Mexico City
Rio de Janeiro	Istanbul	Quito	Kampala	Istanbul	Almaty
Kampala	Rio de Janeiro	Bogota	Nairobi	Buenos Aires	Mumbai
Mumbai	Buenos Aires	Mumbai	Buenos Aires	Kampala	Buenos Aires
Buenos Aires	Kampala	Delhi	Mumbai	Delhi	Delhi
Nairobi	Mumbai	Rio de Janeiro	Rio de Janeiro	Mumbai	Kampala
Almaty	Delhi	Kampala	Delhi	Rio de Janeiro	Nairobi
Delhi	Johannesburg	Nairobi	Almaty	Nairobi	Rio de Janeiro
Johannesburg	Nairobi	Johannesburg	Johannesburg	Johannesburg	Johannesburg

both the leaving ranking flow and entering ranking flow of Shenzhen are higher than those of Shanghai in Scenario 5, making the two cities incomparable.

Increasing the preference threshold makes one city 'harder' to be strictly preferred to another one (in this situation, it is 'harder' for the value of preference functions on each criterion to equal 1). This would decrease both the values of the leaving ranking flow and entering ranking flow. In Scenario 6 (5%/50%), the leaving ranking flow of Shenzhen falls from 0.426 to 0.292 and its entering ranking flow falls from 0.161 to 0.124. At the same time, the leaving ranking flow of Shanghai falls from 0.386 to 0.276 and its entering ranking flow falls from 0.170 to 0.117. Although the ranking flows of both cities decrease, the decrement of entering ranking flow of Shanghai is greater than that of Shenzhen. Therefore, both the leaving ranking flow and entering ranking flow of Shenzhen are higher than those of Shanghai in Scenario 6, making the two cities incomparable.

Therefore, due to the aspect of incomparability, Shenzhen is not necessarily more sustainable. Nevertheless, the two cities are still considered the most sustainable among these Global South cities.

5.2 Sustainable Global South cities and their characteristics

This analysis is based on the eight Global South cities (Shenzhen, Shanghai, Beijing, Lima, Bogota, São Paulo, Quito and Mexico City) which are considered relatively more sustainable than other cities in the dataset. The results show that the cities with high rankings have better performance in unemployment rate, Gini coefficient, number of homicides per 100,000 population, life expectancy at birth, underground stations per 100,000 population and GDP per capita, while they have relatively poorer performance in green area, domestic water consumption and CO_2 emissions. It must be acknowledged that the results are a summary of the characteristics shared by these eight Global South cities, not indicating that cities with poorer performance in environmental indicators are more sustainable. This means the best-ranked cities, Shenzhen and Shanghai, still need to improve in these environmental aspects.

It would be interesting to discuss the reasons for cities having high performance in some social criteria, such as unemployment rate and GDP per capita, but poor performance in other environmental criteria. Population density could be one of the main reasons. It has a close relationship with sustainable development, but also diverse impacts on different dimensions in sustainability.

Shenzhen and Shanghai are among the top twenty most crowded cities in the world. Shenzhen, with a population density of 17,150 people per square kilometer, ranks first in China and fifth globally (Chinadaily, 2010). High population density can reduce the per capita cost of offering urban service and infrastructure (Cohen, 2006). Utilizing public services like health care and sanitation, life expectancy would improve. More schools and universities could also expand the population with tertiary educational attainment. The population density could also make public transportation economically viable. High population density also means there is a sufficient labor force for city development which increases the total GDP. However, the high density would have negative impacts on environmental aspects. Developing the city for a dense population often leads to more carbon emissions. It may also lead to land use problems and less space for green area.

However, population density is just one of the reasons that Shanghai and Shenzhen achieves high ranking among the global south cities; it is not necessarily true for all cities. For example, Mumbai ranks first in population density in the world with 29,650 people per square kilometer (Chinadaily, 2010), but does not achieve a high level of sustainability. This may be due to its low economic development. Mumbai, with only US$6,992 per capita, ranks third last among all 17 Global South cities on this criterion. Therefore, when urban population exceeds the city's carrying capacity, an expanding population would have negative impacts on urban sustainability. Considering the difficulty in defining the thresholds of a city's carrying capacity and its diverse impacts on sustainability, population density is not included as a sustainability criterion.

It is also necessary to point out the fact that Shenzhen and Shanghai have higher rankings than the South American cities only means the two cities have higher overall sustainability but does not mean the two cities are more sustainable than others in each dimension. For example, Quito is much more sustainable than Shenzhen and Shanghai in terms of air quality and green area. That Shenzhen and Shanghai are less sustainable in green space can be explained by their high population density. To explain the air quality of the two cities, it needs to be acknowledged that the sustainability indicators can reflect the sustainability level of a city, but that these indicators cannot fully reflect the efforts the city has made in improving that criterion. For example, air quality is easily influenced by its surrounding area. The cities with relatively higher performance in educational attainment and inflation among the eight are all in China; however, China's air quality is worse than the South American cities among the eight. This could also explain the conflicts between air quality and population with tertiary educational attainment. It must be considered that the analysis is only based on the data of eight Global South cities. As such, the conflict between these criteria is only the statistical results and does not mean that increasing tertiary educational attainment and decreasing inflation would have negative impacts on air quality.

5.3 The sustainable Global South cities vs. sustainable Global North cities

Both Shenzhen and Shanghai published their 13th Five-Year (2016–2020) Plan for Economic and Social Development in 2016 (Policy Research Office of Shenzhen Municipal People's Government, 2016; Shanghai.gov, 2016). The 13th Five-Year Plan for Chinese cities covers all aspects of development goals to achieve by 2020. By comparing the polices from Global South cities and Global North cities, the difference between sustainable development strategies can be seen.

Shmelev (2017c) summarized the sustainability strategies applied in Singapore. Singapore aims to reduce the domestic water consumption to 140 liters per day per person. In Shanghai's Five-Year Plan, there is not a clear goal to achieve in domestic water consumption. Instead of reducing domestic water consumption, Shanghai mainly focuses on reducing industrial water consumption and aims to reduce water consumption per 10,000 yuan in GDP by 25% (Shanghai.gov, 2016). Shenzhen aims to reduce water consumption per 10,000 yuan in GDP by 18.5% in 2020. However, Shenzhen achieved 43% in water consumption per 10,000 yuan in GDP in 2015 compared with 2010 (Policy Research Office of Shenzhen Municipal People's Government, 2016). The difference between water consumption targets in Singapore, Shanghai and Shenzhen indicates that Shanghai and Shenzhen still have significant capacity to improve their industrial water utilization technology.

In terms of air quality, Singapore aims to reduce $PM_{2.5}$ concentration to 12 mg/m^3 by 2020. The targets of $PM_{2.5}$ concentration are much lower for Shanghai and Shenzhen to achieve, according to the 13th Five-Year (2016–2020) Plan. Shanghai and Shenzhen aim to reduce annual $PM_{2.5}$ concentration 42 mg/m^3 and 28 mg/m^3, respectively, by 2020 (Policy Research Office of Shenzhen Municipal People's Government, 2016; Shanghai.gov, 2016). Shanghai's 13th Five-Year (2016–2020) Plan also mentions its main sources of $PM_{2.5}$ are vehicle emissions, power plant burning and other industrial process emissions. The corresponding policies to reduce these emissions are limiting the heavy-polluting vehicles and replacing coal-fired boilers with boilers that use clean energy by 2020 (National Energy Administration, 2013; Shanghai Environmental Protection Bureau, 2017).

In 2007, 79% of Singapore's electricity was generated by natural gas. Since renewable energy such as wind or hydropower is not suitable for Singapore, it aims to improve energy efficiency by promoting energy-efficient technologies and resource-efficient buildings (Shmelev, 2017c). Developing renewable energy is also not a priority for Shanghai. In Shanghai's 13th Five-Year Plan for energy, it aims to increase the proportion of natural gas as energy source from 10% to 12%, and decrease the proportion of coal from 36% to 33%, while increasing its renewable energy rate to approximately 10% in 2020. Waste incineration is viewed as a source of renewable energy in the plan. Other renewable sources of generating electricity for Shanghai are wind and solar. (Shanghai.gov, 2016) Shenzhen is much more sustainable than Shanghai in terms of energy sources. Coal accounted for only 6.4% of its energy in 2015. Other energy sources are mainly oil, natural gas, nuclear and hydropower. Shenzhen also aims to decrease the proportion of coal to 4.6% and increase the proportion of natural gas to 17.2% as energy source (Policy Research Office of Shenzhen Municipal People's Government, 2016).

Singapore reduced its carbon emissions with a large investment in natural gas as an energy source from 2000 to 2007 (Shmelev, 2017c). In Shanghai's 13th Five-Year Plan, it aims to reduce the consumption of coal to under 125 billion tonnes per year and reduce carbon emissions to 250 billion tonnes per year.

It also aims to expand its proportion of clean energy such as wind, solar and natural gas. Boosting the use of new energy vehicles (NEV) is one of the targets in the 13th Five-Year plan (Shanghai.gov, 2016). Shenzhen aims to reduce its carbon emission per 10,000 yuan in GDP by 21% by 2020 and improve energy efficiency by promoting energy-saving industry and green buildings (Policy Research Office of Shenzhen Municipal People's Government, 2016). Promoting NEVs is also the main target of Shenzhen's 3th Five-Year Plan. Although the three cities do not set renewable energy investment as their primary target, improving energy efficiency and reducing carbon emissions play important roles in their governments' plans.

Singapore achieved a 61% recycling rate in 2011, and set the goal of achieving 70% by 2030 (Shmelev, 2017c). However, since the waste classification and recycling policies have not been well established in China, most waste will go to incineration or landfill. According to the 13th Five-Year Plan, Shanghai will promote waste classification and recycling but there is no quantitative goal on this task. This is also mentioned in Shenzhen's 13th Five-Year Plan. It specifically aims to build an electronic waste recycling system, but there is again no quantitative goal on the total recycling rate in this plan. This shows that there is a large gap between cities in China and Singapore regarding waste recycling and treatment.

In terms of reducing unemployment, Singapore provides subsidies for workers to improve their skills through the Continuing Education and Training (CET) system (Shmelev, 2017c). For Shanghai and Shenzhen, as well as creating more jobs, both cities encourage start-ups to reduce unemployment. This could also be a factor in the rapid economic growth of both cities (Policy Research Office of Shenzhen Municipal People's Government, 2016; Shanghai.gov, 2016).

5.4 Conclusion

This chapter focuses on identifying the sustainable Global South cities and analyzing the reasons behind their sustainability. Due to the multidimensional nature of urban sustainability, 17 global south cities (Istanbul, Mumbai, Delhi, Shanghai, Beijing, Shenzhen, Almaty, Johannesburg, Kampala, Nairobi, Mexico City, Bogota, Buenos Aires, Lima, São Paulo, Rio de Janeiro and Quito) are selected and ranked by the MCDA method, PROMETHEE. The ranking is based on the 15 criteria selected by comparing three sets of sustainability indicators: (1) Sustainable Development Goals Index, (2) The UNECE–ITU Smart Sustainable Cities Indicators, and (3) ISO 37120:2014 Sustainable Development of Communities. The PROMETHEE II is used to provide the complete ranking of cities in terms of sustainability, while PROMETHEE I provides the information on incomparability between cities. The PROMETHEE-GAIA offers a graphic representation of the results and a visualization of the relationship between the preferences and criteria. Considering the difference between strong and weak sustainability, the weighted-sum method is used to provide the ranking based on compensation allowance among criteria. Six scenarios, which change weights on different dimensions or change the preference threshold are applied in this chapter to test the robustness of the ranking results. The results show that Shenzhen and Shanghai rank at the top among the Global South cities in most cases. However, Shenzhen and Shanghai are incomparable in Scenarios 5 and 6 based on PROMETHEE II. With respect to the characteristics of Shenzhen and Shanghai, population density plays an important role in their sustainability. The sustainable development strategies of the two cities are also compared with Singapore's policies to illustrate differences and similarities between sustainable Global South and North cities.

References

Albadvi, A., Chaharsooghi, S.K., & Esfahanipour, A., 2007. Decision making in stock trading: An application of PROMETHEE. *European Journal of Operational Research, 177*, pp. 673–683.

Al-Shemmeri, T., Al-Kloub, B., & Pearman, A., 1997. Model choice in multicriteria decision aid. *European Journal of Operational Research, 97*(3), pp. 550–560.

Ayres, R., 2007. On the practical limits to substitution. *Ecological Economics, 61*, pp. 115–128.

Babic, Z., & Plazibat, N., 1998. Ranking of enterprises based on multicriterial analysis. *International Journal of Production Economics, 56–57*, pp. 29–35.

Banos, R., Manzano-Agugliaro, F., Montoya, F.G., Gil, C., Alcayde, A., & Gómez, J., 2011. Optimization methods applied to renewable and sustainable energy: A review. *Renewable and Sustainable Energy Reviews, 15*(4), pp. 1753–1766.

Behzadian, M., Kazemzadeh, R.B., Albadvi, A., & Aghdasi, M., 2010. PROMETHEE: A comprehensive literature review on methodologies and applications. *European Journal of Operational Research, 200*(1), pp. 198–215.

Belton, V., & Stewart, T., 2002. *Multiple Criteria Decision Analysis: An Integrated Approach*. Springer Science & Business Media.

Benoit, V., & Rousseaux, P., 2003. Aid for aggregating the impacts in life cycle assessment. *International Journal of Life Cycle Assessment, 8*, pp. 74–82.

Beynon, M. J., & Wells, P., 2008. The lean improvement of the chemical emissions of motor vehicles based on their preference ranking: A PROMETHEE uncertainty analysis. *OMEGA – International Journal of Management Science, 36*(3), pp. 384–394.

Bilsel, R.U., Buyukozkan, G., & Ruan, D., 2006. A fuzzy preference-ranking model for a quality evaluation of hospital web sites. *International Journal of Intelligent Systems, 21*, pp. 1181–1197.

Bouyssou, D., 1990. Building criteria: A prerequisite for MCDA. In: Bana e Costa, C.A. (ed.), *Readings in Multiple Criteria Decision Aid*. Berlin: Springer-Verlag, pp. 58–80.

Brans, J.P., & Mareschal, B., 1994. The PROMCALC & GAIA decision support system for multi-criteria decision aid. *Decision Support Systems, 12*, pp. 297–310.

Brans, J.P., & Mareschal, B., 2005. PROMETHEE methods. In: Figueira, J., Greco, S., & Ehrgott, M. (eds), *Multiple Criteria Decision Analysis: State of the Art Surveys*. Springer Science + Business Media, Inc., pp. 163–196.

Brans, J.P., & Vincke, P., 1985. Note – a preference ranking organisation method: (The PROMETHEE Method for Multiple Criteria Decision-Making). *Management Science, 31*(6), pp. 647–656.

Brans, J.P., Vincke, P., & Mareschal, B., 1986. How to select and how to rank projects: The PROMETHEE method. *European Journal of Operational Research, 24*, pp. 228–238.

Chang, Y.J., Chu, C.W., & Lin, M.D., 2012. An economic evaluation and assessment of environmental impact of the municipal solid waste management system for Taichung City in Taiwan. *Journal of the Air & Waste Management Association, 62*(5), pp. 527–540.

Chinadaily, 2010. *Shenzhen ranks 5th most crowded city in the world.* www.chinadaily.com.cn/china/2010-02/25/content_9504088.htm [Accessed 13 August 2017].

Choo, E.U., Schoner, B., & Wedley, W.C., 1999. Interpretation of criteria weights in multicriteria decision making. *Computers & Industrial Engineering, 37*(3), pp. 527–541.

Cinelli, M., Coles, S.R., & Kirwan, K., 2014. Analysis of the potentials of multi criteria decision analysis methods to conduct sustainability assessment. *Ecological Indicators, 46*, pp. 138–148.

Cohen, B., 2006. Urbanization in developing countries: Current trends, future projections, and key challenges for sustainability. *Technology in Society, 28*(1), pp. 63–80.

Dağdeviren, M., 2008. Decision making in equipment selection: An integrated approach with AHP and PROMETHEE. *Journal of Intelligent Manufacturing, 19*(4), pp. 397–406.

Daly, H., 1999. *Ecological Economics and the Ecology of Economics*. Cheltenham: Edward Elgar Publishers.

De Montis, A., De Toro, P., Droste-Franke, B., Omann, I., & Stagl, S., 2005. Criteria for quality assessment of MCDA methods. In: Getzner, M., Spash, C., & Stagl, S. (eds), *Alternatives for Environmental Evaluation.* London: Routledge, pp. 99–133.

Diakoulaki, D., Georgiou, P., Tourkolias, C., Georgopoulou, E., Lalas, D., Mirasgedis, S., & Sarafidis, Y., 2007. A multicriteria approach to identify investment opportunities for the exploitation of the clean development mechanism. *Energy Policy, 35*(2), pp. 1088–1099.

Diakoulaki, D., & Karangelis, F., 2007. Multi-criteria decision analysis and cost-benefit analysis of alternative scenarios for the power generation sector in Greece. *Renewable and Sustainable Energy Reviews, 11*(4), pp. 716–727.

Dombi, J., & Zsiros, A., 2005. Learning multicriteria classification models from examples: Decision rules in continuous space. *European Journal of Operational Research, 160*(3), pp. 663–675.

Figueira, J., de Smet, Y., & Brans, J.P., 2004. MCDA methods for sorting and clustering problems: Promethee TRI and Promethee CLUSTER, Université Libre de Bruxelles. *Service de Mathématiques de la Gestion*, Working Paper 2004/02. www.ulb.ac.be/polytech/smg/indexpublications.htm.

Figueira, J.R., Greco, S., Roy, B., & Słowiński, R., 2013. An overview of ELECTRE methods and their recent extensions. *Journal of Multi-Criteria Decision Analysis, 20*, pp. 61–85.

Figueira, J.R., Mousseau, V., & Roy, B., 2005. ELECTRE methods. In: Figueira, J.R., Greco, S., & Ehrgott, M. (eds) *Multiple Criteria Decision Analysis: State of the Art Surveys*. New York: Springer, pp. 133–162.

Georgopoulou, E., Sarafidis, Y., & Diakoulaki, D., 1998. Design and implementation of a group DSS for sustaining renewable energies exploitation. *European Journal of Operational Research, 109*, pp. 483–500.

Gilliams, S., Raymaekers, D., Muys, B., & Van Orshoven, J., 2005. Comparing multiple criteria decision methods to extend a geographical information system on afforestation. *Computers and Electronics in Agriculture, 49*(1), pp. 142–158.

Goletsis, Y., Psarras, J., & Samouilidis, J.E., 2003. Project ranking in the Armenian energy sector using a multicriteria method for groups. *Annals of Operations Research*, 120, pp. 135–157.

Greco, S., Ehrgott, M., & Figueira, J., 2005. *Multiple Criteria Decision Analysis*. Springer International Series in Operations Research & Management Science, New York: Springer.

Greco, S., Słowiński, R., Figueira, J., & Mousseau, V., 2010. Robust ordinal regression. In: Ehrgott, M., Figueira, J., & Greco, S. (eds) *Trends in Multiple Criteria Decision Analysis*. New York: Springer, pp. 241–283.

Guitouni, A., & Martel, J.M., 1998. Tentative guidelines to help choosing an appropriate MCDA method. *European Journal of Operational Research, 109*(2), pp. 501–521.

Hall, P., 2013. *Good Cities, Better Lives: How Europe Discovered the Lost Art of Urbanism*. New York: Routledge.

Hall, P.G., & Pain, K. (eds), 2006. *The Polycentric Metropolis: Learning From Mega-city Regions in Europe*. New York: Routledge.

Hämäläinen, R.P., & Salo, A.A., 1997. The issue is understanding the weights. *Journal of Multi-Criteria Decision Analysis, 6*(6), pp. 340–343.

Hyde, K., Maier, H.R., & Colby, C., 2003. Incorporating uncertainty in the PROMETHEE MCDA method. *Journal of Multi-Criteria Decision Analysis, 12*(4–5), pp. 245–259.

Hyde, K.M., Maier, H.R., & Colby, C.B., 2004. Reliability-based approach to multicriteria decision analysis for water resources. *Journal of Water Resources Planning and Management, 130*(6), pp. 429–438.

ISO, 2014. *ISO 37120:2014 – Sustainable development of communities – Indicators for city services and quality of life*. www.iso.org/standard/62436.html [Accessed 7 July 2017].

Keeney, R.L., 1972. Utility functions for multiattributed consequences. *Management Science, 18*(5-part-1), pp. 276–287.

Kheireldin, K., & Fahmy, H., 2001. Multi-criteria approach for evaluating long term water strategies. *Water International, 26*(4), pp. 527–535.

Krajnc, D., & Glavič, P., 2005. A model for integrated assessment of sustainable development. *Resources, Conservation and Recycling, 43*(2), pp. 189–208.

Macharis, C., Springael, J., De Brucker, K., & Verbeke, A., 2004. PROMETHEE and AHP: The design of operational synergies in multicriteria analysis: Strengthening PROMETHEE with ideas of AHP. *European Journal of Operational Research, 153*(2), pp. 307–317.

Madlener, R., & Stagl, S., 2005. Sustainability-guided promotion of renewable electricity generation. *Ecological Economics, 53*(2), pp. 147–167.

Mareschal, B., & Brans, J.P., 1988. Geometrical representations for MCDA: The GAIA module. *European Journal of Operational Research, 34*(1), pp. 69–77.

Martinez-Alier, J., Munda, G., & O'Neill, J., 1998. Weak comparability of values as a foundation for ecological economics. *Ecological Economics, 26*, pp. 277–286.

Maxim, A., 2014. Sustainability assessment of electricity generation technologies using weighted multi-criteria decision analysis. *Energy Policy, 65*, pp. 284–297.

Mitropoulos, L.K., & Prevedouros, P.D., 2016. Urban transportation vehicle sustainability assessment with a comparative study of weighted sum and fuzzy methods. *Journal of Urban Planning and Development, 142*(4), p.04016013.

Mousseau, V., Figueira, J., Dias, L., Gomes da Silva, C., & Climaco, J., 2003. Resolving inconsistencies among constraints on the parameters of an MCDA model. *European Journal of Operational Research, 147*(1), pp. 72–93.

Müller, E., Hilty, L.M., Widmer, R., Schluep, M., & Faulstich, M., 2014. Modeling metal stocks and flows: A review of dynamic material flow analysis methods. *Environmental Science & Technology, 48*(4), pp. 2102–2113.

Munda, G., 1995. *Multicriteria Evaluation in a Fuzzy Environment. Theory and Applications in Ecological Economics*, Contributions to Economics Series. Heidelberg: Physica-Verlag.

Munda, G., 2005. 'Measuring sustainability': A multi-criterion framework. *Environment, Development and Sustainability, 7*(1), pp. 117–134.

National Energy Administration, 2013. Shanghai targets to replace coal by clean energy as the source of all small to medium sized coal burning boiler in 2015. www.nea.gov.cn/2013-06/25/c_132485053.htm [Accessed 7 July 2017].

Ozelkan, E.C., & Duckstein, L., 1996. Analysing water resources alternatives and handling criteria by multi criterion decision techniques. *Journal of Environmental Management, 48*(1), pp. 69–96.

Polatidis, H., Haralambopoulos, D.A., Munda, G., & Vreeker, R., 2006. Selecting an appropriate multi-criteria decision analysis technique for renewable energy planning. *Energy Sources, Part B, 1*(2), pp. 181–193.

Policy Research Office of Shenzhen Municipal People's Government, 2016. *13th Five-Year Plan for Economic and Social Development for Shenzhen.* www.drc.sz.gov.cn/ztxx/sswgh/201605/t20160509_3619131.htm [Accessed 7 July 2017].

Ringuest, J.L., 1997. LP-metric sensitivity analysis for single and multi-attribute decision analysis. *European Journal of Operational Research, 98*(3), pp. 563–570.

Rosado, L., Niza, S., & Ferrão, P., 2014. A material flow accounting case study of the Lisbon metropolitan area using the urban metabolism analyst model. *Journal of Industrial Ecology, 18*(1), pp. 84–101.

Rowley, H., Peters, G., Lundie, S., & Moore, S., 2012. Aggregating sustainability indicators: Beyond the weighted sum. *Journal of Environmental Management, 111*, pp. 24–33.

Roy, B., 1968. Classement et choix en présence de points de vue multiples. *Revue française d'informatique et de recherche opérationnelle, 2*(8), pp. 57–75.

Roy, B., 1973. How outranking relations help multicriteria decision making. In: Cochran, J.L., & Zeleny, M. (eds), *Multiple Criteria Decision Making.* Columbia: University of South Carolina Press, pp. 179–201.

Roy, B., 1977a. Conceptual framework for a normative theory of decision aid. In: Starr, M.K. & Zeleny, M. (eds), *Multiple Criteria Decision Making.* Amsterdam: North-Holland, pp. 179–210.

Roy, B., 1977b. Partial preference analysis and decision aid: The fuzzy outranking concept. In: Bell, D.E., Keeney, R.L., & Raifa, H. (eds), *Conflicting Objectives in Decisions.* New York: Wiley, pp. 40–75.

Roy, B., 1978. ELECTRE III: Un algorithme de rangement fondé sur une representation floue des preferences en présence de critères multiples, *Cahiers Centre Etudes Rech. Oper., 20*(1), pp. 3–24.

Roy, B., 1996. *Multicriteria Methodology for Decision Aiding.* Boston: Kluwer Academic Publishers.

Roy, B., Brans, J.P., & Vincke, P., 1975. Aide à la décision multicritère. *Revue Belge de Statistique, d'Informatique et de Recherche Opérationnelle, 15*(4), pp. 23–53.

Roy, B., Figueira, J.R., & Almeida-Dias, J., 2014. Discriminating thresholds as a tool to cope with imperfect knowledge in multiple criteria decision aiding: Theoretical results and practical issues. *Omega, 43*, pp. 9–20.

Roy, B., & Hugonnard, J.C., 1982. Ranking of suburban line extension projects on the Paris metro system by a multicriteria method. *Transportation Research Part A: General, 16*(4), pp. 301–312.

Roy, B., & Słowiński, R., 2013. Questions guiding the choice of a multicriteria decision aiding method. *EURO J. Decis. Process., 1*(69). https://doi.org/10.1007/s40070-013-0004-7

Salminen, P., Hokkanen, J., & Lahdelma, R., 1998. Comparing multicriteria methods in the context of environmental problems. *European Journal of Operational Research, 104*, pp. 485–496.

Scott, A.J., 2012. *A World in Emergence: Cities and Regions in the 21st Century*. Cheltenham: Edward Elgar Publishing.

Sdgindex, 2016. *Reports | SDG Index & Dashboards*. http://sdgindex.org/reports/ [Accessed 7 July 2017].

Shanghai.gov., 2016. *13th Five-Year Plan for Economic and Social Development for Shanghai*. www.shanghai.gov.cn/nw2/nw2314/nw2319/nw22396/nw39378/u21aw1101146.html [Accessed 7 July 2017].

Shanghai Environmental Protection Bureau, 2017. Implementation of the fifth stage emission standard for heavy diesel vehicles informed by Shanghai Municipal Environmental Protection Bureau and Municipal Public Security Bureau. www.sepb.gov.cn/fa/cms/shhj/shhj2098/shhj2099/2017/03/95496.htm [Accessed 7 July 2017].

Shmelev, S.E. (ed.), 2017a. *The Green Economy Reader: Lectures in Ecological Economics and Sustainable Development*. Dordrecht & New York: Springer.

Shmelev, S.E., 2017b. Multidimensional assessment of sustainability: Harmony vs. the turning point. In: Shmelev, S.E. (ed.), *Green Economy Reader*. Dordrecht & New York: Springer International Publishing, pp. 67–98.

Shmelev, S.E., 2017c. Multidimensional sustainability assessment for megacities. In: Shmelev, S.E. (ed.), *Green Economy Reader*. Dordrecht & New York: Springer International Publishing, pp. 205–236.

Shmelev, S.E., & Rodríguez-Labajos, B., 2009. Dynamic multidimensional assessment of sustainability at the macro level: The case of Austria. *Ecological Economics*, *68*(10), pp. 2560–2573.

Shmelev, S.E., & Shmeleva, I.A., 2009. Sustainable cities: Problems of integrated interdisciplinary research. *International Journal of Sustainable Development*, *12*(1), pp. 4–23.

Shmelev S.E., & Shmeleva, I.A., 2018. Global urban sustainability assessment: A multidimensional approach, *Sustainable Development*, *26*(6), pp. 904–920.

Shmelev. S.E., & Shmeleva, I.A., 2019. Multidimensional sustainability benchmarking for smart megacities, *Cities*, *92*, pp. 134–163.

Sun, Z., & Han, M., 2010. Multi-criteria decision making based on PROMETHEE method. In: *Proceedings of the 2010 International Conference on Computing, Control and Industrial Engineering*. Los Alamitos: IEEE Computer Society Press, pp. 416–418.

Triantaphyllou, E., & Sanchez, A., 1997. A sensitivity analysis approach for some deterministic multi-criteria decision-making methods. *Decision Science*, *28*(1), pp. 151–194.

UNDESA/PD, 2012. *World Urbanisation Prospects: The 2011 Revision*. United Nations Department of Economic and Social Affairs. Population Division. www.un.org/en/development/desa/population/publications/pdf/urbanization/WUP2011_Report.pdf [Accessed 15 July 2019].

UNECE, 2015. *The UNECE-ITU Smart Sustainable Cities Indicators*. www.unece.org/housing/committee76thsession.html#/ [Accessed 7 July 2017].

United Nations, 2015. *Sustainable Development Goals*. https://sustainabledevelopment.un.org/?menu=1300 [Accessed 7 August 2017].

Vaillancourt, K., & Waaub, J.P., 2004. Equity in international greenhouse gases abatement scenarios: A multicriteria approach. *European Journal of Operational Research*, *153*(2), pp. 489–505.

Van Huylenbroeck, G., 1995. The conflict analysis method bridging the gap between ELECTRE, PROMETHEE and ORESTE. *European Journal of Operational Research*, *82*, pp. 490–502.

Vego, G., Kučar-Dragičević, S., & Koprivanac, N., 2008. Application of multi-criteria decision-making on strategic municipal solid waste management in Dalmatia, Croatia. *Waste management*, *28*(11), pp. 2192–2201.

Von Neumann, J., & Morgenstern, O., 1945. *Theory of Games and Economic Behavior*. Princeton, NJ: Princeton University Press.

Wang, J.J., Wei, C.M., & Yang, D., 2006. Decision method for vendor selection based on AHP/PROMETHEE/GAIA. *Dalian Ligong Daxue Xuebao/Journal of Dalian University of Technology*, *46*(6), pp. 926–931.

Wang, J.J., & Yang, D.L., 2007. Using a hybrid multi-criteria decision aid method for information systems outsourcing. *Computers & Operations Research*, *34*, pp. 3691–3700.

WCED, 1987. *World Commission on Environment and Development, Our Common Future*. Oxford: Oxford University Press.

Yoon, K.P., & Hwang, C.L., 1995. *Multiple Attribute Decision Making: An Introduction* (Vol. 104). Thousand Oaks, CA: Sage Publications.

6 Sustainability assessment of megacities using environmentally extended input–output analysis and network theory

The case of Singapore

Harrison Brook and Stanislav E. Shmelev

Introduction

Sustainable development and cities

As both the world's economy and human population continue to grow, and the corresponding impact on the surrounding natural environment has become more apparent, sustainable development is a concern of ever-increasing importance to policy makers at all levels of government. Sustainable development does not have a rigid, singular definition, but a generally accepted definition was proposed in the Brundtland report (WCED, 1987) as 'meeting the needs of the present without compromising the ability of future generations to meet their needs'. In 2015, the United Nations proposed a set of 17 Sustainable Development Goals (SDGs) that contained specific targets to be achieved in the next 15 years. Since the SDGs' release, others have emphasized the need to refine the goals and align them more towards moral imperatives such as satisfying human needs, ensuring social equity, and respecting environmental limits (Holden et al., 2017), as well as further considering the goals' interrelationships and interactions with one another (Nilsson et al., 2016).

Sustainable development strategy should take into account all relevant economic, environmental, and demographic factors/trends in order to be tailored in the most effective way possible. An important global phenomenon over the last several decades has been increasing urbanization. The proportion of the world's human population that lives in cities has risen from 34.0% in 1960 to 54.0% in 2014, with that figure expected to rise to 60.0% by 2030 (Rosenberg et al., 2016). UN-Habitat also estimates that cities are the source of 70.0% of the world's GHG emissions (UN-Habitat, 2011). More urbanized countries generally have higher CO_2 emissions per capita (World Bank, 2017), as is illustrated in Figure 6.1. Several factors are causal to this relationship, including a higher concentration of motor vehicle pollution in cities (Gately et al., 2015), environmentally impactful urban lifestyle patterns (Weisz & Steinberger, 2010), and the fact that highly developed, more industrialized nations tend to be more urbanized (Roberts et al., 2017).

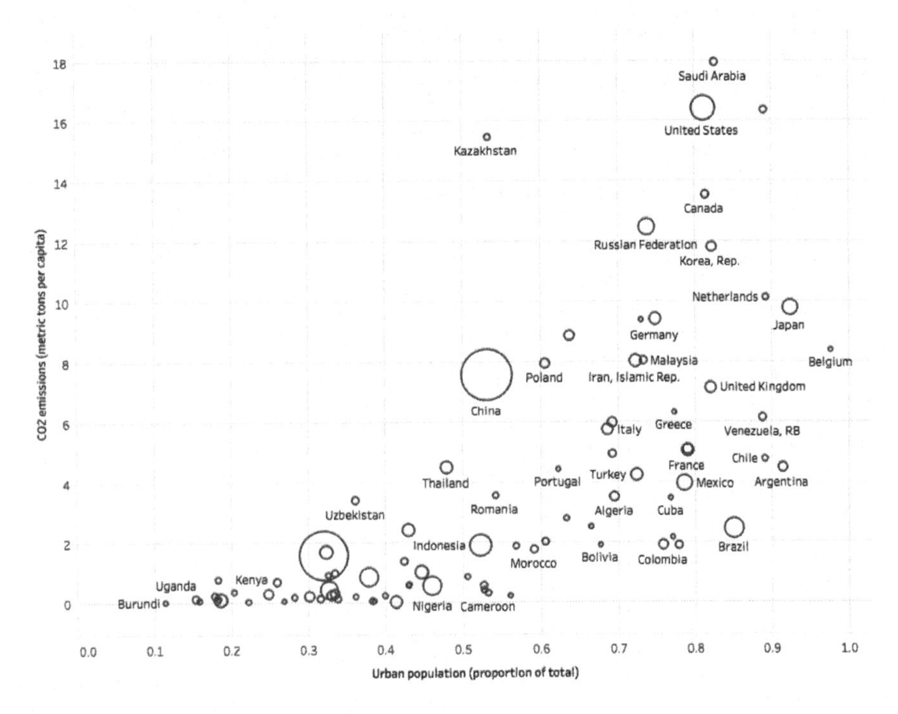

Figure 6.1 CO_2 emissions per capita vs. urban population as a percentage of the total, by country

Note: The size of the circles encodes national population total and only countries with populations over 10,000,000 are included.

Source: World Bank (2017)

Thus, cities must be a key area of focus for sustainable development strategies in the future. Viewing sustainable development through an urban lens leads one to address a specific set of challenges and characteristics, many of which are distinct from those seen at the national and international scales. Shmelev and Shmeleva (2009) identified the following interrelated key problem areas in urban sustainable development: Sustainable energy; Quality of life and health; Psychology of interaction with the environment; Green space and biodiversity; Preservation of natural and cultural heritage; Landscape architecture, eco-design and modernization; Material flows and waste management; Sustainable transport; Environmental conscience and behavior; and Democratic participation.

More complete and accurate understandings of how an urban economy, natural environment, and society function and interact are essential for formulating effective and dynamic sustainable development policy. This understanding can be attained by analyzing the internal structure of an economy and its surroundings using quantitative methods such as environmentally extended input-output analysis

(EE I–O), multi-criteria decision analysis (MCDA) and network analysis. Below, both methods are used to identify key sustainable sectors, linkages and intervention points in the Singapore economy. The purpose of this chapter is to illustrate the usefulness of this methodological framework in the case of Singapore, validating its use as a policy formulation support tool that can be applied to other cities and regions around the world.

Singapore: background

Singapore was selected to illustrate this methodology for several reasons: it is a sovereign city-state; there is high public availability of economic, social, and environmental data; its government has demonstrated leadership in sustainable urban development (Singapore Ministry of the Environment and Water Resources, 2014); and its key role as an international financial and trade hub are interesting dynamics to analyze. Previous studies have found Singapore to be among the most sustainable cities in the world. In Shmelev (2017), Singapore is compared to eleven other global megacities by various sustainability criteria, with Singapore performing the strongest especially under environmental priorities (Shmelev & Shmeleva, 2019). In the 2016 Arcadis Sustainable Cities Index, Singapore is ranked second only to Zurich for overall sustainable and is ranked first overall as measured by purely economic indicators (Arcadis, 2016).

Although Singapore only accounts for 0.12% of global GHG emissions and is already heralded as a leading global city in the field of sustainability (Shmelev, 2017), the country's government is strongly committed to further sustainable development into the future and has outlined their implementation plans for doing so, which include reducing GHG emissions intensity by 36.0% below 2005 levels by 2030 and aiming for total emissions to peak in 2030 (National Environment Agency, 2016). This coincides with another objective of making 80% of buildings in Singapore 'green' by 2030. A number of economic and social improvement initiatives are also being undertaken over the next decade including major public transportation, airport, rail, and container terminal infrastructure development projects (Arcadis, 2016). Applying the aforementioned analytical techniques (EE I–O, MCDA, and network theory) to Singapore is expected to elicit interesting results that could help to further refine its sustainable development strategy in the future and also serve as an example for other cities who wish to follow suit.

Material and methods

Collection of the input–output table and corresponding resource accounts

The data required for this analysis are a symmetric input–output table for Singapore that is disaggregated to a reasonably granular sectoral level, as well as corresponding sectoral final demand, CO_2 emission, and employment vectors. For this study, the 71-sector input–output table for the year 2013 released by Singapore's statistical agency Singstat was used for both the I–O table and the final demand vector. Although there are several databases that aggregate and standardize international input–output data, such including the OECD Input–Output Tables (OECD, 2015), the Eora MRIO Database (Lenzen et al., 2012, 2013), the Global Trade Analysis Project (Aguiar et al., 2016), the World Input–Output Database (Timmer et al., 2012), and EXIOPOL (Tukker et al., 2013), this analysis does not require international or intertemporal comparison, and the data provided by Singstat is sufficiently detailed, accurate, and up-to-date.

The Singapore government releases CO_2 emissions and employment data, but they are not disaggregated into the same sectoral categories as the corresponding I–O table. Therefore, further disaggregation was required to construct those vectors for this analysis. The methodology employed for doing so is described below.

For the employment data, a 15-industry vector was first obtained from Singstat. Then, data from the 152-sector Eora MRIO Database I–O table for Singapore was aggregated to the sectoral level of the 71-sector Singstat I–O table. The categories in the 15-sector table was then mapped to the new 71-sector table, and each of the values in the 71-sector table were adjusted so that their totals added up to each figure in the initial 15-industry Singstat vector. This resulted in a 71-sector employment vector that had both the same number of employees as the Singstat data, but was much more detailed based on the proportions calculated in the Eora database.

A similar but modified technique was used for disaggregating CO_2 emissions data. The initial figures were obtained from a four-industry vector released in Singapore's 'Second biennial update report under the United Nations Framework Convention on Climate Change' (National Environment Agency, 2016). Eora data for CO_2 emissions were again mapped to the Singstat I–O table, and then those figures were adjusted to be proportionally identical to those in the four-industry vector; however, this was only for primary and secondary sectors. For tertiary sectors, the CO_2 emissions figures were set to be proportionally identical to the number of employees in each sector. The resulting 71-sector CO_2 emissions is both commensurable with the Singstat I–O tables and is more accurate than the Eora data considering it is based upon figures released directly by the Singapore government, and not roughly calculated estimates, which is the case for the initial Eora figures. Further discussion of aggregation and disaggregation techniques are described in Lenzen (2011).

Identifying key sectors through linkage coefficients

The method used to calculate the forward and backward linkage coefficients for final demand, CO_2 emissions, and employment in the Singapore economy follows the methodology introduced by Manfred Lenzen in his paper 'Environmentally important paths, linkages and key sectors in the Australian economy' (Lenzen, 2003). The key sectors are those for which the forward linkage > 1 and the backward linkage > 1, implying that the sector has above-average influence on economy-wide use/production for the resource/environmental account in question. For example, if a sector has a forward linkage of 2 and a backward linkage of 0.5, it means that its forward linkage is twice as strong as the average of all sectors and its backwards linkage is half of the average of all sectors.

Calculating sector sustainability with ELECTRE III

To bring together the forward and backward linkage coefficients as criteria to evaluate which sectors in Singapore are the most sustainable, an ELECTRE III MCDA calculation was used with each sector as a decision alternative. A sector is considered more sustainable if its final demand and employment linkages are higher, and its CO_2 emissions linkages are lower. Forward and backward linkages were allocated the same weight in the MCDA, but different weighting schemes were used to emphasize economic, environmental, and social objectives (Table 6.2).

The indifference and preference threshold values (Table 6.1) were set relative to the difference between the minimum and maximum linkages for each criteria. The indifference threshold was set at 1.0% of the difference, and the preference threshold was set at 1.1% of the difference. The veto thresholds were set at 80.0% of the difference for final demand and employment, and at 70.0% of the difference for CO_2 emissions. These threshold settings follow the methodology suggested by Rogers and Bruen (1998a).

Creating networks and calculating sector centrality

As an additional analytical measure, a network analysis was carried out using the aforementioned input–output and sectoral resource data. Adapting the input–output data for use in network analysis is intuitive because of the input–output table's existing resemblance to a network of financial flows (edges) and between sectors (nodes/vertices). Because there is a financial flow between every possible pair of sectors in the Singapore economy, the network is fully connected. Therefore, edge weights had to be taken into account when calculating centrality measures. The method for doing so in this study followed the methodology set out in Blöchl et al. (2011).

Because these are directed networks, centrality was calculated for both directions in the network. This reveals which are the most central 'producing' sectors (measured by their outgoing resource flows) and the most central 'consuming' sectors (measured by their incoming resource flows). To calculate centrality in terms of CO_2 emissions and employment, a novel technique was employed where the outgoing edges from each node were multiplied by the corresponding resource intensity of the node to create a weighted edge value. This produced vastly different networks of financial flows, embodied CO_2 emissions, and employment (Figures 6.7, 6.8, and 6.9).

Table 6.1 Criteria thresholds and goal settings for Singapore ELECTRE III MCDA calculation

Criteria	*Max linkage coefficient*	*Min linkage coefficient*	*Difference (Max - min)*	*Indifference threshold (i)*	*Preference threshold (p)*	*Veto threshold (v)*	*Goal*
Final demand, forward linkage	8.68	0.02	8.67	0.09	0.10	6.936	Maximize
Final demand, backward linkage	6.15	0.18	5.98	0.06	0.07	4.784	Maximize
CO_2 emissions, forward linkage	27.17	0.00	27.17	0.27	0.30	19.019	Minimize
CO_2 emissions, backward linkage	17.43	0.06	17.36	0.17	0.19	12.152	Minimize
Employment, forward linkage	7.33	0.00	7.33	0.07	0.08	5.864	Maximize
Employment, backward linkage	5.88	0.12	5.76	0.06	0.06	4.608	Maximize

Table 6.2 Weighting schemes for the Singapore ELECTRE III MCDA calculation

Criteria	*Equally weighted*	*Final demand-weighted*	*CO_2 emissions-weighted*	*Employment-weighted*
Final demand, forward linkage	16.7%	25.0%	12.5%	12.5%
Final demand, backward linkage	16.7%	25.0%	12.5%	12.5%
CO_2 emissions, forward linkage	16.7%	12.5%	25.0%	12.5%
CO_2 emissions, backward linkage	16.7%	12.5%	25.0%	12.5%
Employment, forward linkage	16.7%	12.5%	12.5%	25.0%
Employment, backward linkage	16.7%	12.5%	12.5%	25.0%

Three types of centrality were calculated: degree, closeness, and betweenness. Because the network is fully connected, in-degree and out-degree centrality are simply the sum of the incoming and outgoing edge weights from each node, respectively. Closeness centrality in a fully connected network was calculated using a random walk based on edge-weight probability distributions between every possible combination of nodes to determine which node took the least number of steps to reach, on from every other node in the network. Betweenness was also calculated using a random walk, and is simply the sum of the amount of times each node was passed through on the random walks between every possible combination of nodes. Self-loops were included, and sectors with high self-loop rates tended to have higher betweenness. This methodology is explained in further detail in Blöchl et al. (2011) and the Python code for simulating the random walk can be found at https://pastebin.com/i96zjk3N.

Theory/calculation

Environmentally extended input–output analysis

Input–output analysis (I–O) is a method of quantitative economic analysis based on the relationships and interactions between intermediate sectors in an economy (Leontief, 1941, 1986; Miller & Blair, 2009). The primary component of I–O is a matrix representation of resource flows (usually financial) between different economic sectors in a region over a given period of time (usually one year). Coefficients can be calculated from this matrix that describe a sector's degree of interlinkage with other sectors and the upstream and downstream effects of smaller-scale economic activity at the macro level. Building upon Leontief's work, notable advances in the study of I–O include Rasmussen (1956) and Hirschman (1958), who proposed the idea of forward and backward linkage coefficients and used them to identify key sectors, respectively. Key sectors are those with forward and backward linkage coefficients greater than 1, meaning that they have above-average upstream and downstream influence on macro-scale economic activity. Empirical studies of the actual influence of key sectors in stimulating economic growth have been explored in some detail in past literature (Panchamukhi, 1975; Bulmer-Thomas, 1982; Hewings, 1982; Hewings et al., 1984; Clements & Rossi, 1991).

Environmentally extended input–output analysis (EE I–O) expands the Leontief I–O model to include impact on environmental and other resources (Leontief, 1970; Shmelev, 2012). EE I–O takes into account sectoral resource use and emissions data in order to calculate the indirect, upstream, and downstream environmental/social effects of economic activity. It is a robust way of evaluating the relationships between economic activity, trade, and consumption on environmental impacts, including natural resource extraction and degradation (Kitzes, 2013). EE I–O can be used to assess hidden and indirect impacts on the use of any resource indicator for which data is disaggregated on a sectoral level, such as GHG emissions, land use, water use, employment, and final demand. EE I–O was first developed by Leontief and has seen important many important contributions since (Førsund & Strøm, 1976; Gay & Proops, 1993; Aroche-Reyes, 2003; Peters & Hertwich, 2006). In 2003, Manfred Lenzen introduced the concept of environmentally weighted forward and backward linkages (Lenzen, 2003), which has been further applied by others including Shmelev (2010), and is a primary methodology used in this study. EE I–O has been influential on fields such as industrial ecology, material flow analysis, and life-cycle assessment, all of which have been employed in the sustainability assessment of cities (Blazek et al., 1999; Bai, 2007; Kennedy et al., 2007; Banar et al., 2009).

Network theory

Network theory is the study of complex interacting systems structured as nodes/vertices linked together by edges (Newman, 2010). The economy is an intuitive subject for network analysis, given its dynamic internal characteristics of sectors (nodes) connected by material flows (edges). An illustration of a simple network is shown in Figure 6.2. Various measures of node centrality, or importance to the network, can be elicited from mathematical network models (Scott, 2017). Loenhard Euler is regarded as the first person to conceptualize and write about network theory in his famous problem 'The bridges of Konigsberg' (Gribkovskaia et al., 2007). It has since been developed and applied to myriad contexts, including the social sciences (Borgatti et al., 2009), finance (Iori et al., 2008; Allen & Babus, 2008), engineering, and computer science (Deo, 2017), and economic input–output analysis (Chen & Chen, 2013; Blöchl et al., 2011; McNerney et al., 2013).

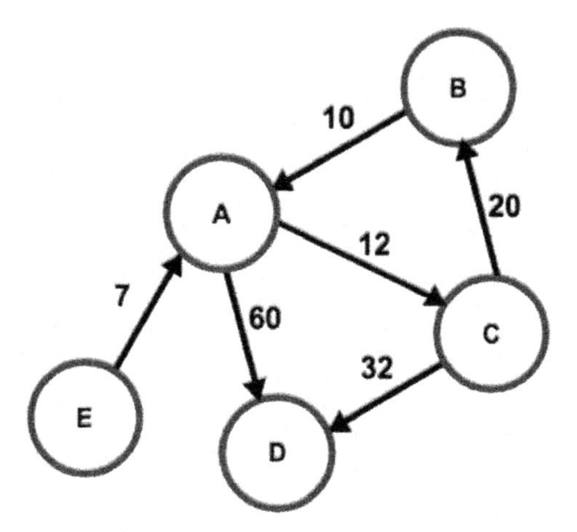

Figure 6.2 Example of a weighted, directed network

Note: Nodes (circles) and connected by directed, weighted edges (black lines).

This chapter will propose a novel use and extension of conventional network analysis, or graph theory, to include EE I–O techniques, with the result being networks of embodied resource use flows between sectors in Singapore's economy.

Multi-criteria decision analysis (MCDA)

In order to determine the most sustainable sectors according to EE I–O linkage coefficients, MCDA (more specifically ELECTRE III) was employed. MCDA is a broad set of decision-making tools that can be used to evaluate and rank alternatives by sets of often-conflicting and incommensurable criteria (Roy, 1968; Vincke, 1992). These criteria are often not commensurable in nature and units of measurement. MCDA allows for a methodical way to structure decision making of this nature, and many different MCDA techniques have been developed over the last several decades to address different types of problems. Applications of MCDA to environmental sustainability have been carried out in a number of categories including renewable energy (Shmelev & van den Bergh, 2016), dynamic macro sustainability assessment (Shmelev & Rodríguez-Labajos, 2009; Shmelev, 2010), and environmental and social impact assessment (ESIA) (Janssen, 2001).

For this study, an outranking approach (one in which preference for alternatives are shown as rankings in relation to one another) called ELECTRE III is used. Several versions of ELECTRE exist; I, II, III, and IV, but ELECTRE III is the most sophisticated and suitable for this study. The ELECTRE (*ELimination Et Choix Traduisant la REalité*) family of MCDA methods was devised by Bernard Roy in response to the problems faced by decision makers using the weighted-sum method (Roy, 1968). ELECTRE III involves two stages: establishing an outranking framework using pairwise comparisons, and undertaking a comparison analysis based on the recommendations in the first phase (Hokkanen & Salminen, 1997).

ELECTRE III is differentiated from other MCDA methods in that it makes use of uncertainty parameters to more closely emulate real-life decision making. These parameters are as follows (from Rogers & Bruen, 1998a):

1. Indifference threshold (q), beneath which the decision maker is indifferent to two project option valuations.
2. Preference threshold (p), above which the decision maker shows a clear strict preference of one project over another.
3. Veto threshold (v), where a 'discordant' difference in favor of one option greater than this value will require the decision maker to negate any possible outranking relationship indicated by other thresholds.

Criteria weights are also employed to further model subjective input into uncertain decision making in ELECTRE III. Guidance for setting threshold levels and criteria weightings in ELECTRE are further discussed in Rogers and Bruen (1998a, 1998b), and Figueira and Roy (2002), among others.

Urban sustainability

Cities were first widely and officially recognized as important focal points of sustainable development by the United Nations following the 1992 Earth Summit in Rio de Janeiro (UN General Assembly, 1992). A plethora of literature exists that addresses the multidimensional nature of sustainability problems at the urban level, which typically are related to issues such as those emphasized by London's sustainable development strategy. These factors largely extend to other urban areas and include air quality, ambient noise, biodiversity, energy, municipal and business waste management, economic development, culture, transport, food, housing, and water (Shmelev & Shmeleva, 2009).

Select contributions to urban sustainability literature include methods used in order to monitor the success of urban sustainability initiatives (Maclaren, 1996), critical analysis of the role of cities in climate change (Bulkeley & Betsill, 2005), and urban carbon mapping (Wiedmann et al., 2016). Applications of input–output analysis have included applications of industrial ecology and metabolism to urban areas (Newman, 1999; Warren-Rhodes & Koenig, 2001) and analysis of intersectoral energy flows in cities (Chen & Chen, 2015). The ISO has also set out international standards for sustainable city services and quality of life (ISO, 2014). Singapore itself has been the subject of studies on making the local construction supply chain more environmentally sustainable (Ofori, 2000); the effect of restricting car traffic on travel behavior and urban sustainability (Olszewski, 2007), and Singapore's urban ecosystem (Tan & bin Abdul Hamid, 2014).

Results

Key sectors and linkages

Figures 6.3, 6.4, and 6.5 show the forward and backward linkage coefficients of the 71 sectors in the Singapore input–output table, weighted by final demand, CO_2 emissions, and employment. Key sectors (those that have forward and backward linkages greater than one) are colored lighter. The full linkage coefficient results can be found in Appendix 6.1.

In 2013 Singapore exported 45.8% of its total industrial and commercial output (Department of Statistics Singapore, 2017a). As can be seen above in Figure 6.3, sectors which have a large proportion of their goods and services exported tend to be more deeply linked final demand sectors. Wholesale trade, water transport and supporting services, semiconductors, construction of buildings and civil engineering, banking and finance, and petrochemicals and petrochemical products are sectors whose growth stimulates the largest upstream and downstream effects on the economy as a whole, reflecting Singapore's strong

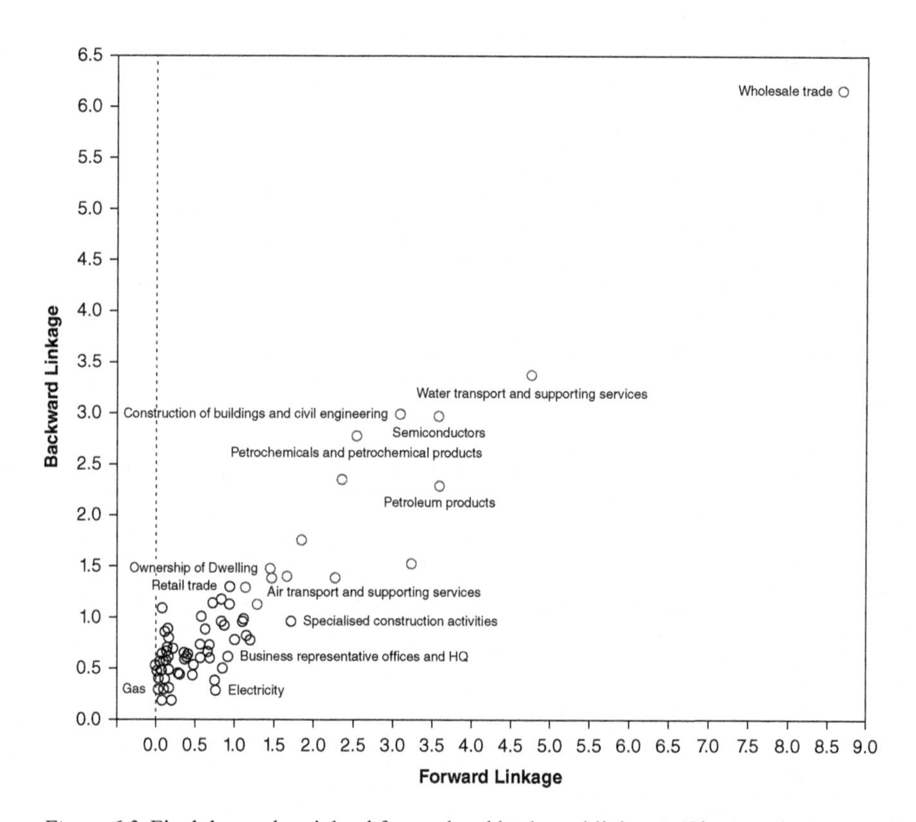

Figure 6.3 Final demand-weighted forward and backward linkages (Singapore)

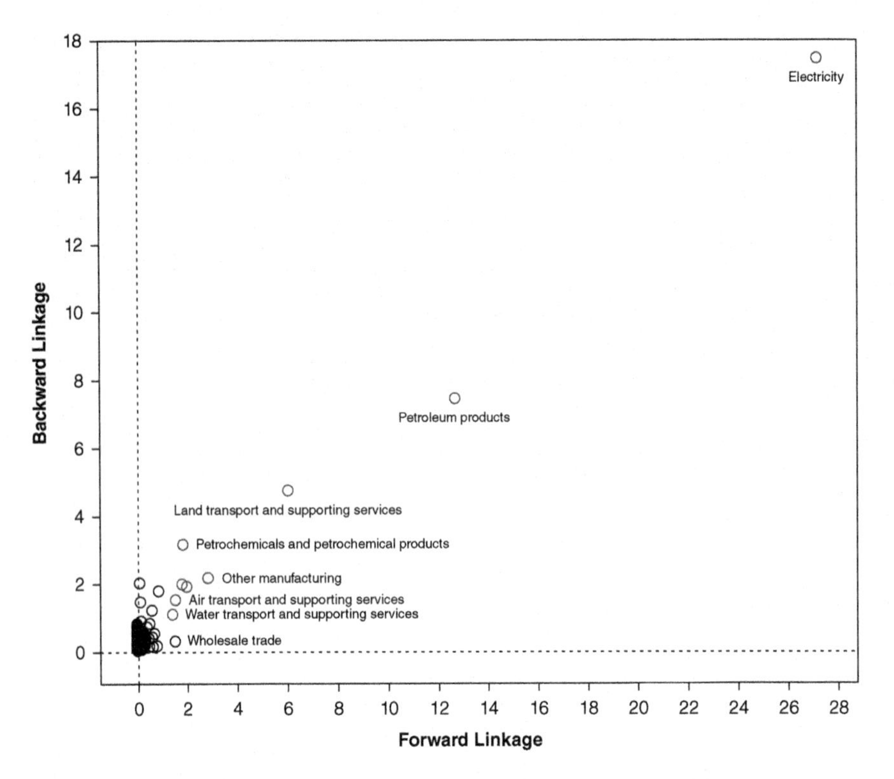

Figure 6.4 CO_2 emissions-weighted forward and backward linkages (Singapore)

economic focus on international trade, financial services, and the manufacture of computer parts.

Weighting linkage coefficients by CO_2 emissions results in a very different set of sectors being highlighted, as can be seen in Figure 6.4. Electricity is by far the most deeply linked sector in this case, largely because of Singapore's reliance on the combustion of imported liquid natural gas LNG for 90% of its domestic electricity generation (Kannan et al., 2007). Singapore has expanded its LNG transmission infrastructure by constructing a new terminal which commenced operations in 2013 and is expected to contribute greatly to the country's energy security (SLNG, 2014). Other fossil-fuel related industries such as petroleum products, land transport and supporting services, and petrochemicals and petrochemical products comprise the most deeply-linked CO_2 sectors in Singapore.

The service sectors employed 72.1% of Singapore's workforce in 2013 (Department of Statistics Singapore, 2017b). The key employment sectors as highlighted above in Figure 6.5 show public administration and defense, wholesale trade, construction of buildings and civil engineering, food and

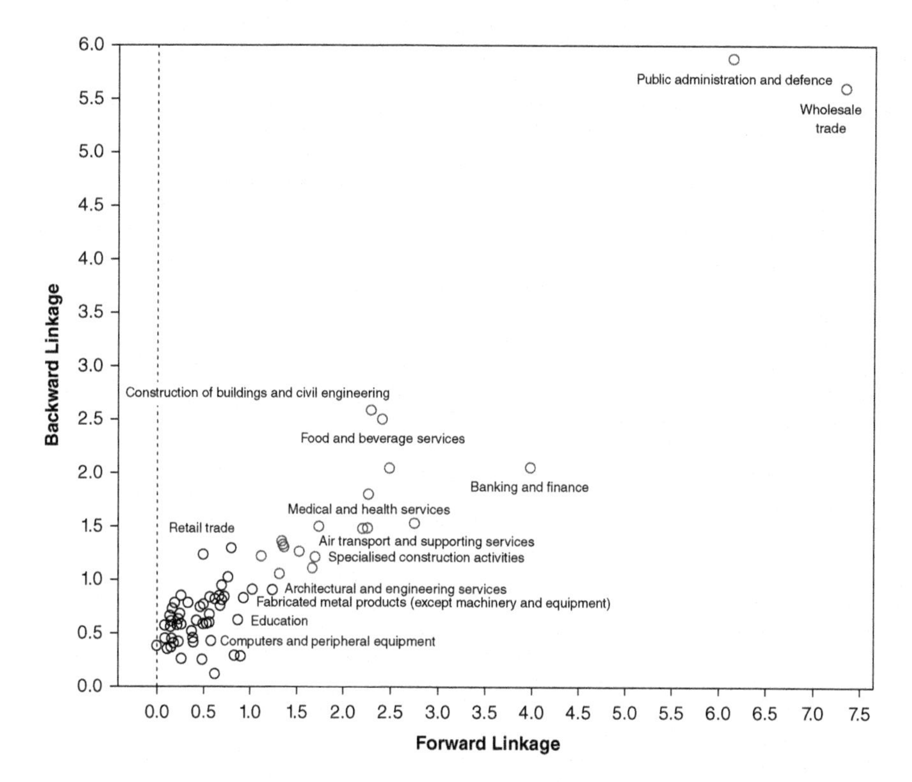

Figure 6.5 Employment-weighted forward and backward linkages (Singapore)

beverage services, and banking and finance as being the sectors that have the strong upstream and downstream linkage effects on employment across the Singaporean economy as a whole. This is in line with Singapore's major economic focus on trade and financial services, and also with its particularly strong focus on the public sector and the military, with military spending in Singapore comprising 3.4% of GDP compared to the global average of 2.2% (World Bank, 2017).

Figure 6.6 shows the trade-off of the average of the forward and backward linkage coefficients for final demand, CO_2 emissions, and employment in Singapore. Sectors that cluster towards the top-left and have large circles can be considered generally more sustainable because they have high final demand and employment linkages, while also having low CO_2 emissions linkages. These sectors include wholesale trade, construction of buildings and civil engineering, and water transport and supporting services. Sectors at the bottom-right of the chart with smaller circles can be considered relatively less sustainable. These include electricity, land transport and supporting services, and petroleum products.

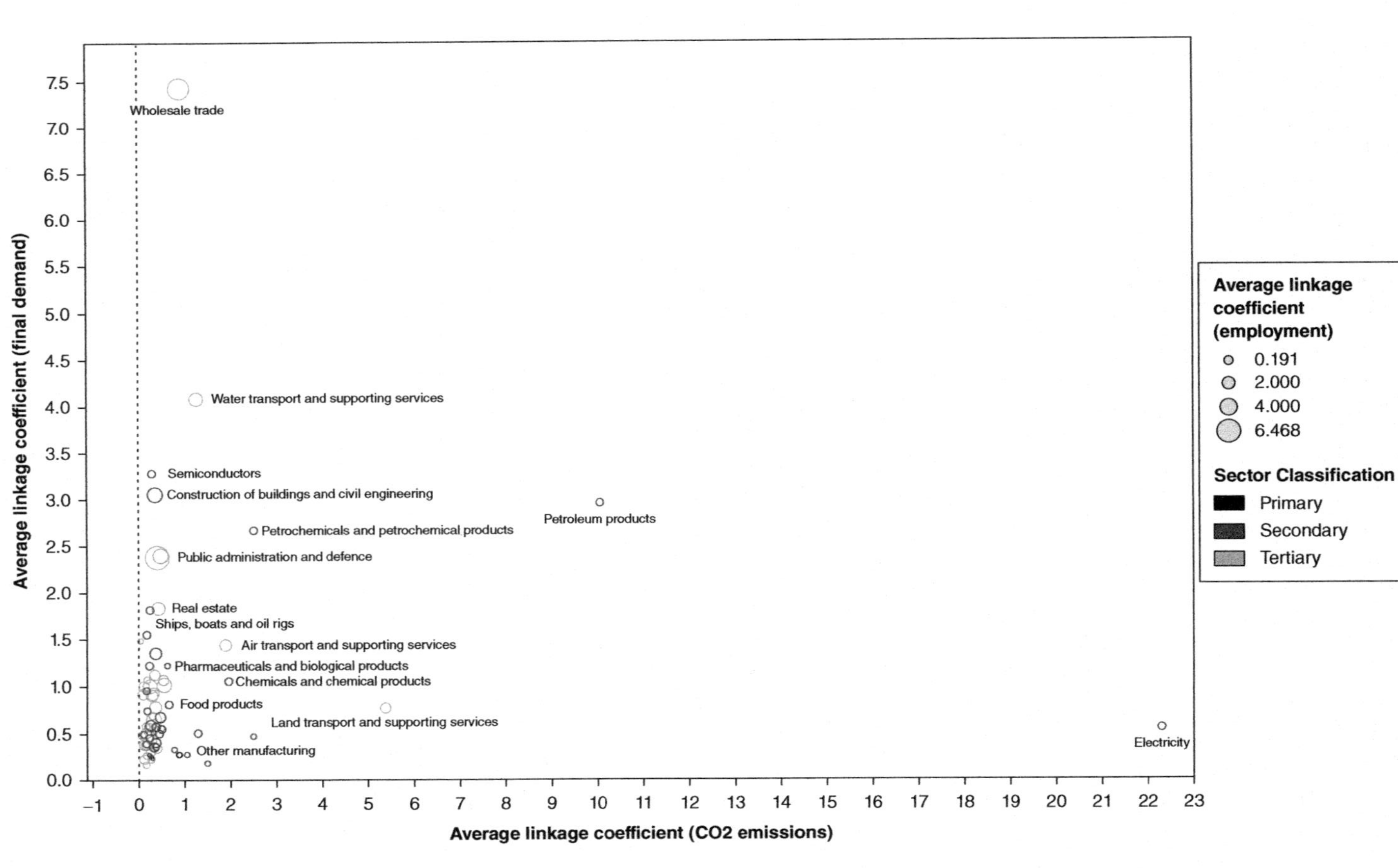

Figure 6.6 Trade-off of average forward and backward linkage coefficients (Singapore)

Network analysis

Figures 6.7, 6.8, and 6.9 are visualizations of financial (unweighted edges), environmental (CO_2 emissions-weighted edges), and social (employment-weighted edges) economic networks in Singapore. For illustrative purposes, only the top 500 edges by weight for each network are shown. The node size represents out-degree centrality. The edge thickness/transparency encodes edge weighting.

Figure 6.7 shows the unmodified input–output table for Singapore represented as a network. The network largely clusters around sectors that facilitate trade, which is intuitive, as it has deeper links with the rest of the economy. The largest five material flows (excluding self-loops) are those from specialized construction activities to construction of buildings and civil engineering ($14,740,971,800), water transport and supporting services to wholesale trade ($7,776,360,600), banking and finance to real estate ($5,205,811,200), petroleum products to petrochemicals and petrochemical products ($4,998,345,700), and medical and health services to public administration and defense ($3,612,958,100). Some sectors, although large by out-degree centrality, are not particularly central in the network according to random walk centrality including petrochemicals and petrochemical products, specialized construction activities, and construction of buildings and civil engineering.

Figure 6.8 is a network visualization of CO_2 emissions embodied in financial flows throughout the Singaporean economy. Its structure is much different than the unweighted network in Figure 6.7, and a clear hub-and-spoke-like pattern forms around the CO_2 emissions-intensive electricity sector, which is responsible for 45.9% of all CO_2 emissions in Singapore (National Environment Agency, 2016). It also has the second-highest CO_2 emissions intensity in the Singapore economy at 1770.6 tons per million dollars of output (second only to agriculture and fishing which has a CO_2 emissions intensity of 1,777.7 tons per million dollars of output). The largest intersectoral flows of embodied CO_2 emissions occur as a result of the financial flows from electricity to semiconductors (1,706,595 tons), electricity to petrochemicals and petrochemical products (1,422,140 tons), petroleum products to petrochemicals and petrochemical products (888,154 tons), electricity to real estate (886,456 tons), and electricity to public administration and defense (848,328 tons). Few sectors come close to being as central producers in this network as electricity. As far as embodied CO_2 consumption as measured by in-degree centrality, the most important sectors are petrochemicals and petrochemical products (2,676,772 tons), electricity (2,393,592 tons), semiconductors (1,807,477 tons), wholesale trade (1,477,628 tons), and petroleum products (1,375,610 tons).

The network in Figure 6.9 above shows the number of employees required to provide the annual amount of goods and services that flow between each pair of sectors. The structure is relatively similar to the unweighted financial

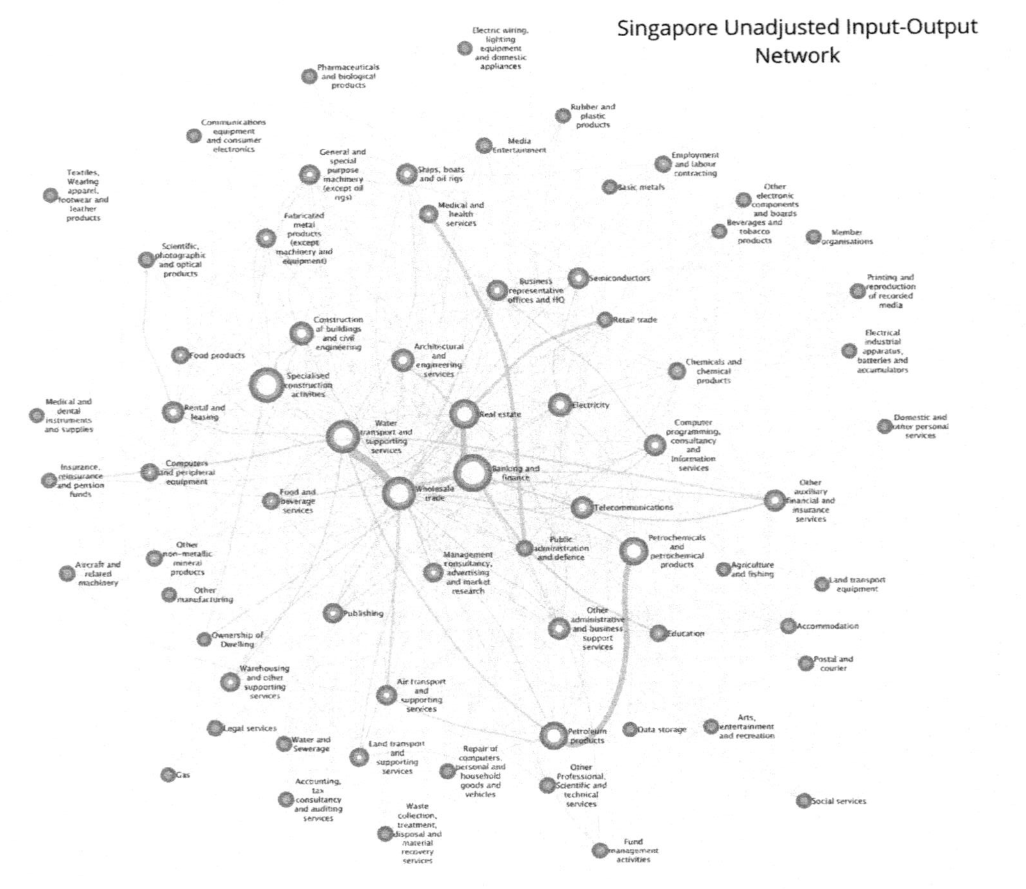

Figure 6.7 Network of intersectoral monetary flows in Singapore

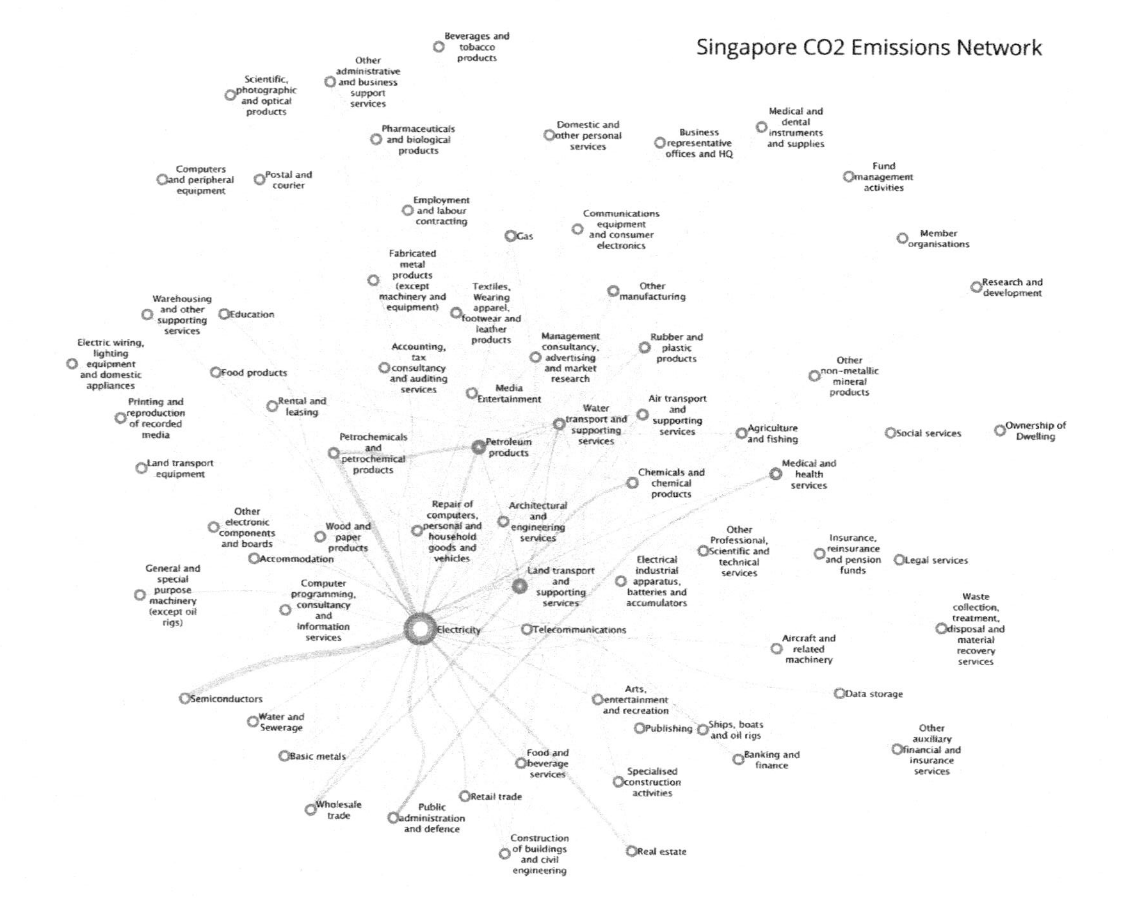

Figure 6.8 Network of intersectoral embodied CO_2 emissions in Singapore

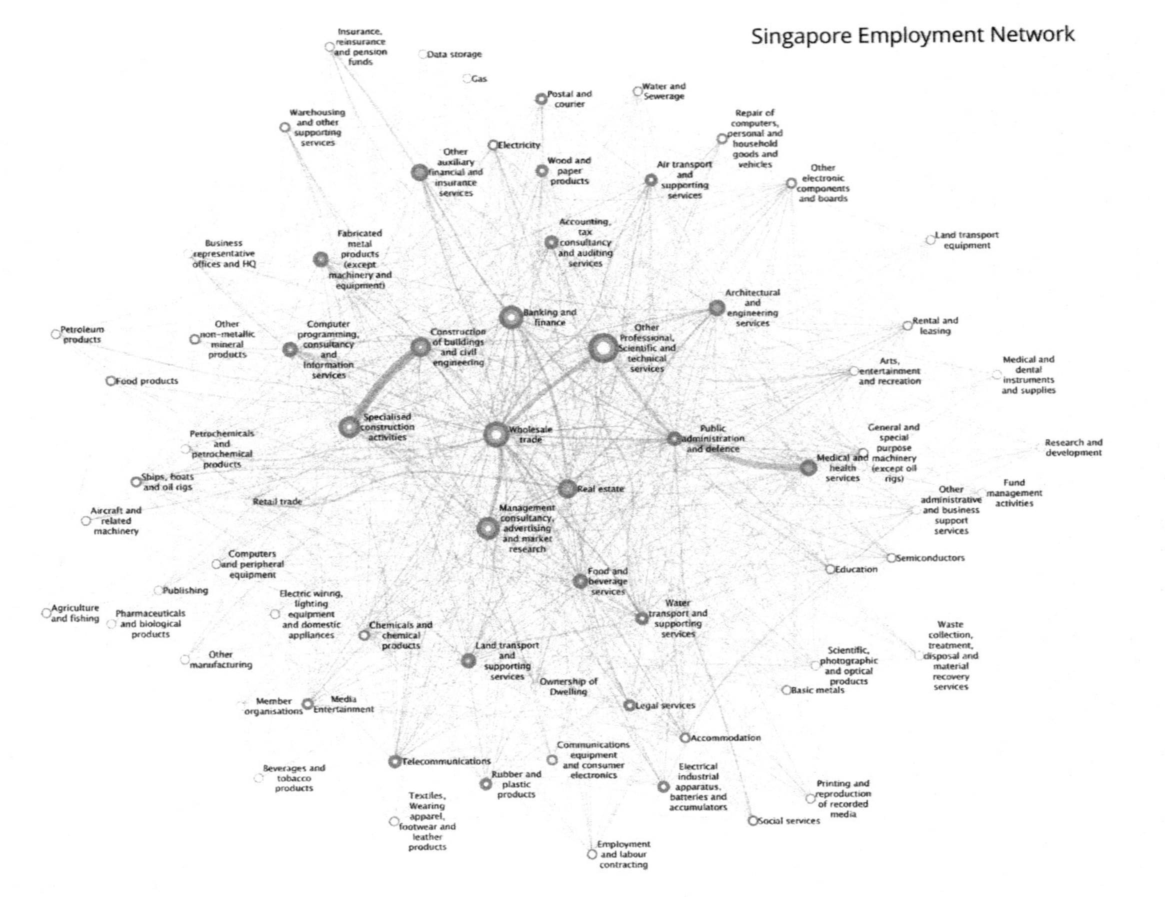

Figure 6.9 Network of intersectoral embodied employment in Singapore

flow network, and it should be noted that node size (out-degree centrality) is not representative of total number of employees in Singapore, but rather the total number of employees required to only serve the intermediate demand in the economy. The largest embodied flows of employment between sectors are from medical and health services to public administration and defense (19,468 employees); specialized construction activities to construction of buildings and civil engineering (14,158 employees), banking and finance to real estate (9,392 employees), other administrative and business support services to wholesale trade (8,727 employees), and management consultancy, advertising, and market research to wholesale trade (8,609 employees). Sectors with the highest in-degree centrality, or those that require the highest number of employees in other sectors to produce their inputs, are wholesale trade (77,117 employees), public administration and defense (55,163 employees), construction of buildings and civil engineering (43,331 employees), water transport and supporting services (36,016 employees), and banking and finance (35,919 employees).

Node centrality

Table 6.3 shows, for each edge weighting scheme (final demand, CO_2 emissions, and employment), the five most central nodes by three different centrality measures (node strength, closeness, and betweenness), for both directions in each network. Measures with a direction of 'in' describe the most central consumers of the resource in question, and those with a direction of 'out' are the most central producers of each resource.

MCDA results: most sustainable sectors

Figures 6.10–6.13 show the sectors at the top of the four weighted ELECTRE III calculations. The first ten layers of the hierarchy are shown, and the full ranking of sectors, including those that are the least sustainable, can be found in Appendix 6.2.

When economic, environmental, and social weightings are equal across forward and backward linkages, public administration and defense emerges as the most sustainable sector in Singapore due to its strong links to final demand growth, employment, and its relatively low CO_2 emissions intensity. Interestingly, the most sustainable sectors overall include several secondary manufacturing industries such as semiconductors, and ships, boats, and oil rigs, because of their major contribution to exports. Sectors that are highly economically important but are not at the very top of the hierarchy because of their environmental impact include petroleum products, petrochemicals, and petrochemical products, and air transport and supporting services. The bottom of the sustainability hierarchy is occupied by industries with high CO_2 emissions-weighted linkages and that are relatively insignificant in terms of final demand and employment. These sectors

Table 6.3 Five most central sectors in Singapore by edge weighting type, centrality measure, and direction

Centrality measure	*Direction*	*Edge weighting type*		
		Unweighted	*CO_2 emissions*	*Employment*
Node strength (degree)	In	Wholesale trade	Petrochemicals and petrochemical products	Wholesale trade
		Construction of buildings and civil engineering	Electricity	Public administration and defense
		Water transport and supporting services	Semiconductors	Construction of buildings and civil engineering
		Petrochemicals and petrochemical products	Wholesale trade	Water transport and supporting services
		Public administration and defense	Petroleum products	Banking and finance
	Out	Banking and finance	Electricity	Other administrative and business support services
		Specialized construction activities	Petroleum products	Banking and finance
		Wholesale trade	Land transport and supporting services	Wholesale trade
		Water transport and supporting services	Gas	Management consultancy, advertising, and market research
		Real estate	Other manufacturing	Computer programming, consultancy, and information services
Random walk (closeness)	In	Wholesale trade	Public administration and defense	Public administration and defense
		Construction of buildings and civil engineering	Petrochemicals and petrochemical products	Wholesale trade
		Public administration and defense	Wholesale trade	Construction of buildings and civil engineering
		Specialized construction activities	Construction of buildings and civil engineering	Specialized construction activities
		Water transport and supporting services	Food and beverage services	Real estate

	Out	Wholesale trade	Electricity	Other administrative and business support services
		Banking and finance	Gas	Wholesale trade
		Real estate	Petroleum products	Computer programming, consultancy, and information services
		Computer programming, consultancy, and information services	Land transport and supporting services	Real estate
		Petroleum products	Chemicals and chemical products	Management consultancy, advertising, and market research
Random walk (counting betweenness)	In	Construction of buildings and civil engineering	Public administration and defense	Construction of buildings and civil engineering
		Specialized construction activities	Wholesale trade	Specialized construction activities
		Petrochemicals and petrochemical products	Construction of buildings and civil engineering	Public administration and defense
		Wholesale trade	Specialized construction activities	Wholesale trade
		Semiconductors	Real estate	Ships, boats and oil rigs
	Out	Banking and finance	Electricity	Other administrative and business support services
		Wholesale trade	Gas	Employment and labor contracting
		Telecommunications	Petroleum products	Postal and courier
		Electricity	Land transport and supporting services	Computer programming, consultancy, and information services
		Petroleum products	Chemicals and chemical products	Telecommunications

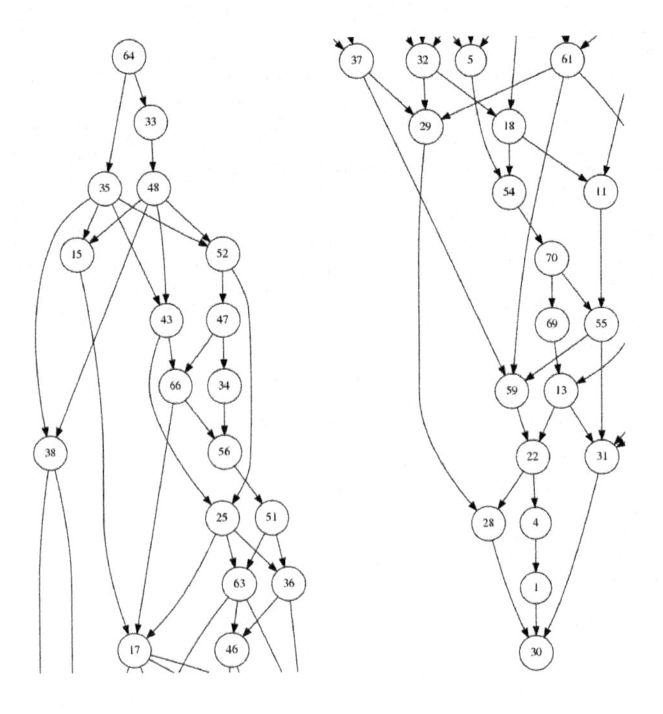

Sector ID	*Sector name*	*Sector ID*	*Sector name*
64	Public administration and defense	37	Land transport and supporting services
33	Construction of buildings and civil engineering	32	Waste collection, treatment, disposal and material recovery services
35	Wholesale trade	5	Wood and paper products
48	Banking and finance	61	Rental and leasing
15	Semiconductors	29	Electricity
52	Real estate	18	Data storage
43	Food and beverage services	54	Legal services
47	Computer programming, consultancy and information services	11	Rubber and plastic products
66	Medical and health services	70	Repair of computers, personal and household goods and vehicles
34	Specialized construction activities	69	Member organizations
38	Water transport and supporting services	55	Accounting, tax consultancy and auditing services
56	Business representative offices and HQ	59	Research and development

25	Ships, boats, and oil rigs	13	Basic metals
51	Other auxiliary financial and insurance services	22	Electric wiring, lighting equipment, and domestic appliances
63	Other administrative and business support services	31	Water and sewerage
36	Retail trade	28	Other manufacturing
17	Computers and peripheral equipment	4	Textiles, wearing apparel, footwear, and leather products
46	Telecommunications	1	Agriculture and fishing
		30	Gas

Figure 6.10 The most (a) and least (b) sustainable sectors in Singapore (equal-weighted)

include gas (which appears at the bottom of the sustainability hierarchy for every weighting scheme); agriculture and fishing; textiles, wearing apparel, footwear, and leather products, and other manufacturing. Electricity, which has extremely high CO_2 forward and backward linkages of 27.2 and 17.4, respectively, is not at the very bottom of the hierarchy as one might expect, because of its relatively strong contributions to final demand and employment, although it is not a key sector by either of those two criteria.

Figure 6.11 shows the results of the ELECTRE III calculation under an environmental weighting scheme emphasizing the influence of CO_2 emissions forward and backwards linkage criteria. This results in a new order of sector sustainability with construction of buildings and civil engineering at the top of the hierarchy, rather than public administration and defense. Public administration and defense is still near the top of the hierarchy; however, wholesale trade and banking and finance are no longer in the top ten levels and semiconductors has been pushed farther down in rank. Sectors which have moved significantly higher in the hierarchy under this weighting include education, domestic and other personal services, and management consultancy, advertising, and market research. Gas is still the least sustainable sector under the environmental weighting, with other non-metallic mineral products moving significantly further down the hierarchy.

The employment-weighted ELECTRE III calculation results in Figure 6.12 show public administration and defense, wholesale trade, construction of buildings and civil engineering, and banking and finance at the top of the sustainability hierarchy. Compared to the equally weighted MCDA, manufacturing industries which have relatively small numbers of employees such as semiconductors do not rank as high on the MCDA hierarchy. Sectors which move higher up in the hierarchy compared to the equal weighting include member organizations, legal services, and other administrative and business support services. Sectors which are relatively less sustainable by the employment weighting include ownership of dwelling, pharmaceuticals and biological products, and computers and peripheral equipment.

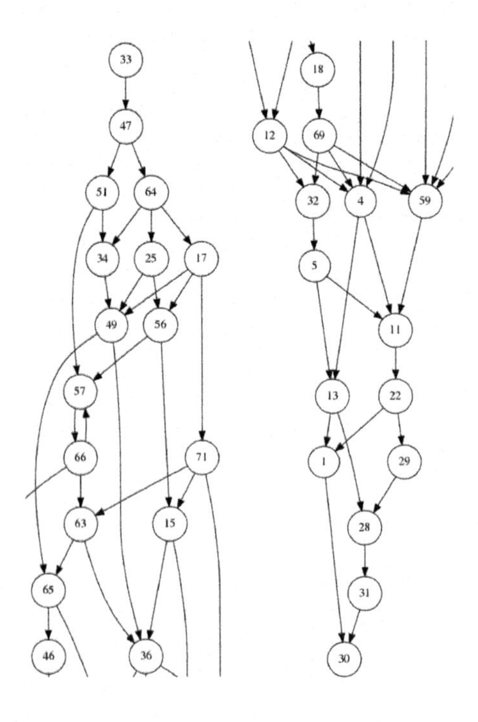

Sector ID	*Sector name*	*Sector ID*	*Sector name*
33	Construction of buildings and civil engineering	18	Data storage
47	Computer programming, consultancy, and information services	12	Other non-metallic mineral products
51	Other auxiliary financial and insurance services	69	Member organizations
64	Public administration and defense	32	Waste collection, treatment, disposal, and material recovery services
34	Specialized construction activities	4	Textiles, wearing apparel, footwear, and leather products
25	Ships, boats, and oil rigs	59	Research and development
17	Computers and peripheral equipment	5	Wood and paper products
49	Insurance, reinsurance, and pension funds	11	Rubber and plastic products
56	Business representative offices and HQ	13	Basic metals
57	Management consultancy, advertising, and market research	22	Electric wiring, lighting equipment, and domestic appliances

66	Medical and health services	1	Agriculture and fishing
71	Domestic and other personal services	29	Electricity
63	Other administrative and business support services	28	Other manufacturing
15	Semiconductors	31	Water and sewerage
65	Education	30	Gas
46	Telecommunications		
36	Retail trade		

Figure 6.11 The most (a) and least (b) sustainable sectors in Singapore (CO_2 emissions-weighted)

The final demand-weighted hierarchy is shown in Figure 6.13, where once again public administration and defense is at the top of the hierarchy. Sectors that have moved up in ranking compared to the unweighted hierarchy include food products, research and development, and other non-metallic mineral products. Sectors that are relatively less sustainable in this ranking include employment and labor contracting, Postal and courier, and media entertainment.

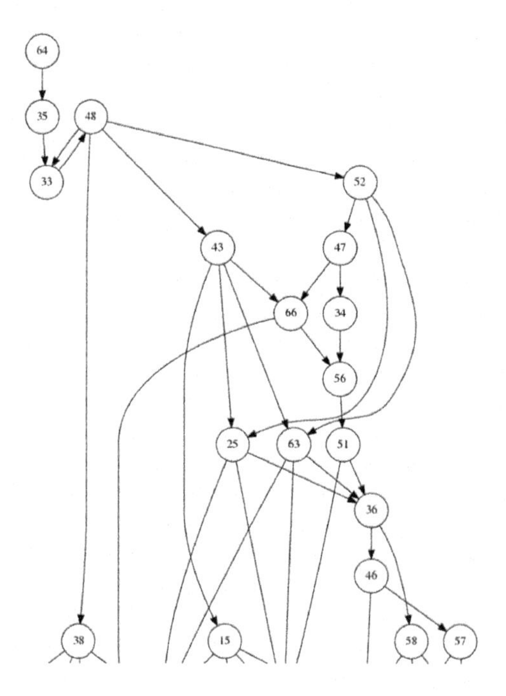

Figure 6.12 (continued)

(continued)

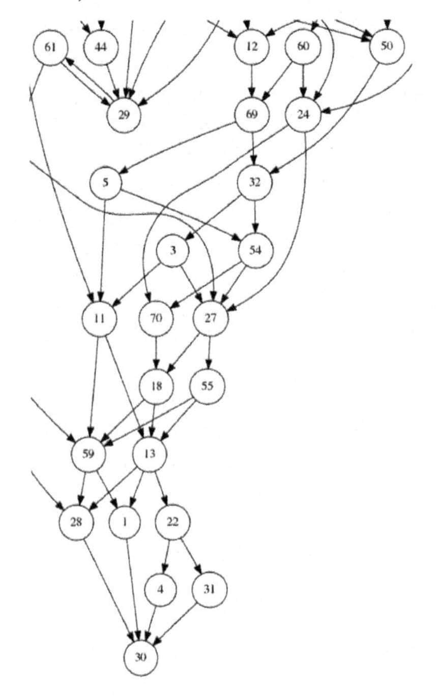

Sector ID	*Sector name*	*Sector ID*	*Sector name*
64	Public administration and defense	61	Rental and leasing
35	Wholesale trade	44	Publishing
33	Construction of buildings and civil engineering	12	Other non-metallic mineral products
48	Banking and finance	60	Other professional, scientific, and technical services
52	Real estate	50	Fund management activities
43	Food and beverage services	29	Electricity
47	Computer programming, consultancy, and information services	69	Member organizations
66	Medical and health services	24	Land transport equipment
34	Specialized construction activities	5	Wood and paper products
56	Business representative offices and HQ	32	Waste collection, treatment, disposal, and material recovery services
25	Ships, boats, and oil rigs	3	Beverages and tobacco products
63	Other administrative and business support services	54	Legal services
51	Other auxiliary financial and insurance services	11	Rubber and plastic products
36	Retail trade	70	Repair of computers, personal and household goods, and vehicles

46	Telecommunications	27	Medical and dental instruments and supplies
38	Water transport and supporting services	18	Data storage
15	Semiconductors	55	Accounting, tax consultancy, and auditing services
58	Architectural and engineering services	59	Research and development
57	Management consultancy, advertising, and market research	13	Basic metals
		28	Other manufacturing
		1	Agriculture and fishing
		22	Electric wiring, lighting equipment, and domestic appliances
		4	Textiles, wearing apparel, footwear, and leather products
		31	Water and sewerage
		30	Gas

Figure 6.12 The most (a) and least (b) sustainable sectors in Singapore (employment-weighted)

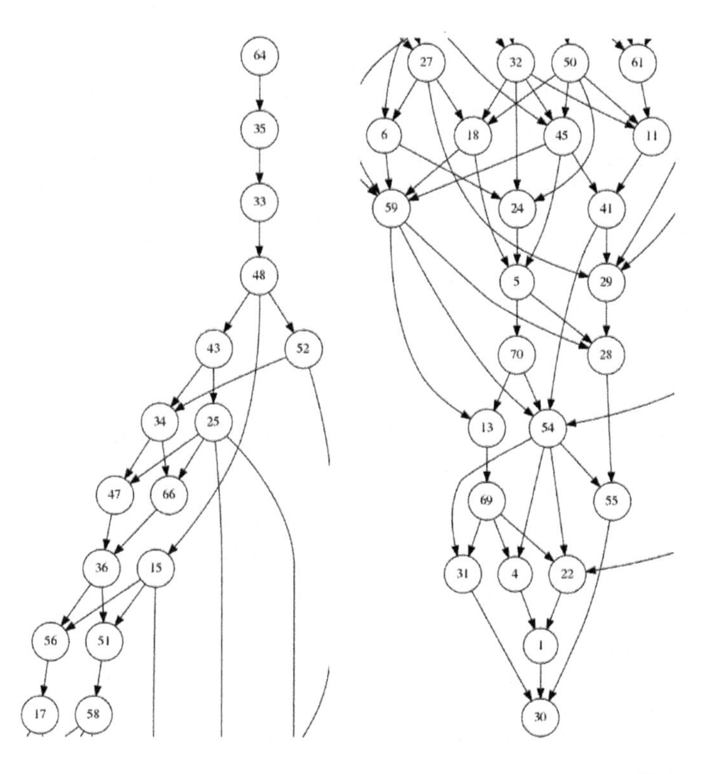

Figure 6.13 (continued)

(continued)

Sector ID	Sector name	Sector ID	Sector name
64	Public administration and defense	27	Medical and dental instruments and supplies
35	Wholesale trade	32	Waste collection, treatment, disposal, and material recovery services
33	Construction of buildings and civil engineering	50	Fund management activities
48	Banking and finance	61	Rental and leasing
43	Food and beverage services	6	Printing and reproduction of recorded media
52	Real estate	18	Data storage
34	Specialized construction activities	45	Media entertainment
25	Ships, boats, and oil rigs	11	Rubber and plastic products
47	Computer programming, consultancy, and information services	59	Research and development
66	Medical and health services	24	Land transport equipment
36	Retail trade	41	Postal and courier
15	Semiconductors	5	Wood and paper products
56	Business representative offices and HQ	29	Electricity
51	Other auxiliary financial and insurance services	70	Repair of computers, personal and household goods, and vehicles
17	Computers and peripheral equipment	28	Other manufacturing
58	Architectural and engineering services	13	Basic metals
		54	Legal services
		69	Member organizations
		55	Accounting, tax consultancy, and auditing services
		31	Water and sewerage
		4	Textiles, wearing apparel, footwear, and leather products
		22	Electric wiring, lighting equipment, and domestic appliances
		1	Agriculture and fishing
		30	Gas

Figure 6.13 The most (a) and least (b) sustainable sectors in Singapore (final demand-weighted)

Discussion

Policy implications in Singapore

The above results are significant in that they elicit insights about the underlying structure of the Singaporean economy and its interactions with the environment and society through CO_2 emissions and employment. These insights can be useful to policy makers in Singapore in further identifying ways in which to grow and manage the local economy in the most effective way possible.

Singapore is particularly forward-thinking and progressive regarding the future direction of the city-state and its inhabitants. The book *Singapore 2065: Leading insights on economy and environment from 50 Singapore icons and beyond* (Quah, 2016) outlines a 50-year vision for Singapore that specifically addresses the city's past, present, and speculative future relationship between its economy and natural environment. In order to thoughtfully formulate and achieve policy goals that can bring the city towards the vision outlined in *Singapore 2065*, evidence-based analytical approaches should be taken, as has been explored in other cases and studies (Allen et al., 2010; Anderson et al., 2005; Finco & Nijkamp, 2001; While et al., 2004). The results of this study could be directly applicable in constructing a road map of economic stimulus and technology development initiatives, focusing on the most sustainable sectors as calculated above in the MCDA, which can be tailored according to different priorities and weighting schemes.

Network analysis is particularly useful in generating actionable insights for sustainable management of the Singapore economy. For example, modeling the CO_2 emissions network results in clear intervention points that fall into three categories: the sectors whose output should be decreased, the sectors whose CO_2 intensity should be decreased, and sectors whose consumption should be decreased because that consumption contains high levels of embodied CO_2 emissions. The quantification of these sectors using the methods describe above is shown in Tables 6.4, 6.5, and 6.6.

Table 6.4 shows the sectors that emit the largest total amounts of direct CO_2, along with corresponding total output of these sectors. This is useful in determining the major sources of CO_2 and act as a starting point for policy guidance to reduce total output in these industries.

A more optimized approach would take into consideration the CO_2 intensity of each sector, or how many tons of CO_2 are emitted per million dollars of total output. Table 6.5 shows the top 20 sectors in the Singaporean economy by CO_2 intensity. Many of them are essential to the city's infrastructure and their presence and expansion are necessary. Therefore, policy makers should look to employ new and innovative technology to reduce CO_2 intensity from these industries rather than decrease total output.

Table 6.6 show the most CO_2-intensive intersectoral relationships (including feedback self-loops) in the Singaporean economy. This information could be used by policy makers to encourage lessened consumption by the input sectors of the

Table 6.4 Sectors in Singapore that emit the highest amounts of CO_2

Sector	*Total CO_2 emissions (tons)*	*Total output (\$)*
Electricity	20,003,560	11,297,614,800
Petroleum products	9,035,239	50,848,450,400
Land transport and supporting services	4,660,757	9,248,349,100
Other manufacturing	2,315,659	3,572,220,300
Chemicals and chemical products	1,429,048	13,705,659,100
Air transport and supporting services	1,270,669	20,988,194,000
Petrochemicals and petrochemical products	771,085	35,991,680,100
Water transport and supporting services	763,869	67,278,073,900
Gas	627,949	848,706,300
Rubber and plastic products	591,480	2,180,054,600
Pharmaceuticals and biological products	558,133	18,406,608,500
Agriculture and fishing	447,046	251,472,400
Other non-metallic mineral products	259,117	2,529,478,100
Food products	244,132	8,556,609,400
Fabricated metal products (except machinery and equipment)	183,099	9,731,296,100
Textiles, wearing apparel, footwear, and leather products	142,411	823,016,200
Electrical industrial apparatus, batteries and accumulators	101,722	2,373,078,400
Other electronic components and boards	87,607	5,477,175,900
Communications equipment and consumer electronics	85,578	1,503,692,600
Ships, boats, and oil rigs	81,377	26,426,064,000

Table 6.5 Sectors in Singapore that have the highest CO_2 emissions intensity per unit of total output

Industry name	*Total output (\$)*	*CO_2 intensity (tons/\$1M output)*
Agriculture and fishing	251,472,400	1,778
Electricity	11,297,614,800	1,771
Gas	848,706,300	740
Other manufacturing	3,572,220,300	648
Land transport and supporting services	9,248,349,100	504
Rubber and plastic products	2,180,054,600	271
Petroleum products	50,848,450,400	178
Textiles, wearing apparel, footwear, and leather products	823,016,200	173
Chemicals and chemical products	13,705,659,100	104
Other non-metallic mineral products	2,529,478,100	102
Basic metals	1,049,194,700	71
Electric wiring, lighting equipment, and domestic appliances	653,947,700	62

Air transport and supporting services	20,988,194,000	61
Communications equipment and consumer electronics	1,503,692,600	57
Electrical industrial apparatus, batteries, and accumulators	2,373,078,400	43
Wood and paper products	1,618,062,100	38
Pharmaceuticals and biological products	18,406,608,500	30
Food products	8,556,609,400	29
Beverages and tobacco products	2,452,247,100	25
Petrochemicals and petrochemical products	35,991,680,100	21

Table 6.6 Intersectoral relationships in Singapore that result in the highest levels of embodied CO_2 emissions

Output sector	*Input sector*	*Annual monetary flow between sectors ($)*	*Embodied CO_2 emissions (tons)*
Electricity	Electricity	1,234,601,100	2,185,985
Electricity	Semiconductors	963,851,200	1,706,595
Electricity	Petrochemicals and petrochemical products	803,196,500	1,422,140
Petroleum products	Petrochemicals and petrochemical products	4,998,345,700	888,154
Electricity	Real estate	500,652,900	886,456
Electricity	Public administration and defense	479,118,600	848,328
Electricity	Petroleum products	471,814,300	835,395
Electricity	Chemicals and chemical products	378,043,400	669,364
Electricity	Wholesale trade	342,506,200	606,442
Electricity	Food and beverage services	320,634,600	567,716
Land transport and supporting services	Wholesale trade	1,119,629,100	564,243
Electricity	Retail trade	280,007,000	495,780
Electricity	Accommodation	218,451,000	386,789
Electricity	Arts, entertainment, and recreation	214,139,200	379,155
Petroleum products	Water transport and supporting services	2,112,614,700	375,390
Petroleum products	Petroleum products	2,091,386,900	371,618
Electricity	Education	209,702,700	371,300
Electricity	Water and sewerage	193,405,800	342,444
Land transport and supporting services	Water transport and supporting services	668,654,300	336,972
Petroleum products	Air transport and supporting services	1,690,351,600	300,358

particular good coming from the output sectors. Again, these are not exact, empirical figures but rather estimates that are proportional to the output sectors' CO_2 emissions and the portion of the monetary output that is consumed by the corresponding input sector. More efficient use and lessened consumption of electricity by the electricity sector itself, petrochemicals and petrochemical products, real estate, and public administration and defense sectors would make a proportionally significant impact on embodied CO_2 emissions within the economic network. Considering the government and military's consumption of electricity is one of the largest embodied CO_2 flows in Singapore, its own operations should be a primary focus of new energy-efficient policy making.

The potential application of the above methodology to other cities could result in more effective urban sustainability around the globe, and contribute in some part to national emissions reduction targets as well as sustainable development strategy as a whole. As long as the required data could be obtained (input–output tables with corresponding sectoral resource vectors), the same insights could be discovered for any city, region, or country of application.

Conclusions

Summary

In this study, 71 economic sectors in Singapore have been assessed for their sustainability, as measured by their upstream and downstream economic, environmental, and social impacts effects. EE I–O, MCDA, and network theory were used to form a framework of identifying key intervention points for sustainable development at the urban level. The most sustainable sectors, by an equally weighted MCDA using the forward and backward linkages as criteria, are public administration and defense, construction of buildings and civil engineering, wholesale trade, banking and finance, and real estate. Under different criteria weightings, certain sectors become more or less desirable. For example, food products, Research and development, and other non-metallic mineral products become more desirable under final demand weighting. Domestic and other personal services, legal services, and fund management activities become more desirable under CO_2 emissions weighting. Member organizations, legal services, and other administrative and business support services become more desirable under employment weighting. Network centrality was also calculated and visualized, with the most central sectors, by outgoing random walk centrality (closeness) being wholesale trade (unweighted edges), electricity (CO_2 emissions-weighted edges, and other administrative and business support services (employment-weighted edges).

This novel analysis reveals insights that could be highly beneficial to policy makers in Singapore aiming to advance sustainable development in the future. Actionable insights include which sectors to focus on for economic stimulus (those which rank higher in the key linkage MCDA), sectors to focus on optimizing resource use/emissions intensity (through network analysis), and sectoral linkages that should be expanded/reduced to either minimize emissions and resource use or maximize employment and other social benefit in the most efficient way possible (through analyzing edges weighted by different criteria in the network analysis). This framework can easily be applied to other cities using data that is readily available for many developed areas of the work. All that is required is a symmetric input–output table and sector-disaggregated resource vectors.

Limitations of study

A major limitation of this study was having to disaggregate the CO_2 and employment vectors manually, which was necessary since the figures for all of the individual 71 sectors weren't included in Singstat's input–output tables. Having full, empirical data would make the results empirically more reliable and robust.

A higher sector resolution would have also been advantageous given the highly granular nature of assessing sustainability at the urban level. Although a slightly higher sector resolution is available with EORA, the input–output data made available directly from the Singaporean government lent robustness and credibility to the analysis.

Directions for further research

As far as the network component of this study goes, there are clear directions for further research in order to better understand the nature and significance of these networks. Extending this analysis to a multi-region input–output (MRIO) model would allow one to better understand the entirety of direct and indirect resource use in the Singaporean economy considering the importance of international trade to the nation's economy. 'Zooming out' and conceptualizing an international network of domestic trade networks of CO_2 emissions embodied in both domestic and international trade could further allow one to uncover previously unknown intervention points in the economy. Comparing Singapore to other cities, and comparing progress over time would likely elicit other key insights that could help shape sustainable development economic policy formulation in the future.

References

Aguiar, A., Narayanan, B., McDougall, R., 2016. An overview of the GTAP 9 data base. *Journal of Global Economic Analysis* 1, 181–208. doi:10.21642/JGEA.010103AF

Allen, F., Babus, A., 2008. Networks in finance. Wharton Financial Institutions Center Working Paper No. 08–07.

Allen, J., Browne, M., Holguin-Veras, J., 2010. Sustainability strategies for city logistics. In: *Green Logistics: Improving the Environmental Sustainability of Logistics*, 282–305.

Anderson, S., Allen, J., Browne, M., 2005. Urban logistics: How can it meet policy makers' sustainability objectives? *Journal of Transport Geography* 13, 71–81.

Arcadis, 2016. *Sustainable Cities Index 2016.* www.arcadis.com/en/global/our-perspectives/sustainable-cities-index-2016/

Aroche-Reyes, F., 2003. A qualitative input–output method to find basic economic structures. *Papers in Regional Science* 82, 581–590. doi:10.1007/s10110-003-0149-z

Bai, X., 2007. Industrial ecology and the global impacts of cities. *Journal of Industrial Ecology* 11, 1–6.

Banar, M., Cokaygil, Z., Ozkan, A., 2009. Life cycle assessment of solid waste management options for Eskisehir, Turkey. *Waste Management* 29, 54–62.

Blazek, M., Rhodes, S., Kommonen, F., Weidman, E., 1999. Tale of two cities: Environmental life cycle assessment for telecommunications systems: Stockholm, Sweden and Sacramento, CA. In: *Electronics and the Environment, 1999. ISEE-1999. Proceedings of the 1999 IEEE International Symposium*. IEEE, pp. 76–81.

Blöchl, F., Theis, F.J., Vega-Redondo, F., Fisher, E.O., 2011. Vertex centralities in input–output networks reveal the structure of modern economies. *Physical Review E* 83, 46127.

Borgatti, S.P., Mehra, A., Brass, D.J., Labianca, G., 2009. Network analysis in the social sciences. *Science* 323, 892–895. doi:10.1126/science.1165821

Bulkeley, H., Betsill, M.M., 2005. *Cities and Climate Change: Urban Sustainability and Global Environmental Governance*. Psychology Press.

Bulmer-Thomas, V., 1982. *Input–output Analysis in Developing Countries: Sources, Methods and Applications*. Wiley.

Chen, S., Chen, B., 2015. Urban energy consumption: Different insights from energy flow analysis, input–output analysis and ecological network analysis. *Applied Energy* 138, 99–107.

Chen, Z.-M., Chen, G.Q., 2013. Demand-driven energy requirement of world economy 2007: A multi-region input–output network simulation. *Communications in Nonlinear Science and Numerical Simulation* 18, 1757–1774. doi:10.1016/j.cnsns.2012.11.004

Clements, B.J., Rossi, J.W., 1991. Interindustry linkages and economic development: The case of Brazil reconsidered. *The Developing Economies* 29, 166–187.

Deo, N., 2017. *Graph Theory With Applications to Engineering and Computer Science*. Courier Dover Publications.

Department of Statistics Singapore, 2017a. *Singapore Supply, Use and Input–output Tables 2013.* www.singstat.gov.sg/publications/publications-and-papers/economy (accessed August 15, 2017).

Department of Statistics Singapore, 2017b. *Labour, Employment, Wages and Productivity*. www.singstat.gov.sg/statistics/browse-by-theme/labour-employment-wages-and-productivity (accessed August 15, 2017).

Figueira, J., Roy, B., 2002. Determining the weights of criteria in the ELECTRE type methods with a revised Simos' procedure. *European Journal of Operational Research, EURO XVI: O.R. for Innovation and Quality of Life* 139, 317–326. doi:10.1016/S0377-2217(01)00370-8

Finco, A., Nijkamp, P., 2001. Pathways to urban sustainability. *Journal of Environmental Policy and Planning* 3, 289–302.

Førsund, F.R., Strøm, S., 1976. The generation of residual flows in Norway: An input–output approach. *Journal of Environmental Economics and Management* 3, 129–141. doi:10.1016/0095-0696(76)90027-9

Gately, C.K., Hutyra, L.R., Sue Wing, I., 2015. Cities, traffic, and CO_2: A multidecadal assessment of trends, drivers, and scaling relationships. *Proceedings of the National Academy of Sciences* 112, 4999–5004. doi:10.1073/pnas.1421723112

Gay, P.W., Proops, J.L.R., 1993. Carbon dioxide production by the UK economy: An input–output assessment. *Applied Energy, Energy and Environmental Management* 44, 113–130. doi:10.1016/0306-2619(93)90057-V

Gribkovskaia, I., Halskau, Ø., Laporte, G., 2007. The bridges of Königsberg: A historical perspective. *Networks* 49, 199–203. doi:10.1002/net.20159

Hewings, G.J.D., 1982. The empirical identification of key sectors in an economy: A regional perspective. *The Developing Economies* 20(2), 173–195.

Hewings, G.J.D., Merrifield, J. and Schneider, J., 1984. A regional test of the linkage hypothesis. *Revue d'Economie Regionale et Urbaine* 25, 275–290.

Hirschman, A.O., 1958. *The Strategy of Economic Development.* Yale University Press.

Hokkanen, J., Salminen, P., 1997. ELECTRE III and IV decision aids in an environmental problem. *Journal of Multi-Criteria Decision Analysis* 6, 215–226. doi:10.1002/(SICI)1099-1360(199707)6:4<215::AID-MCDA139>3.0.CO;2-P

Holden, E., Linnerud, K., Banister, D., 2017. The imperatives of sustainable development. *Sustainable Development* 25, 213–226. doi:10.1002/sd.1647

Iori, G., De Masi, G., Precup, O.V., Gabbi, G., Caldarelli, G., 2008. A network analysis of the Italian overnight money market. *Journal of Economic Dynamics and Control* 32, 259–278.

ISO, 2014. ISO 37120:2014 – Sustainable development of communities – Indicators for city services and quality of life. www.iso.org/standard/62436.html

Janssen, R., 2001. On the use of multi-criteria analysis in environmental impact assessment in The Netherlands. *Journal of Multi-Criteria Decision Analysis* 10, 101–109. doi:10.1002/mcda.293

Kannan, R., Leong, K.C., Osman, R., Ho, H.K., 2007. Life cycle energy, emissions and cost inventory of power generation technologies in Singapore. *Renewable and Sustainable Energy Reviews* 11, 702–715.

Kennedy, C., Cuddihy, J., Engel-Yan, J., 2007. The changing metabolism of cities. *Journal of Industrial Ecology* 11, 43–59.

Kitzes, J., 2013. An introduction to environmentally-extended input–output analysis. *Resources* 2, 489–503. doi:10.3390/resources2040489

Lenzen, M., 2003. Environmentally important paths, linkages and key sectors in the Australian economy. *Structural Change and Economic Dynamics* 14, 1–34. doi:10.1016/S0954-349X(02)00025-5

Lenzen, M., 2011. Aggregation versus disaggregation in input–output analysis of the environment. *Economic Systems Research* 23, 73–89. doi:10.1080/09535314.2010.548793

Lenzen, M., Kanemoto, K., Moran, D., Geschke, A., 2012. Mapping the structure of the world economy. *Environmental Science & Technology* 46, 8374–8381. doi:10.1021/es300171x

Lenzen, M., Moran, D., Kanemoto, K., Geschke, A., 2013. Building eora: A global multi-region input–output database at high country and sector resolution. *Economic Systems Research* 25, 20–49. doi:10.1080/09535314.2013.769938

Leontief, W., 1970. Environmental repercussions and the economic structure: An input–output approach. *Review of Economics and Statistics* 52, 262–271. doi:10.2307/1926294

Leontief, W.W., 1941. *The Structure of American Economy, 1919–1929: An Empirical Application of Equilibrium Analysis*. Harvard University Press.

Leontief, W.W., 1986. *Input–output Economics*. Oxford University Press.

Maclaren, V.W., 1996. Urban sustainability reporting. *Journal of the American Planning Association* 62, 184–202.

McNerney, J., Fath, B.D., Silverberg, G., 2013. Network structure of inter-industry flows. *Physica A: Statistical Mechanics and its Applications* 392, 6427–6441.

Miller, R.E., Blair, P.D., 2009. *Input–output Analysis: Foundations and Extensions*, 2nd ed. Cambridge University Press.

National Environment Agency, 2016. *Singapore's Second Biennial Update Report under the United Nations Framework Convention on Climate Change*. Singapore.

Newman, M.E.J., 2010. *Networks: An Introduction*. Oxford University Press.

Newman, P.W., 1999. Sustainability and cities: Extending the metabolism model. *Landscape and Urban Planning* 44, 219–226.

Nilsson, M., Griggs, D., Visbeck, M., 2016. Map the interactions between sustainable development goals: Mans Nilsson, Dave Griggs and Martin Visbeck present a simple way of rating relationships between the targets to highlight priorities for integrated policy. *Nature* 534, 320–323.

OECD, 2015. *Input–output Tables*. http://stats.oecd.org/Index.aspx?DataSetCode=IOTS (accessed July 3, 2017).

Ofori, G., 2000. Greening the construction supply chain in Singapore. *European Journal of Purchasing & Supply Management* 6, 195–206.

Olszewski, P.S., 2007. Singapore motorisation restraint and its implications on travel behaviour and urban sustainability. *Transportation* 34, 319.

Panchamukhi, V.R. (1975) Linkages in industrialisation: A study of selected developing countries in Asia. *United Nations Journal of Development Planning*, 8.

Peters, G.P., Hertwich, E.G., 2006. Pollution embodied in trade: The Norwegian case. *Global Environmental Change* 16, 379–387. doi:10.1016/j.gloenvcha.2006.03.001

Quah, E. (ed.), 2016. *Singapore 2065: Leading Insights on Economy and Environment from 50 Singapore Icons and Beyond*. World Scientific.

Rasmussen, P.N., 1956. *Studies in Intersectoral Relations*. North-Holland.

Roberts, M., Blankespoor, B., Deuskar, C., Stewart, B.P., 2017. Urbanization and development: Is Latin America and the Caribbean different from the rest of the world? *Policy Research* Working Paper.

Rogers, M., Bruen, M., 1998a. Choosing realistic values of indifference, preference and veto thresholds for use with environmental criteria within ELECTRE. *European Journal of Operational Research* 107, 542–551.

Rogers, M., Bruen, M., 1998b. A new system for weighting environmental criteria for use within ELECTRE III. *European Journal of Operational Research* 107, 552–563. doi:10.1016/S0377-2217(97)00154-9

Rosenberg, P., World Health Organization, Centre for Health Development, United Nations Human Settlements Programme, 2016. Global report on urban health: Equitable, healthier cities for sustainable development. https://unhabitat.org/.../global-report-on-urban-health-equitable-healthier-cities-for-su... (accessed July 14, 2019).

Roy, B., 1968. Classement et choix en présence de points de vue multiples. *RAIRO – Operations Research – Recherche Opérationnelle* 2, 57–75.

Roy, B., 1978. ELECTRE III: Un algorithme de classement fondé sur une représentation floue des préférences en présence de critères multiples. *Cahiers du CERO* 20, 3–24.
Scott, J., 2017. *Social Network Analysis*. Sage.
Shmelev, S., 2012. *Ecological Economics: Sustainability in Practice*. Springer.
Shmelev, S., 2017. Multidimensional sustainability assessment for megacities. In S. Shmelev (ed.), *Green Economy Reader: Lectures in Ecological Economics and Sustainable Development*. Springer.
Shmelev, S.E., 2010. Environmentally extended input–output analysis of the UK economy: Key sector analysis. https://core.ac.uk/download/pdf/6307360.pdf (accessed July 14, 2019).
Shmelev, S.E., Rodríguez-Labajos, B., 2009. Dynamic multidimensional assessment of sustainability at the macro level: The case of Austria. *Ecological Economics* 68, 2560–2573. doi:10.1016/j.ecolecon.2009.03.019
Shmelev, S.E., Shmeleva, I.A., 2009. Sustainable cities: Problems of integrated interdisciplinary research. *International Journal of Sustainable Development* 12, 4–23.
Shmelev, S.E., Shmeleva, I.A., 2019. Multidimensional sustainability benchmarking for smart megacities. *Cities*, 92, 134–163.
Shmelev, S.E., van den Bergh, J.C., 2016. Optimal diversity of renewable energy alternatives under multiple criteria: An application to the UK. *Renewable and Sustainable Energy Reviews* 60, 679–691.
Singapore Ministry of the Environment and Water Resources, 2014. Our home, our environment, our future: Sustainable Singapore blueprint 2015. Singapore: Ministry of the Environment and Water Resources: Ministry of National Development, Singapore.
SLNG, 2014. Importance of LNG to Singapore www.slng.com.sg/website/content.aspx?wpi=Importance+of+LNG+to+Singapore&mmi=27&smi=117 (accessed June 23, 2017).
Tan, P.Y., bin Abdul Hamid, A.R., 2014. Urban ecological research in Singapore and its relevance to the advancement of urban ecology and sustainability. *Landscape and Urban Planning* 125, 271–289.
Timmer, M., Erumban, A.A., Gouma, R., Los, B., Temurshoev, U., Vries, G.J. de, Arto, I., Genty, V.A.A., Neuwahl, F., Rueda Cantuche, J.M., Francois, J., Pindyuk, O., Pšschl, J., Stehrer, R., Streicher, G., 2012. The World Input–output Database (WIOD): Contents, sources and methods (No. 20120401), IIDE Discussion Papers. Institute for International and Development Economics.
Tukker, A., Koning, A. de, Wood, R., Hawkins, T., Lutter, S., Acosta, J., Cantuche, J.M.R., Bouwmeester, M., Oosterhaven, J., Drosdowski, T., Kuenen, J., 2013. Exiopol: Development and illustrative analyses of a detailed global Mr Ee Sut/Iot. *Economic Systems Research* 25, 50–70. doi:10.1080/09535314.2012.761952
UN General Assembly, 1992. *Rio Declaration on Environment and Development*. www.un.org/documents/ga/conf151/aconf15126-1annex1.htm
UN-Habitat (ed.), 2011. *Cities and Climate Change, Global Report on Human Settlements*. Earthscan.
Vincke, P., 1992. *Multicriteria Decision-aid*. Wiley.
Warren-Rhodes, K., Koenig, A., 2001. Escalating trends in the urban metabolism of Hong Kong: 1971–1997. *AMBIO: A Journal of the Human Environment* 30, 429–438.
WCED, 1987. *Our Common Future*. UN World Commission on Environment and Development.

Weisz, H., Steinberger, J.K., 2010. Reducing energy and material flows in cities. *Current Opinion in Environmental Sustainability* 2, 185–192. doi:10.1016/j.cosust.2010.05.010

While, A., Jonas, A.E., Gibbs, D., 2004. The environment and the entrepreneurial city: Searching for the urban 'sustainability fix' in Manchester and Leeds. *International Journal of Urban and Regional Research* 28, 549–569.

Wiedmann, T.O., Chen, G., Barrett, J., 2016. The concept of city carbon maps: A case study of Melbourne, Australia. *Journal of Industrial Ecology* 20, 676–691. doi:10.1111/jiec.12346

World Bank, 2017. DataBank | The World Bank. http://databank.worldbank.org/data/home.aspx (accessed June 22, 2017).

1.8 Appendix 6.1: Complete forward and backward linkage results for Singapore

Key sectors where the forward and backward linkage are both greater than 1 are highlighted in grey.

Table A.6.1 Complete forward and backward linkage coefficients for Singapore

Sector ID	*Sector name*	*Final demand*		*CO_2 emissions*		*Employment*	
		Fwd	*Bckwd*	*Fwd*	*Bckwd*	*Fwd*	*Bckwd*
1	Agriculture and fishing	0.02	0.53	0.54	1.22	0.09	0.57
2	Food products	0.61	1.00	0.47	0.83	0.56	0.82
3	Beverages and tobacco products	0.17	0.72	0.10	0.35	0.15	0.67
4	Textiles, wearing apparel, footwear, and leather products	0.06	0.40	0.19	0.32	0.17	0.45
5	Wood and paper products	0.12	0.64	0.17	0.58	0.25	0.68
6	Printing and reproduction of recorded media	0.20	0.61	0.08	0.65	0.49	0.77
7	Petroleum products	3.62	2.29	12.67	7.44	0.83	0.29
8	Petrochemicals and petrochemical products	2.56	2.77	1.85	3.15	0.50	0.58
9	Chemicals and chemical products	0.97	1.12	1.99	1.92	0.70	0.81
10	Pharmaceuticals and biological products	1.31	1.12	0.69	0.54	0.26	0.26
11	Rubber and plastic products	0.16	0.85	0.81	1.78	0.47	0.74

12	Other non-metallic mineral products	0.18	0.88	0.39	0.69	0.26	0.84
13	Basic metals	0.07	0.57	0.12	1.47	0.15	0.55
14	Fabricated metal products (except machinery and equipment)	0.69	0.66	0.38	0.58	1.03	0.91
15	Semiconductors	3.60	2.96	0.11	0.49	0.52	0.59
16	Other electronic components and boards	0.39	0.64	0.14	0.74	0.68	0.75
17	Computers and peripheral equipment	1.69	1.39	0.18	0.15	0.58	0.42
18	Data storage	0.42	0.62	0.06	0.57	0.18	0.41
19	Communications equipment and consumer electronics	0.11	1.09	0.12	0.42	0.50	1.24
20	Scientific, photographic, and optical products	0.77	1.14	0.07	0.29	0.22	0.57
21	Electrical industrial apparatus, batteries, and accumulators	0.17	0.57	0.16	0.54	0.69	0.94
22	Electric wiring, lighting equipment, and domestic appliances	0.05	0.46	0.07	0.39	0.09	0.45
23	General and special purpose machinery (except oil rigs)	1.16	1.29	0.15	0.34	0.42	0.61
24	Land transport equipment	0.18	0.61	0.08	0.27	0.22	0.64
25	Ships, boats, and oil rigs	1.88	1.75	0.19	0.33	0.67	0.85
26	Aircraft and related machinery	0.71	0.72	0.10	0.27	0.39	0.46
27	Medical and dental instruments and supplies	0.39	0.58	0.03	0.27	0.23	0.42
28	Other manufacturing	0.25	0.69	2.85	2.15	0.17	0.61
29	Electricity	0.80	0.29	27.17	17.43	0.91	0.29
30	Gas	0.06	0.29	1.50	1.51	0.10	0.35
31	Water and sewerage	0.15	0.39	0.09	2.03	0.25	0.43

(continued)

Table A.6.1 (continued)

Sector ID	*Sector name*	*Final demand*		*CO_2 emissions*		*Employment*	
		Fwd	*Bckwd*	*Fwd*	*Bckwd*	*Fwd*	*Bckwd*
32	Waste collection, treatment, disposal, and material recovery services	0.19	0.79	0.08	0.77	0.17	0.73
33	Construction of buildings and civil engineering	3.11	2.98	0.29	0.45	2.30	2.59
34	Specialized construction activities	1.73	0.96	0.28	0.44	1.70	1.21
35	Wholesale trade	8.68	6.15	1.50	0.33	7.33	5.60
36	Retail trade	0.96	1.29	0.09	0.60	0.81	1.30
37	Land transport and supporting services	0.66	0.87	6.02	4.75	1.36	1.33
38	Water transport and supporting services	4.78	3.36	1.42	1.11	2.25	1.48
39	Air transport and supporting services	1.49	1.38	1.80	1.96	1.74	1.50
40	Warehousing and other supporting services	0.60	0.60	0.28	0.51	0.56	0.60
41	Postal and courier	0.13	0.29	0.13	0.30	0.73	0.84
42	Accommodation	0.43	0.63	0.07	0.80	1.12	1.22
43	Food and beverage services	0.85	1.16	0.16	0.91	2.42	2.50
44	Publishing	0.79	0.37	0.19	0.08	0.62	0.12
45	Media entertainment	0.31	0.44	0.06	0.26	0.56	0.67
46	Telecommunications	0.94	0.60	0.32	0.40	1.68	1.11
47	Computer programming, consultancy, and information services	1.17	0.82	0.33	0.22	2.21	1.48
48	Banking and finance	3.26	1.52	0.80	0.16	3.98	2.05
49	Insurance, reinsurance, and pension funds	0.90	0.91	0.11	0.06	0.63	0.81
50	Fund management activities	0.48	0.48	0.03	0.08	0.38	0.45

51	Other auxiliary financial and insurance services	1.22	0.77	0.14	0.10	1.24	0.90
52	Real estate	2.30	1.37	0.45	0.41	2.75	1.53
53	Ownership of dwelling	1.48	1.46	0.00	0.07	0.00	0.38
54	Legal services	0.19	0.30	0.05	0.19	0.38	0.51
55	Accounting, tax consultancy, and auditing services	0.12	0.18	0.05	0.14	0.39	0.42
56	Business representative offices and HQ	1.03	0.78	0.36	0.16	1.53	1.26
57	Management consultancy, advertising, and market research	0.51	0.56	0.16	0.25	1.36	1.31
58	Architectural and engineering services	1.14	0.98	0.62	0.42	1.32	1.06
59	Research and development	0.29	0.45	0.05	0.37	0.16	0.36
60	Other professional, scientific and technical services	0.18	0.66	0.19	0.38	0.33	0.78
61	Rental and leasing	0.88	0.50	0.51	0.16	0.48	0.25
62	Employment and labor contracting	0.23	0.18	0.08	0.10	0.93	0.83
63	Other administrative and business support services	0.69	0.60	0.25	0.46	2.49	2.04
64	Public administration and defense	2.38	2.35	0.24	0.55	6.15	5.88
65	Education	1.13	0.95	0.04	0.36	0.87	0.63
66	Medical and health services	0.87	0.95	0.06	0.49	2.27	1.80
67	Social services	0.19	0.48	0.02	0.81	0.76	1.02
68	Arts, entertainment, and recreation	0.61	0.73	0.03	0.57	0.50	0.75
69	Member organizations	0.11	0.49	0.02	0.51	0.20	0.77
70	Repair of computers, personal and household goods, and vehicles	0.13	0.56	0.17	0.38	0.26	0.57
71	Domestic and other personal services	0.31	0.44	0.03	0.24	1.35	1.36

1.9 Appendix 6.2: Full MCDA results summary

Table A.6.2 Complete sector sustainability rankings in Singapore by weighting scheme, by number of other sectors outranked

Sector ID	*Sector name*	*Overall rank by weighting scheme (1 = most sustainable)*			
		Equal	*Final demand*	*CO_2 emissions*	*Employment*
1	Agriculture and fishing	68	68	69	67
2	Food products	43	33	49	37
3	Beverages and tobacco products	40	38	42	47
4	Textiles, wearing apparel, footwear, and leather products	66	65	60	67
5	Wood and paper products	52	51	60	46
6	Printing and reproduction of recorded media	42	48	45	39
7	Petroleum products	57	58	55	51
8	Petrochemicals and petrochemical products	55	51	55	53
9	Chemicals and chemical products	50	42	55	47
10	Pharmaceuticals and biological products	47	42	49	61
11	Rubber and plastic products	57	54	64	55
12	Other non-metallic mineral products	43	35	53	41
13	Basic metals	63	59	66	59
14	Fabricated metal products (except machinery and equipment)	27	27	31	24
15	Semiconductors	11	8	12	16
16	Other electronic components and boards	34	30	42	31
17	Computers and peripheral equipment	13	16	4	27
18	Data storage	50	49	49	55
19	Communications equipment and consumer electronics	25	25	21	23
20	Scientific, photographic, and optical products	27	24	24	32
21	Electrical industrial apparatus, batteries, and accumulators	31	33	34	30
22	Electric wiring, lighting equipment, and domestic appliances	65	65	65	63
23	General and special purpose machinery (except oil rigs)	22	18	19	25
24	Land transport equipment	47	50	38	47
25	Ships, boats, and oil rigs	9	6	5	8
26	Aircraft and related machinery	32	30	27	36

27	Medical and dental instruments and supplies	43	45	39	52
28	Other manufacturing	68	65	68	67
29	Electricity	66	64	67	63
30	Gas	71	71	71	71
31	Water and sewerage	68	68	69	67
32	Waste collection, treatment, disposal, and material recovery services	47	40	59	45
33	Construction of buildings and civil engineering	2	3	1	3
34	Specialized construction activities	10	9	11	11
35	Wholesale trade	3	2	29	2
36	Retail trade	16	12	15	14
37	Land transport and supporting services	60	63	60	55
38	Water transport and supporting services	29	23	34	27
39	Air transport and supporting services	46	38	45	41
40	Warehousing and other supporting services	39	36	44	38
41	Postal and courier	37	56	31	34
42	Accommodation	24	25	30	22
43	Food and beverage services	6	5	21	6
44	Publishing	52	46	47	62
45	Media entertainment	35	46	25	34
46	Telecommunications	17	20	18	15
47	Computer programming, consultancy, and information services	6	10	2	7
48	Banking and finance	3	4	21	3
49	Insurance, reinsurance, and pension funds	22	17	12	19
50	Fund management activities	38	40	26	44
51	Other auxiliary financial and insurance services	13	13	8	13
52	Real estate	5	6	15	5
53	Ownership of dwelling	33	32	27	59
54	Legal services	54	59	41	47
55	Accounting, tax consultancy, and auditing services	60	68	55	55
56	Business representative offices and HQ	11	15	6	12
57	Management consultancy, advertising, and market research	18	22	9	18
58	Architectural and engineering services	18	14	36	16
59	Research and development	64	54	63	63

(continued)

Table A.6.2 (continued)

Sector ID	Sector name	Overall rank by weighting scheme (1 = most sustainable)			
		Equal	*Final demand*	*CO_2 emissions*	*Employment*
60	Other professional, scientific, and technical services	40	42	40	40
61	Rental and leasing	57	51	49	63
62	Employment and labor contracting	26	59	20	25
63	Other administrative and business support services	15	18	12	8
64	Public administration and defense	1	1	3	1
65	Education	20	21	15	21
66	Medical and health services	8	10	9	8
67	Social services	35	37	37	32
68	Arts, entertainment, and recreation	29	28	31	29
69	Member organizations	60	62	53	43
70	Repair of computers, personal and household goods, and vehicles	55	56	47	53
71	Domestic and other personal services	20	28	6	19

7 Multidimensional assessment of sustainability of Taipei and Almaty

Zhanar M. Kadyrkhanova, Stanislav E. Shmelev, Rimma K. Sagiyeva, Yelena Y. Chzhan and Irina A. Shmeleva

1. Introduction

Active and dynamic processes of globalization, international interaction, scientific and technical breakthroughs and the introduction of digital technologies in all spheres of the life of society exacerbate the contradictions between the economic and environmental trends in the development of regions and cities.

The challenge of building a green economy in Kazakhstan is magnified by the strong and established status of Kazakhstan as a resource-exporting country. Kazakhstan is the 52nd largest exporter in the world. The country's volume of trade amounted to $55.4 billion in 2017. Exports increased by 31% to $34.5 billion due to the price increase for energy resources and products of metallurgy. The supplies of ferroalloys, petroleum products, copper, rolled metal, propane, butane and zinc increased. The share of processed goods in total exports is 32%, which is 7% higher than in 2012, according to total.kz. In tonnage, exports increased by 4%. Over the past five years, the qualitative structure of exports has improved significantly, while a third of the exported products falls to the non-primary sector of the economy (MKKZ, 2017). According to Zhenis Kasymbek, the Minister for Investment and Development of the Republic of Kazakhstan, the country exports electric locomotives to Turkmenistan, Tajikistan, Azerbaijan, Kyrgyzstan and Ukraine; assembles cars for export to Tajikistan, Uzbekistan, Kyrgyzstan and China; and exports electric batteries to the Middle East, the Caucasus and Belarus.

Dynamics of the economy in Kazakhstan is unfortunately accompanied by deterioration in the environmental quality in regions of the country. An unsuitable ecological situation complicates the problem of stable territorial development. The high level of atmospheric air pollution in the cities of Kazakhstan with such pollutants as nitrogen dioxide, carbon monoxide, sulfur dioxide, hydrogen sulfide is caused by congestion of roads by urban transport, dissipation of emissions from industrial enterprises and low aeration in cities (Kazhydromet, 2017).

Accounting for environmental factors in the development of regional programs, the rationale for investment policy, and the resolution of resource-saving issues is becoming increasingly important, since there is a close interconnection between the economic and environmental spheres that are the main components of sustainable territorial development.

The chapter will be structured as follows: Section 2 offers a conceptual framework of green economy in the urban sustainability context; Section 3 covers data and methods used; Section 4 presents the analysis and results of the study; Section 5 offers a discussion, and Section 6 concludes.

2. Conceptual framework

After the global financial crisis of 2008–2009, the United Nations Environment Programme issued a proposal to use green economy as a key paradigm for economic reform, not only tackling the environmental issues through the development of renewable energy, circular economy and sustainable cities, but also stimulating economic progress through increased investment (UNEP, 2011; Shmelev, 2017).

The UN Conference on Sustainable Development in Rio de Janeiro, held in 2012, considered the following areas: the "green" economy in the context of sustainable development, and institutional framework for sustainable development. The final document of the Rio+20 conference sets out the basis for a new model of green economic development that will promote social progress, sustainable consumption and production. Thus, the green economy concept, that has spread after the adopted "green economy approach to the sustainable development and poverty eradication" of the UN Environment Program (UNEP) in 2011, is positioned as a way to achieve sustainable development within global systems (UNEP, 2011). The transition to a green economy involves the use of clean energy and green technologies in production, reducing the negative impact on the environment and climate, and at the same time eradicating poverty.

The concept of sustainable development was even more specific in the Agenda for Sustainable Development until 2030, adopted by the UN member states on September 25, 2015 (The Sustainable Development Goals (SDGs), officially known as "Transforming Our World: The 2030 Agenda for Sustainable Development until 2030"). Seventeen goals were identified and expressed quantitatively in a number of indicators, which should be achieved within 15 years. It clarified that joint efforts of governments, the private sector, civil society and the world's inhabitants are needed to achieve these Sustainable Development Goals.

Since there has been unprecedented urban growth in the world in recent decades and with about 54% of the world's population living there, there are also attendant problems of rapid urbanization: increasing number of slum dwellers, increasing air pollution, industrial and solid domestic waste, water pollution, etc. Therefore, Goal 11 on the Agenda for Sustainable Development until 2030 was defined as "Ensuring the openness, security, resilience and environmental sustainability of cities and human settlements" (UN, 2017). The urban slum population in developing countries declined from 39% in 2000 to 30% in 2014. However, despite some progress, the absolute number of urban residents living in slums continued to increase, partly due to the acceleration of urbanization, the growth of the total population and the lack of sound land and housing policies. In 2014, it is estimated that 880 million urban dwellers lived in slums, while in 2000 the number of such residents was 792 million.

As more and more people move to urban areas, cities' geographical boundaries tend to expand to accommodate new residents. Between 2000 and 2015, the expansion of urban areas in all regions of the world was faster than the urban population growth. As a result, as cities grow, population density decreases, while unplanned urban sprawl complicates the introduction of more sustainable urban development models. As of May 2017, 149 countries were engaged in the development of urban policies at the national level.

One of the most important environmental services in cities is the safe disposal of solid waste. Unsuccessful solid waste disposal blocks gutters, causing flooding and can lead to the spread of diseases transmitted through water. Data from cities in 101 countries for the period from 2009 to 2013 indicate that municipal waste collection systems covered only 65% of the urban population.

One of the main environmental risk factors for health is air pollution. In 2014, nine out of ten people living in cities breathed air that did not meet the WHO's safety standard. The total area of the world's cities is only 3% of the earth's land area, but they account for 60–80% of energy consumption and 75% of carbon dioxide emissions. Rapid urbanization rates have a negative impact on freshwater supplies, sewerage systems, habitat and public health systems. The relatively high density of cities can contribute to improving the efficiency of the economy and the introduction of technological innovations while reducing consumption of resources and energy (UN, 2017).

Thus, SDGs openly acknowledged the need to take into account the interrelation between ecological and economic parameters of the development of modern cities, which has also been emphasized in the UN Indicators of Sustainable Development (UN, 2007).

Table 7.1 provides an overview of the literature on sustainable development and the green economy.

Thus, "sustainable development" is a multifaceted and broadly interpreted category that integrates economic, environmental and social components.

The "green" economy according to UNEP is the economy that leads to "improved human well-being and social equity, significantly reducing environmental risks and environmental deficits." In a new compendium, *Green Economy Reader* (Shmelev, 2017), the green economy is shown to be methodologically underpinned by the developments in ecological economics, an interdisciplinary field focusing on economy-environment interactions in the broadest sense. Ecological economics uses a wide spectrum of analytical tools to support decision making for sustainable development.

Thus, the achievement of the goal of openness, security, vitality and environmental sustainability of cities and settlements assumes a comprehensive assessment of their environmental, social and economic development in comparison with the best world examples.

Table 7.1 Review of interdisciplinary literature on sustainable urban development and the green economy

Author and year	*Research focus (city, country)*	*Method*	*Indicators*	*Conclusions*
(Mayer & Knox, 2006)	Waldkirch, Hersbruck (Germany)	Urban regime theory	City budget, number of restaurants, unemployment, number of farmers, trees planted, indigenous recipes and ingredients preserved, population affected	The article explores two distinct examples of the slow city and slow food movements in Germany in Hersbruck and Waldkirch. Using the urban regime theory, the article illustrates how alternative development ideas can generate local community-based processes promoting the local economy and environmentally sensitive development strategies.
(Shmelev & Shmeleva, 2009)	London (UK), St Petersburg (Russia)	Indicator-based multidimensional urban sustainability assessment	Sustainable energy, sustainable transport, material flows and waste management, quality of life, health, psychology of interaction with the environment, green space, biodiversity, preservation of natural and cultural heritage, landscape architecture, ecodesign and democratic participation.	The article offers a holistic approach to urban sustainability incorporating the diversity of perspectives: the overview of the relevant problem areas to identify knowledge gaps and the analysis of interdisciplinary linkages among different sustainability dimensions to reveal the problem complexity and hidden connections. The article performs a comparative indicator-based sustainability assessment of London and St Petersburg to highlight differences in sustainability performance and policies used in two cities.
(Beddoe et al., 2009).	Global	Institutional analysis	Well-being, genuine progress indicator (GPI), GDP, raw materials, energy use, public goods, life satisfaction, education, infrastructure	The authors suggest an integrated set of worldviews, institutions and technologies to stimulate and seed the evolutionary redesign of the current socio-ecological regime to achieve global sustainability.

(Shen et al., 2011)	PET recycling system	Life-Cycle Assessment	Electricity, heat, CO_2, sulfuric acid, solid waste, number of recycling trips	The article offers an analysis of PET recycling systems. It shows possible ways to reduce environmental impacts and illustrates how impacts could be reduced by 20–30% depending on the configuration of the recycling system. The article concludes that the system's environmental impact can be optimized by maximizing the amount of recycled PET and using bio-based polymers.
(Lorek & Fuchs, 2011).	Global	Sustainable consumption governance	De-growth, energy efficiency, eco-efficiency, resource scarcity, distribution, mass consumption, social innovation, NGOs, governance	This article frames the necessary changes in consumption patterns required to achieve sustainability in the context of the de-growth discourses, paying particular attention to resource scarcity and distributional effects and emphasizing the role of social innovation and NGO strategies for sustainability governance.
(Yu, Pagani & Huang, 2012)	Nanjing, Changzhou, Wuxi, Hangzhou (China), Athens, Barcelona, Bologna, Brussels, Frankfurt, Geneva, Glasgow, Hamburg, Helsinki, London, Ljubljana, Madrid, Naples, Oslo, Paris, Porto, Prague, Rotterdam, Stockholm, Stuttgart, Turin Veneto (EU)	CO_2 emissions analysis	CO_2 emissions for residential sector, tertiary sector, industrial sector, public transport sector, private and commercial sector.	This paper conducts a thorough assessment of CO_2 emissions for the four urban centers in China: Nanjing, Changzhou, Wuxi and Hangzhou, and compares these results with emissions inventories of European cities. The paper concludes that Chinese cities in highly urbanized areas contribute a much higher per-capita emissions than their European competitors. Several policy suggestions considering regional disparities are provided aiming to reduce the CO_2 emissions of highly urbanized areas in China.

(continued)

Table 7.1 (continued)

Author and year	*Research focus (city, country)*	*Method*	*Indicators*	*Conclusions*
(Beatley & Newman, 2013)	Brisbane (Australia), New York (US)	Institutional analysis of biophilic cities	Biophilic city design elements (green roofs, urban trees, urban forests, community gardens, greens schools, green transport corridors). Biophilic conditions and infrastructure (percentage of population within a few hundred feet or meters of a park or green space, percentage of city area covered by vegetation, number of green design features, extent of natural images, shapes and forms in urban architecture, extent of flora and fauna species found in cities). Biophilic behaviors, patterns in lifestyles (average proportion of day spent outside, visitation rates for city parks, percent of trips made by walking, membership in local conservation clubs). Biophilic attitudes and knowledge (percent of residents who express care and concern for nature, percent of residents who could identify common species of flora and fauna). Biophilic institutions and governance (priority to nature conservation in local government, design and planning regulations supporting green cities, presence of educational institutions, number of nature organizations).	The article argues that a transition to make cities greener, more natural, more *biophilic*, will also help to make them more resilient in the long run, ecologically, economically and socially. The authors offer a system of conditions for developing a biophilic city to foster social and landscape resilience, in the face of climate change, natural disasters, economic uncertainty and various other shocks that cities will face in the future.

(Höjer & Wangel, 2014)	Stockholm (Sweden)	Life cycle assessment	Natural resources and energy, transportation and mobility, buildings, living, government, economy and people.	The article argues that a Smart Sustainable City is a city that meets the needs of its present inhabitants without compromising the ability for other people or future generations to meet their needs, and thus, does not exceed local or planetary environmental limitations and where this process is supported by ICT. The example of Stockholm is considered at length to illustrate this point.
(Cioca et al., 2015)	Romania	CO_2 emissions analysis	Energy sector, transport sector, waste sector, industry sector.	The article focuses on Romania's CO_2 emissions in the transport and energy sectors, industries and, especially, waste management sector. It argues for a transition towards the use of renewable resources at the national level.
(David & Fistung, 2015)	Romania	Transport research policy analysis	ICT, energy, environment, physical and fundamental sciences, health, agriculture, food safety and security, biotechnology, materials, processes and innovative products, space and security, socio-economic and humanistic dimensions.	The article considers the prospects for developing a modern transport infrastructure in Romania based on sustainable principles. The role of the EU in enabling such a transition is emphasized.
(Imai, 2015)	Japan	Patent data analysis	International patent applications, technological innovation, green technologies, competitiveness.	This article argues that Japan is losing its international technological advantage in the field of environmental technologies. Based on patent data from 2006 and 2012, it found that the international market competitiveness of Japanese environmental technologies is lower than in the EU and the US. The situation is attributed to the significant loss of momentum in technological innovation in Japan and noticeable progress in China's technological development.

(continued)

Table 7.1 (continued)

Author and year	*Research focus (city, country)*	*Method*	*Indicators*	*Conclusions*
(Hemani & Punekar, 2015)	India	Sustainable design	Broad design parameters: experience of arriving at the station complex, enjoying the station complex, moving in and around the station complex. Detailed design parameters: key functional areas.	The article makes the case for design education for sustainability in the rapidly urbanizing Indian context and concludes that design education should include approaches and methods that sensitize the students to the parameters of sustainability which are contextually relevant, inclusive and socially acceptable.
(Mukhtarova & Zhidebekkyzy, 2015)	Kazakhstan	Analysis of legislation	Energy efficiency, share of fossil fuels, share of renewable energy, CO_2 emissions per capita, urban pollution, exhaustion of natural resources, satisfaction with the actions for the protection of the environment.	The research shows that despite Kazakhstan's intentions to update and develop much of its infrastructure over the coming 20 years, inefficient use of resources is currently observed in every sector. It is necessary to encourage scientists and entrepreneurs to invent and commercialize new green technologies, which could be the basis for a successful transition to green economy.
(Mukhtarova, Trifilova & Zhidebekkyzy, 2016)	US, UK, China, India, Canada, Germany, Australia	Secondary literature review	Technology and industry; key variables of successful commercialization of green technologies, profitability and competitiveness of green technology, research of green technology commercialization issues.	The study identifies the most relevant research on commercialization of green technologies. It forms a report on the existing practices and positive experience in responsible leadership for policy makers. The study shows which subject areas create profitable growth in compliance with environmental sustainability and good corporate citizenship.

(Shmelev, 2017)	London, New York, Hong Kong, Los Angeles, São Paolo, Rio de Janeiro, Paris, Berlin, Singapore, Shanghai, Sydney, Tokyo	Multi-criteria decision aid (MCDA): ELECTRE III, NAIADE and APIS.	Top twelve world cities were assessed on eleven sustainability indicators: GRP, income per capita, higher education level, unemployment, Gini, CO_2, PM_{10}, green space, water use, waste generation, recycling rates.	The research has applied multi-criteria decision aid (MCDA) to assess sustainability of the major global cities. The results have shown that Singapore dominates sustainability rankings in most multi-criteria applications, showing particular strength in economic and environmental dimensions and a slightly less strong performance in the social dimension. Overall, Singapore, London and Sydney are leading and Shanghai, Rio de Janeiro and Los Angeles are still lagging behind in terms of sustainability. Learning from best practices and worst cases in this context provides an invaluable insight for policy reform to create smarter, greener, more compact, socially diverse, economically strong and less polluting cities around the world.
(Kumar et al., 2017)	Renewable energy systems	Multi-criteria decision aid (MCDA).	Environmental, socio-economic, technical and institutional barriers in energy planning	This article develops an insight into various MCDM techniques, progress made by considering renewable energy applications over MCDM methods and future prospects in this area. An extensive review in the sphere of sustainable energy has been performed.
(MCC, 2017).	Global	Global Environmental Assessment	Climate change, ecosystems, resources, biodiversity, governance, stakeholder engagement	The article provides an overview of the Global Environmental Assessments in the fields of climate change, resource use, ecosystems and biodiversity. The role of governance and stakeholder engagement receives particular attention.
(Biermann, Kanie & Kim, 2017).	Global	Sustainable Development Goals	Sustainable Development Goals (17 SDGs and their national adaptation is the measure of progress).	The article reviews the recently adopted Sustainable Development Goals in the context of new global initiatives in the area of sustainable development and environmental policy.

3. Data and methods

It is well known that the ecological situation in the cities of the Republic of Kazakhstan is characterized by unfavorable parameters. In particular, if we look at the official statistics, the cities of the eastern region (Ust-Kamenogorsk, Ridder), the southern (Almaty, Shymkent, Kyzylorda, Taraz) and the central (Karaganda, Temirtau, Zhezkazgan), where the largest metallurgy, chemical and petrochemical, and heat-and-power enterprises are located, fall into the priority list of the cities with the highest pollution level (Ust-Kamennogorsk, 2016). The high level of pollution in these cities has persisted for many years, which indicates the need to strengthen environmental measures to improve the situation in the air.

The level of atmospheric pollution is estimated from the complex index of atmospheric pollution (API-5), which is calculated from five substances (sulfur dioxide, carbon monoxide, nitrogen dioxide, phenol, formaldehyde) with the highest values at maximum permissible concentration (MPC), taking into account their hazard class. Depending on the API value, the level of air pollution is determined as follows: low – less than or equal to 5; increased – 5–7; high – 7–14; very high – greater than or equal to 14 (Kazinform, 2012). According to the Kazgidromet Regional State Enterprise data for 2014, the atmospheric pollution index (API-5) was 10.7 in Shymkent, 10.4 in Ust-Kamenogorsk and 10 in Almaty, which indicates a high level of pollution (see Figure 7.1) (State Fund of Ecological Information, 2015).

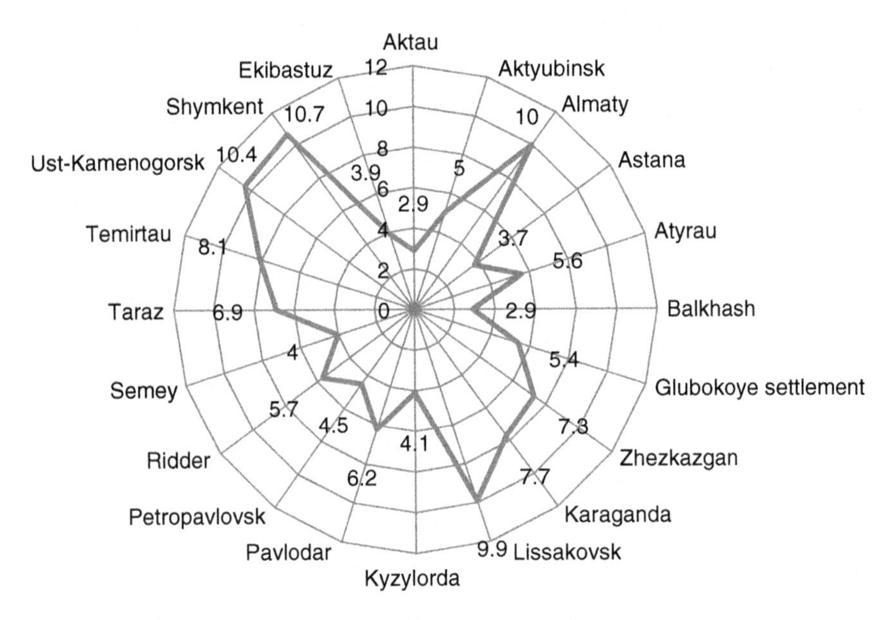

Figure 7.1 Index of air pollution in the cities of Kazakhstan, 2014

Source: "Kazhydromet" RSE

According to API-5 estimates, for the year 2017, the highest pollution levels were observed in the cities of Zhezkazgan, Karaganda, Shymkent, Temirtau, Ust-Kamenogorsk and Glubokoe settlement. The increased level of pollution was found in the cities of Astana, Almaty, Aktobe, Semey, Ridder, Taraz, Aktau and Balkhash. Low pollution levels were detected in the cities of Zyryanovsk, Lissakovsk, Kostanay, Turkestan, Uralsk, Kokshetau, Kyzylorda, Pavlodar, Ekibastuz, Taldykorgan, Atyrau and Petropavlovsk. In comparison with 2016, the level of atmospheric air pollution (according to API-5) increased in the cities of Aktau, Ust-Kamenogorsk and in the village of Glubokoe (Kazhydromet, 2017).

According to the *akimat* (regional government administration) of Almaty, the sources of harmful emissions are: motor transport – 80%, or 231 thousand tons per year; heat and power facilities – 12.5%, or 37 thousand tons; private residential sector – 5.6%, or 11 thousand tons; industrial enterprises – 1.9%, or 6 thousand tons.

The following factors contributing to the high level of environmental pollution in Almaty should also be taken into account: the high ash content (about 40%) of Ekibastuz coal, which is the fuel for the CHPP-2 (Coal Heat Power Plant); new construction carried out without taking into account prevalent winds as expressed by the wind rose in the city, as well as in park and water protection zones; cutting down the "lungs" of the metropolis, capable of absorbing up to 70% of dust, dirt and noise (Pchelyanskaya, 2017).

In order to understand what is the starting point for improving the ecological situation, in particular in Almaty, it is necessary to become familiar with the experience of leading cities, in which development occurs in a more harmonious way, according to the principles of sustainable development, with equal attention given to economic, social and environmental dimensions. Based on the methodological ideas outlined in Shmelev and Shmeleva (2009) and Shmelev (2017), we have adopted a multi-criteria approach to try to determine the pressing urban development problems and sustainability potential of Almaty.

For comparison, we suggest considering the city of Taipei, one of the leading cities in South East Asia, which demonstrates sustainable development, as environmental indicators do not deteriorate with dynamic and green economic growth. The choice is due to the fact that the cities are comparable in population: 2,704,810 inhabitants in Taipei in 2015 (City Population, 2015), and 1,552,349 residents in Almaty (Express-K, 2016). In addition to the constantly growing population within the city limits, Almaty became the core of the Almaty agglomeration with a population of 2,460,400 on January 1, 2015 (Express-K, 2016). At the same time, the area of Taipei is 271.8 km^2 (Taipei, 2013), while the area of Almaty is 682 km^2 (Kozybayev, 1983).

To analyze the ecological and economic situation in selected cities, we use sustainability indicators (Shmelev, 2012; Shmelev & Rodriguez-Labajos, 2009; Shmelev, 2011), which will be conditionally divided into three major sectors:

- economic parameters that demonstrate the dynamics of the economy of these cities;
- social parameters associated with living conditions of the population;
- environmental parameters associated with the consequences of economic activities for the environment.

Table 7.2 presents the economic, social and environmental parameters for Taipei and Almaty for the year 2015.

4. Analysis and results

Comparative analysis of the two cities, Taipei and Almaty, based on their sustainable development performance shows the following trends. The value of Taipei's gross regional product exceeds the GRP of Almaty by almost 1.5 times, which

Table 7.2 Economic, social and environmental parameters for Taipei and Almaty for the year 2015

No	*Indicators*	*Taipei**	*Almaty*
	Economic parameters		
1	Gross regional product	$44,173	$27,658
2	Number of metro stations per million inhabitants	43.26	5.72
3	The share of public transport in the city's total passenger traffic (in percent)	34	32
4	The share of cycling in the city's total passenger traffic (in percent)	4	1
5	Number of cars per 1000 people	46	67
	Social parameters		
6	Average monthly wage (in US$)	1583	568
7	Unemployment rate (in percent)	3.8	5.5
8	Average life expectancy (years)	80	74
9	Patents for scientific research (per 1000 inhabitants)	3.24	0.45
10	Gini index	32.5	27.1
	Environmental parameters		
11	Sulfur dioxide (SO_2, μg/m^3)	3	36
12	Nitrogen dioxide (NO_2, μg/m^3)	21,8	77
13	The amount of PM_{10} (μg/m^3), WHO standard – 20 μg/m^3	41,7	120
14	The use of water (in liters per day)	326	226,5
15	"Green" spaces (in m^2 per capita)	3.62	12
16	Amount of municipal waste generated (kg/per person per year)	101	409
17	Recycling (in percent)	54	5

Sources:
* Taipei City Government. https://english.gov.taipei/
Committee on Statistics of the Ministry of National Economy of the Republic of Kazakhstan. http://stat.gov.kz/

indicates the larger scale of economic development in Taipei with a smaller population. The level of development of the city's infrastructure, which has positive environmental implications, can be judged by such indicators as the number of metro stations per million inhabitants (in Taipei, there are eight times more than in Almaty) and the share of bicycle transport in the cities' overall passenger flow (this figure in Taipei is four times more than Almaty's). At the same time, the share of public transport in the total city passenger traffic in the compared cities is approximately the same, and the number of cars per 1,000 people in Taipei is about 1.5 times less than in Almaty. Thus, the economy of Taipei is developing dynamically, using more environmentally friendly transport, which significantly reduces the amount of harmful emissions in the atmosphere. How does Taipei manage to achieve a much stronger sustainable development performance with an area 2.5 times smaller than that of Almaty?

Turning to social performance indicators, let us examine the average monthly wage level – the main source of income for the population of both cities. In Taipei, this level is 2.7 times higher than that of Almaty. If we focus on real wages, then with an inflation rate of 1.7% in Taipei and 7% in Almaty, we can state that the actual average standard of living of Taipei's population exceeds that of Almaty by a factor of 5. Analysis of labor market statistics for both cities shows a greater number of unemployed in Almaty (5.5%) than in Taipei (3.8%), which also emphasizes the better social and economic well-being of the latter, since a significant part of the economically active population is engaged in creative production. The next criterion of social development is the average life expectancy at birth. In Taipei, it is six years higher than in Almaty, which can also be associated with a higher level of well-being in the population as well as better environmental quality.

If we consider another social indicator, such as the Gini index of income inequality, then in Almaty it is 27.1%, which indicates a more even distribution of the population's income in the megalopolis than in the city of Taipei, where it is 32.5%. Meanwhile, if we look at additional statistics showing the level of incomes of Almaty's population, such as the existence of savings and consumer loans, as well as the number of people owning real estate, then the problem of inequality takes on a special meaning. The monetary policy pursued in Kazakhstan, in turn, influenced the population's incomes (Kondybayeva & Ishuova, 2013). Thus, the proportion of citizens with a high standard of living and those who come in search of work and are severely restricted in funds is high in the city (in comparison with the average republican values). The problem of inequality is aggravated by employment and housing issues, which are often faced by new residents.

Social inequality is especially evident in matters of ownership of real estate. Housing rental makes up the lion's share of the costs for those arriving in Almaty. This phenomenon is also characteristic of a number of developing countries, where "contrast cities" often appear, for example, Rio de Janeiro or Caracas; and also of those countries with a high standard of living, in whose megacities the spatial differentiation of the regions by prestige, housing prices, the composition of the population in terms of "rich" and "poor" are clearly marked. Sociologist V. I. Ilyin in his book *The Theory of Social Inequality* (regarding the structure-construction paradigm) specifically considers physical space in the context of the formation of social inequality. The phenomenon of spatial stratification is also found in Almaty. In this case, a property's location is also an indicator of social inequality in the city. Researchers note that in the former capital of Kazakhstan this is manifested in the development of segregated areas into zones of closed elite housing and areas inhabited mostly by the poor.

"Housing conflicts" and, in general, the problem of "self-construction" also indicate a high level of inequality in Almaty. Many experts believe that housing issues can affect the growth in the potential for conflict among the population (Zhusupova, 2015).

Excessive inequality is a threat to society's stable development, by exacerbating the growth of social tension and deep social stratification; overall, inequality is more dangerous than poverty in its consequences for society. Here we can distinguish three main factors:

- Inequality gives rise, on the one hand, to the growth of apathy and passivity in society, and on the other, the desire to monopolize the sphere of political decision making.
- Inequality entails the marginalization of society, which creates the basis for political radicalism and extremism.
- Inequality contributes to the development of the atmosphere in society that undermines the foundations of social justice.

Now the problem is more acute, which is illustrated by the overpopulated Almaty, which has not had time to adapt to the influx of migrants from rural areas of Kazakhstan (Zhusupova, 2015), though it should be noted that at current official levels, inequality in Almaty is at the level of European cities and lower than in the US, China, or South America.

The next indicator included in the group of social parameters is the number of patents for research work per capita. This indicator not only reflects the "innovation" of the city's economy, but also fulfills the goal of generating new ideas, since the development and implementation of know-how and technologies, in addition to progress, help create new jobs which require a qualitatively high level of skills, solve environmental and infrastructure problems, and reduce the unequal income distribution between the rich and the poor. As can be seen from Table 7.3, the number of patents received for research work in Almaty shows a significant backlog: 0.5 patents per capita, while in Taipei this figure is seven times higher. In fact, the southern capital of Kazakhstan, despite the considerable concentration of the country's scientific potential (41% of scientific-sector employees are concentrated in Almaty), does not realize it effectively.

Thus, summing up the analysis, we can state that in the social sphere, Almaty lags far behind Taipei in accordance with the goals of sustainable development.

The most important component of a city's harmonious development is its ecology. At the same time, to minimize the harm to the environment from resident's growing economic activity, it is important to achieve the goals of sustainable development.

Consider the indicators that can describe a city's environmental performance. First of all, let's pay attention to air quality. Concerning one of the most harmful emissions – sulfur dioxide (SO_2, mg/m^3) – average concentrations of this pollutant in Taipei are twelve times lower than in Almaty; the release of nitrogen dioxide (NO_2, mg/m^3) in Taipei demonstrates a value of three times less than in Almaty. Concentrations of particulate matter or PM_{10} (in μg/m^3, the WHO's norm of 20), which according to WHO studies is directly affecting life expectancy, in Taipei are three times less than in Almaty (WHO, 2010). This disappointing fact is confirmed by the indicator of average life expectancy (in years) in Almaty, which is six years less than in Taipei.

A significant ecological criterion for a city's sustainable development is the indicator of water use (in liters per day), which is less by 100 liters per day in Almaty than in Taipei. However, "economical" water consumption in Almaty is not a result of the use of resource-saving technologies, but the population's lack of access to the central water supply.

The most favorable ecological indicator of the Kazakh city is the "green" spaces (in m^2 per capita), which is four times more in Almaty than in Taipei. This is a great advantage of Almaty, because the presence of green spaces allows mitigation of the city's ecological situation, the spaces acting as a natural filter protecting the city's air.

The two following indicators of cities' ecological well-being – the volume of municipal waste and recycling (in percent) – also show a significant backlog in Almaty. The volume of municipal waste in Almaty is four times higher than in Taipei, with recycling lagging behind by a factor of ten.

Thus, the analysis of Taipei's ecological, economic and sociological parameters reflects its greater compliance with sustainable development goals.

Figure 7.2 is a comparative analysis of sustainability in Almaty and Taipei, as of 2015.

Turning to the history of Taiwan's environmental problems, back in the late 1990s, it was considered to be one of the world's most polluted regions, as the country neglected environmental issues and increased the number of industrial enterprises in an effort to increase employment and encourage exports.

First of all, experts noted particularly acute problems of garbage recycling, and air and water pollution, which was due to the island's high population density; in 1998, this amounted to almost 580 people/km^2 (22 times more than in the US) with a total population of 21.82 million. The number of industrial companies reached 157,000, and their density was 4.4 companies/km^2 (12 times more than in the US). At the same time, enterprises that used and produced pathogenic materials found it difficult to dispose these harmful wastes.

In turn, high population density led to a higher concentration of debris. The total amount of debris produced in Taiwan in 1996 was 8.6 million tons. This means that each person produced more than 1 kilogram of garbage a day – an amount comparable to that produced by residents of OECD (Organization of Economic Cooperation and Development) countries. Therefore, many

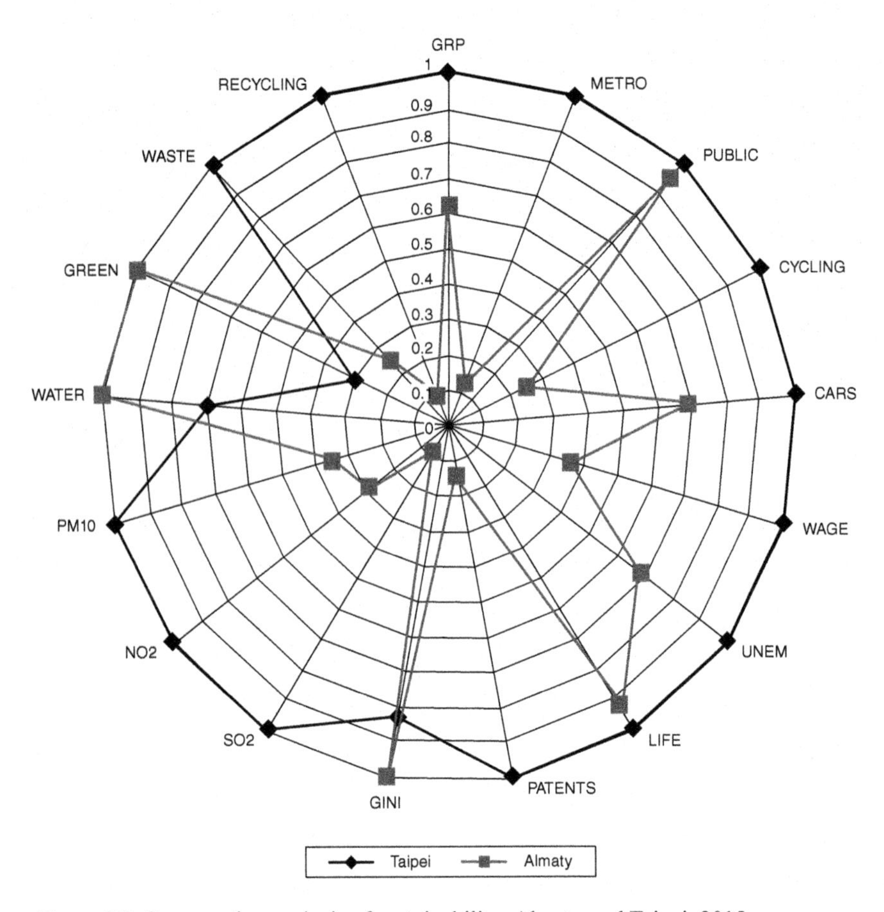

Figure 7.2 Comparative analysis of sustainability: Almaty and Taipei, 2015

Taiwanese cities suffered from the fact that they did not have an efficient waste disposal system.

The total number of vehicles was 15.77 million units, including 5.35 million vehicles and 10.42 million motor scooters and motorcycles, a density of 438 vehicles/km^2 (18 times more than in the US).

River and coastal waters in many parts of Taiwan have been severely contaminated, since industrial, agricultural and residential wastewater often directly merged into rivers. By 1997, 38% of Taiwan's main rivers and 29.5% of secondary rivers were polluted. The primary causes of the pollution problem were the urban areas, mainly because the island did not have a modern sewerage system and wastewater treatment facilities. Even in Taipei, where the sewerage system construction began in 1972, only 40% of the sewage system was connected to residential buildings, while in Kaohsiung, the rate fell to 5.3% (Paley, 2003).

Taiwan's ecological problems were exacerbated by complex natural and climatic conditions, as the island is periodically exposed to such natural disasters as earthquakes, hurricanes, typhoons and droughts.

Thus, it required the consolidation of the efforts of the entire population and the Taiwanese government to combat environmental pollution. Even though Taiwan is not a member of the UN and acceding to the United Nations Framework Convention on Climate Change, it actively supported all international initiatives to protect its environment, including the Paris Agreement on Climate Change, which was due to enter into force in 2020. Taiwan has been actively cooperating with the United States on environmental issues for over twenty years, beginning in 1993, when both parties signed the Agreement on Cooperation in the field of environmental protection. For two decades, the United States and Taiwan have been sharing experiences in addressing issues regarding water quality, chemical safety, soil contamination, electronic waste processing and other areas. One of the main components of the ecological cooperation between Taiwan and the United States is the "International Environmental Partnership" program, which has allowed the development of a number of strategic measures for the water and air purification, waste processing and the research of natural disasters.

Backed by the support of international organizations, the government of Taiwan has taken a number of administrative, legislative and economic measures that have allowed the phased environmental problems to be addressed in stages, implementing "green" programs.

The government of Taiwan established the Environmental Protection Agency, investing significant funds and implementing a number of government programs aimed at addressing the problems of air quality, water pollution and recycling of waste. "Green" energy was one of the six priority industries to be identified, beginning in 2010. The remaining five priority sectors are tourism, biotechnology, health, agriculture, and creative and cultural industries.

In addition to actively participating in all international environmental agreements, Taiwan's environmental reforms were reinforced by the efforts of the business community. The Environmental Protection Agency encourages local enterprises' participation in energy conservation programs, including the provision of technical assistance, the development of the integration of energy resources in industrial zones, as well as assistance in the introduction of power management systems and compliance with the criteria for obtaining "green" certificates. The Environmental Protection Agency also formed working groups with representatives from seven major industry sectors and held 17 meetings to discuss possible solutions and measures to overcome the difficulties in reducing emissions. The voluntary system for CO_2 emissions reduction is active in industries such as cement, steelmaking and petrochemical production, where from 2006 to 2015, more than 7,776 emission reduction measures were introduced. As a result of 42.1 billion NT$ (about US$1.3 billion) spent, the level of CO_2 emissions decreased by 10,220,000 tons, and overall productivity increased. This indicates the effectiveness of interaction between the state and the private sector on the transition to an environmentally friendly production cycle (Radio Taiwan International, 2011).

The government of Taiwan is continuously improving environmental legislation to prevent environmental crime. Taiwan is one of the few countries that have adopted legislation to reduce greenhouse gas emissions. In July 2015, the government published the relevant law. In addition, the government has adopted a plan to reduce emissions by 50% by 2030 to 214 million tons, which would mean an actual reduction of 20%. Since 2017, it has been proposed to toughen penalties for violations in this area by increasing fines, encouraging cooperation in the search for offenders and obliging violators to provide evidence that their activities do not threaten the environment and people. Also at this stage, the Executive Yuan approved the draft law on ecological agriculture, which will later be submitted to the Legislative Yuan. According to the Council for Agriculture, in Taiwan now more than 7,200 hectares of land are allocated for the cultivation of organic products. Over the past ten years, the area of such fields has grown by a factor of 3.5. However, at the moment the Taiwanese legislation regulates only the norms of finished products; the law does not specify the rules for conducting clean agriculture.

The government of Taiwan stimulates the development of "green" innovations:

- Several research institutions are currently active in the "green economy" field in Taiwan: the Research Institute of Industrial Technology, the Taiwan Institute of Renewable Energy and others.
- Taiwan's information center on environmental protection gathered proposals on sharing information on "green" cases. The site presented eight topics related to environmental protection: water resources, resource recycling, ecosystem conservation, green public initiative, green transport, green consumption. Everyone who comes to the site can pledge to act in one of these areas, whether already accomplished or to be enacted soon.
- Taiwan has become a platform for international forums and conferences on environment protection and the development of "green" technologies. For example, in Taipei, the annual international conference on resource processing is organized by the Environmental Protection Agency and discusses issues of sustainable development of urban habitat, and the creation of a "green" economic system for resources use.
- Following the summit of the leaders of APEC countries in Honolulu, the Asia-Pacific region's economies agreed to lower duties for producers of "green" energy products to below 5%; as a result, small Taiwanese companies decided to integrate to strengthen their positions in this market. In 2011, the total productivity of Taiwan's green industries reached NT$390 billion. "Green" trade has become a new trend for the development of small- and medium-sized Taiwanese companies, that is, a course was adopted for trade in goods, created on the basis of energy-saving technologies and free of harmful emissions.
- To promote industrial research and development, the government has created a number of scientific parks and economic zones that provide tax breaks and specialized lending rates to attract investment. The first of these, Hsinchu

Science Park was established in 1980 by the National Science Council with an emphasis on research and development in information and biotechnology.

The government of Taiwan is carrying out activities to consolidate the efforts of the whole society to solve environmental problems. In the implementation of environmental reforms, citizens and non-governmental organizations take an active part in environmental protection. For example, Taipei residents positively perceive the city mayor's proposals to economize on water consumption, reduce catering establishments' use of firewood and coal in residential areas, and prohibit the entry of vehicles into the city center implemented under the project "Provisions on self-government and the creation of favorable life conditions in Taipei City." One of the initiatives of this project was the installation of LED lighting in the streets, including 220,000 street lamps, as well as the distribution of the public bicycle system "YouBike," the use of which is free for the first 30 minutes.

Thus, the consistent and comprehensive implementation of measures by the government of Taiwan, continuing at this stage, has a positive economic, social and environmental effect in Taipei, which is planned to be extended to other regions of the country.

In 2011, according to the influential English news agency Economist Intelligence Unit, for active efforts to improve citizens' quality of life and the urban environment, Taipei took second place after Singapore among 22 Asian cities. The experts' report "Comparison of the 'green' cities in Asia" estimated the following indicators: energy and CO_2 emissions, land use and architecture, transport, waste management, water resources, health infrastructure, air quality and environmental management. Together with Taipei, second place was shared by Tokyo, Osaka, Yokohama, Hong Kong and Seoul. However, other regions of Taiwan have not achieved such excellent indicators, for example, Yunlin County is still the worst among all Taiwan's administrative-territorial units in terms of air pollution (Radio Taiwan International, 2011).

Taiwanese company MiniwizLtd, specializing in the transformation of household waste into materials with high technological characteristics, was recently named "Technology Pioneer 2015" by the World Economic Forum (WEF). This honorary title was first conferred on the company from Taiwan. Miniwiz's revolutionary approach, incorporating proprietary technologies, and its own engineering and architectural designs, is used to create the next generation of resources obtained by recycling end-of-life electronic devices, food waste, plastic and various packaging materials. One of Miniwiz's most famous technological solutions was realized in the "Eco-Ark" – a structure that has become one of the main attractions of the Taipei International Flower Exhibition (Flora-Expo), held in the Taiwanese capital in 2010. This building, the height of a nine-story building, was built from 1.5 million recyclable polyethylene terephthalate bottles. As noted by Huang Qian-chih, Miniwiz's chief executive, this Taipei-based company owes its technological development and rapid growth to the Taiwanese industrial sector's advanced level of development and high efficiency – in particular, support from the local powerful and extensive circulation chain of industrial and domestic waste (Radio Taiwan International, 2011).

Taiwan's National Laboratory of Applied Research first demonstrated unmanned aerial vehicles for the study of typhoons. Six devices ordered in Australia in 2015 at a price of AU$81,5000–120,000, will be used in the research program in 2019. They are equipped with video cameras and sensors to record data on temperature, humidity, air pressure, wind speed and direction. The maximum flight time of unmanned drones is 10–18 hours (Radio Taiwan International, 2011).

Also, Taipei introduces "smart," "green" power management. The company that manages the Taipei 101 skyscraper, which is officially certified as the world's tallest green, i.e. environmentally acceptable, building, signed a memorandum of understanding with Siemens Ltd, Taiwan regarding the implementation of the energy complex management platform based on cloud technologies. This allows this building, which is a symbol and one of Taipei's main attractions, to confidently embark on a path leading to its transformation into one of the planet's not only most "green," but also the most intellectually advanced structures. According to the memorandum, "Taipei 101" will use the Siemens navigator to control the energy supply, energy consumption and operation of the building's entire power system, and also make detailed analytical reports based on the incoming data. All this should improve the building operation's efficiency, reduce costs and promote the development of the strategy for managing the energy complex in the spirit of the concept of sustainable development.

In Kazakhstan, green economy principles are gradually introduced into the development model by means of establishing new green institutions and joining international agreements, which play a very important role. These steps are outlined in Table 7.3. Kazakhstan's active participation in international agreements in the field of environmental protection positively affected the solution of environmental problems.

5. Discussion

Thus, comparing the effect of environmental reforms on the sustainability performance of Taipei and Almaty, the following differences can be noted:

- Commitment to the steady implementation of sustainable development programs in Taiwan was due to very acute environmental and natural conditions (such as lack of territory, and the political and socio-economic situation in Taiwan).
- The government of Taiwan relied on technological, financial and institutional international cooperation with developed countries, both in the development and implementation of the strategy for sustainable development.
- The government of Taiwan has provided various support for the transition of industrial enterprises to more environmentally friendly and energy-efficient technologies.
- The government of Taipei has connected both the intellectual potential of the country and ordinary citizens, as well as non-governmental organizations, to the implementation of green projects.

Table 7.3 Significant milestones in the transition towards the green economy in the Republic of Kazakhstan

No	*Date*	*Document*	*Comments*
1	1997	Long-Term Development Strategy of Kazakhstan until 2030	In 1997, the Long-Term Development Strategy of the country until 2030 was adopted, in which, along with improving the conditions and living standards of the people of Kazakhstan, stabilization of the quality of the environment, conservation of natural resources for future generations was identified as one of the country's main development priorities.
2	1997	Vienna Convention	Kazakhstan acceded to the Vienna Convention on October 30, 1997 (Law of the Republic of Kazakhstan dated October 30, 1997 N 177-1 "On Accession of the Republic of Kazakhstan to the Vienna Convention for the Protection of the Ozone Layer"). The Vienna Convention entered into force (for Kazakhstan and other parties to the Convention) on August 26, 1998.
3	1997	Montreal Protocol	Kazakhstan acceded to the Montreal Protocol on October 30, 1997 (Law of the Republic of Kazakhstan dated October 30, 1997 No. 176 "On Accession of the Republic of Kazakhstan to the Montreal Protocol on Substances that Deplete the Ozone Layer"). The Montreal Protocol entered into force on August 26, 1998.
4	1998	National Action Plan for Environmental Protection	In 1998, the National Action Plan for Environmental Protection was developed, based on the ideology of sustainable development (NAPEP / SD).
5	2001	London Amendment to Montreal Protocol	Kazakhstan acceded to the London Amendment on May 7, 2007 (Law of Kazakhstan No. 191 dated 7 May 2001 "On Accession of the Republic of Kazakhstan to the Amendment to the Montreal Protocol on Substances that Deplete the Ozone Layer, London, 27–29 June 1990"). The amendment entered into force on July 26, 2001.
6	2002	World Summit on Sustainable Development, Johannesburg	In 2002, the Interdepartmental Commission was established to prepare for the UN World Summit on Sustainable Development in Johannesburg and the Kazakhstani Agenda for the twenty-first century.
7	2004	Council for Sustainable Development	In 2004, the Council for Sustainable Development under the Government of the Republic of Kazakhstan was established.
8	2006	The Concept for Transition of the Republic of Kazakhstan to Sustainable Development	On November 14, 2006, by the Decree of the President of the Republic of Kazakhstan, the Concept for Transition of the Republic of Kazakhstan to Sustainable Development for 2007–2024 was approved (the Decree became invalid).
9	2007	Environmental Code	On January 9, 2007, the Environmental Code of the Republic of Kazakhstan was adopted.

(continued)

Table 7.3 (continued)

No	*Date*	*Document*	*Comments*
10	2010	First National Report on the Green Economy	On 6 May 2010, Almaty presented the UN annual report on the ESCAP countries (Economic and Social Commission for Asia and the Pacific), as well as the First National Report on the Green Economy.
11	2011	Copenhagen Amendment to Montreal Protocol	Kazakhstan ratified the Copenhagen Amendment by the Law of the Republic of Kazakhstan dated April 6, 2011 No. 426-IV. The amendment entered into force on June 28, 2011.
12	2012	Strategy Kazakhstan-2050	In 2012, the Message of the President of the Republic of Kazakhstan – the Leader of the Nation N. Nazarbayev to the People of Kazakhstan Strategy “Kazakhstan-2050: New political course of the state.”
13	2012	Green Bridge Partnership Programme at Rio+20	The Partnership Program “Green Bridge” was included in the Declaration Rio+20 in 2012. On June 26, 2012 at the UN Conference on Sustainable Development “Rio+20” in Rio de Janeiro (Brazil), the final declaration “The Future We Want” was adopted, which reflected the initiatives of the President of the Republic of Kazakhstan N. Nazarbayev – Partnership program “Green Bridge” and Global Energy and Ecological Strategy.
14	2013	Green Economy implementation measures for 2013–2020	On May 30, 2013, the Decree of the President of the Republic of Kazakhstan approved the “Concept for the transition of the Republic of Kazakhstan to Green Economy. Plan of measures for the implementation of the Concept for 2013–2020.”
15	2014	Beijing Amendment to Montreal Protocol	Kazakhstan ratified the Beijing Amendment to the Montreal Protocol by the Law of the Republic of Kazakhstan, dated April 23, 2014 No. 198-V.
16	2015	Order N.214 of the Energy Minister	Order No. 214 of the Energy Minister of the Republic of Kazakhstan dated March 18, 2015 “On Approval of the Rules for Conducting Verification of the Completeness, Transparency and Reliability of the State Inventory of GHG Emissions and Absorption” registered in the Ministry of Justice of the Republic of Kazakhstan dated May 20, 2015 No. 11090.
17	2016	State Environmental Information Fund	On October 13, 2016, the Resolution of the Government of the Republic of Kazakhstan “On Approval of the Rules for the Maintenance of the State Environmental Information Fund” was adopted.

- Measures to stimulate the development and implementation of both national and foreign environmental innovations played a significant role.
- Legislation in the field of environmental protection has become tougher with respect to industries that cause damage to both the environment and the public.

To improve air quality in Almaty, a set of measures is required, both from the government and from the residents.

The city's main sources of air pollution are the emissions from the CHPP (Coal Heat Power Plant), road transport and the private sector. Here is an outline of several proposed immediate steps that could be taken in the context of Almaty.

- The Government may take the following measures:
  - The CHPP must be switched to natural gas. It is necessary to start using coal with ash content below 39%. Filters that are more efficient can be installed on the pipes. The largest point source of pollution is CHPP-2, which burns more than two million tons of coal annually. The quality of coal is low – Ekibastuz high-ash product – which when burned forms a large amount of useless ash. Some part of it is emitted through a high pipe into the atmosphere.

    To solve the air pollution situation, the government plans to allocate a budget for gas supply in 2019. "Samruk-Kazyna" JSC has started to develop a project on the transfer of CHPP-2 to gas; this is a very labor-consuming and costly process, and these measures would solve only 20% of the pollution problem.
  - Introduce a carbon tax on fuel.
  - Improve the quality of fuel sold in the city.
  - Transfer the private sector to gas heating; this will significantly reduce the amount of emissions in the air. The private sector also uses coal as fuel because it is cheaper than gas, but highly ashy. In addition, in private sector, people do not use filters, thereby contributing to the creation of urban smog.

Residents should think about using environmentally friendly means of transportation, not burning rubbish and leaves. Cars certainly do not make the air cleaner. Eighty percent of emissions comes from traffic; therefore, there is a need to solve the problem of traffic congestion: streamline the car flow, reduce the number of cars. This requires the political will of the city authorities.

It is necessary to constantly monitor the level of emissions, which will predict the increase in permissible concentrations, evaluate the effectiveness of the measures taken, identify new harmful substances, and find the most dangerous and illegal sources of pollution. Kazakhstan's air-monitoring system requires modernization.

To identify the main source of pollution, serious work of scientific institutions is necessary: year-round monitoring, conducting analysis, and compiling the scientific base. Only after that, it will be possible to identify the dynamics of air pollution by different sources.

Authorities should listen to independent organizations and city activists who measure pollution levels. More data on the state of air quality are needed. Kazhydromet is guided by the environmental code, where the rules for the implementation of environmental monitoring are prescribed. For example, all devices must have a certificate of compliance; in addition, all devices are regularly checked to verify serviceability or to set up correct functioning. If after monitoring, we see a tenfold excess of the maximum permissible concentrations, which is called the "case of high pollution", then it is necessary to immediately send the data to the territorial Department of Ecology in order to identify the pollution source and take action.

People are not yet aware of the whole scale of the problem, because they simply do not have the time for that: they need to feed their children and pay their loans, and the issue of ecology is at a completely different level of the needs pyramid. The media should bring awareness of the problem to people; the media need to dig, to conduct investigations, reveal the origins of the problem and who should take responsibility, and inform citizens what they can do.

Almaty has been balancing on the brink of ecological catastrophe for many years. A huge number of citizens come to polyclinics with complaints about respiratory problems. The state in turn spends millions on their treatment, although it would be cheaper to eliminate the source of the problem. Pollution gravely outstrips the maximum allowable concentrations, and the authorities still do not pay proper attention to this problem.

The best thing that Almaty residents can do is to demand from the municipal authorities affordable and practical public transport and developed bicycle infrastructure, as well as convenient infrastructures for pedestrians (Ramazanova, 2017).

Air pollution is the most important problem in both Almaty and many other cities in Kazakhstan. Of course, Almaty has other important problems – poor road and public transport systems, overcrowded kindergartens and schools, criminalization and marginalization of the city, indiscriminate tree-felling and criminal seizure of national park lands and reserves in the city's southern suburbs, replacement of architectural monuments and cinemas with unhealthy fast-food establishments, etc.

But in terms of neglect and scale, the air pollution problem exceeds all the others described above. There cannot be one solution to this issue; a complex of modern and effective measures is needed here. These could include many steps:

- choice in favor of grass, not asphalt;
- choice in favor of gas, not coal;
- choice in favor of eco-friendly transport, not fume-emitting decommissioned Chinese buses;
- choice in favor of accessible galleries, independent cultural and educational centers, kindergartens and schools, and not in favor of shopping centers, gas stations, bookmakers and pubs;
- choice in favor of beautiful, honest and eco-friendly coat of arms with a snow leopard, glaciers, apple and flowers, not a monstrous acid logo. Almaty does

deserve these proud symbols. After all, it was the Almaty land that gave the planet tulips and apple trees;

- expansion of public transport and reduction of the number of toxic vehicles;
- support for environmental art and culture;
- introduction of environmental standards, as well as air and fuel measurements at the European level;
- returning to the values of the green city and trees, and not keeping to the barbarous practice of tree-felling and reducing green zones and reserves;
- refusal to destroy environmentally friendly modes of transport (as was the case with the destruction of the Almaty tram system), in favor of introduction and expansion of clean modes of transport: trams, trolleybuses, bicycling;
- revival of the "lungs" of Almaty – the Kok-Zhailau National Park – rather than corrupt buildings and destruction of the reserve;
- expanding pedestrian alleys and parks, rather than highways and car interchanges;
- pedestrians, not driver;
- ecological activism and ecological tourism.

It is important to deal with real and not imaginary problems and issues; that is, it is better to use the taxpayers' resources and money to ensure the health and safety of city residents, and not burn these funds to rename the streets in the Zimbabwean-North Korean style. Not to pressure and pursue environmental and civil activists, but work together to solve common problems. Introduce environmental lessons from kindergartens, schools and universities to private companies, *akimats* and ministries. Judging by what happens to our cities and regions, to our nature and reserves, the vast majority of the population does not have even basic knowledge of their environment.

To see the real scale of the problem, one should climb the mountains or look at the city and the smog through the porthole of a flying plane. It is important to realize that in these airplanes the number of educated citizens who emigrate permanently from their country has increased recently, not least due to deteriorating environmental situation.

Air pollution is an urgent problem throughout the Republic. One can see different kinds of smog in Kazakhstan's cities: in Karaganda, Astana, Ust-Kamenogorsk, Pavlodar, Temirtau. Surprisingly and sadly, the political capital and the cultural capital compete in terms of air pollution with the level of pollution of industrial and toxic areas and even surpass them. This is not very compatible with expensive pomp and statements about innovation, green economy, the energy of the future and modernization of consciousness. Without clean air, our consciousness will most likely experience harmful mutations, and not transformation towards innovation and spiritual revival. Taking into account this sad fact, it is high time that everyone should understand the real scale of the problem, and start making the right choices to solve it.

6. Conclusions

Kazakhstan's past history clearly illustrates the effect of a shortsighted approach, greed and ignoring ecological values combined with bad management in relation to cities and nature. Examples are abandoned ghost cities in central Kazakhstan, the Semipalatinsk nuclear test site, the desert where the Aral Sea used to splash, and many other places afflicted by polluted air amid a post-apocalyptic landscape. We would like Almaty to move towards Scandinavian environmental values, towards Brooklyn's gentrification, towards Istanbul's public transport system and Amsterdam's bicycle use (Ramazanova, 2017).

Thus, for Almaty's further sustainable development, it is necessary to continue pressing for environmental reforms, stimulating production using green technologies, encouraging foreign and domestic eco-innovations, as well as connecting non-governmental organizations and members of the public to support environmental reforms. Only consistent and systemic steps will help Kazakhstan's megalopolis move towards sustainable development.

References

Beatley, T., & Newman, P. (2013). Biophilic cities are sustainable, resilient cities. *Sustainability*, *5*, 3328–3345. DOI:10.3390/su5083328.

Beddoe, R. et al. (2009). Overcoming systemic roadblocks to sustainability: The evolutionary redesign of worldviews, institutions, and technologies. *PNAS*, *106*(8), 2483–2489.

Biermann, F., Kanie, N., & Kim, R.E. (2017). Global governance by goal-setting: The novel approach of the UN Sustainable Development Goals. *Current Opinion in Environmental Sustainability*, 26–27.

Cioca, L.-I. et al. (2015). Sustainable development and technological impact on CO_2 reducing conditions in Romania. *Sustainability*, *7*, 1637–1650. DOI:10.3390/su7021637.

City Population (2015). *Taiwan: Counties and Cities*. www.citypopulation.de/Taiwan-Cities.html/2015

David, L., & Fistung, F.D. (2015). Scientific research support in developing sustainable transport in Romania. *World Review of Science, Technology and Sustainable Development*, *12*(1), 49–66.

Express-K (2016). The population of Almaty exceeded 1.7 million people. https://express-k.kz/news/lenta_novostey/chislennost_naseleniya_almaty_prevysila_1_7_mln_chelovek-66193/2016/02/03

Hemani, S., & Punekar, R.M. (2015). Design education for sustainability: A case study for an inclusive approach to design in India. *World Review of Science, Technology and Sustainable Development*, *12*(1), 29–48.

Höjer, M., & Wangel, J. (2014). Smart sustainable cities: Definition and challenges. *ICT Innovations for Sustainability: Advances in Intelligent Systems and Computing 310* (pp. 1–15). Springer International Publishing.

Imai, K. (2015). International market competitiveness of Japanese green innovation technologies: An analysis using patent data. *World Review of Science, Technology and Sustainable Development*, *12*(1), 77–94.

Kazhydromet (2017). Information Bulletin on the State of the Environment of the Republic of Kazakhstan for 2017, Ministry of Energy of the Republic of Kazakhstan, 'Kazhydromet' RSE, Department of Environmental Monitoring. https://kazhydromet.kz/upload/pdf/en_1516788286.pdf.

Kazinform (2012). The atmospheric pollution index in Almaty decreased by 2.*8 units*. www.inform.kz/ru/indeks-zagryazneniya-atmosfery-v-almaty-snizilsya-na-2-8-edinicy_a2501100/2012/10/09

Kondybayeva, S., & Ishuova, Z. (2013). The effect of monetary policy on real house price growth in the Republic of Kazakhstan: A vector autoregression analysis. *World Applied Sciences Journal*, *22*(10), 1384–1394.

Kozybayev, M. (1983). *Alma-Ata. Encyclopedia*. Alma-Ata.

Kumar, A. et al. (2017). A review of multi criteria decision making (MCDM) towards sustainable renewable energy development. *Renewable and Sustainable Energy Reviews*, *69*, 596–609.

Lorek, S., & Fuchs, D. (2011). Strong sustainable consumption governance: Precondition for a degrowth path? *Journal of Cleaner Production*, *30*, 1–8. DOI:10.1016/j.jclepro.2011.08.008.

Mayer, H., & Knox, Paul L. (2006). Slow cities: Sustainable places in a fast world. *Journal of Urban Affairs*, *28*(4), 321–334.

MCC (2017). Solution-oriented global environmental assessments: Opportunities and challenges. www.mcc-berlin.net/en/research/cooperation/unep.html. The MCC is an independent social-science research institute in the field of sustainability located in Berlin.

MKKZ (2017). Kazakhstan exports more than 800 kinds of goods to 113 countries. http://mk-kz.kz/articles/2017/11/21/kazakhstan-eksportiruet-bolee-vidov-800-tovarov-v-113-stran-mira.html.

Mukhtarova, K.S., & Zhidebekkyzy, A. (2015). Development of green economy via commercialization of green technologies: Experience of Kazakhstan. *Journal of Asian Finance, Economics and Business*, *2*(4), 21–29. DOI:10.13106/jafeb.2015.vol2.no4.21.

Mukhtarova, K.S., Trifilova, A.A., & Zhidebekkyzy, A. (2016). Commercialization of green technologies: An exploratory literature review. *Journal of International Studies*, *9*(3), 75–87. DOI:10.14254/2071-8330.2016/9-3/6

Paley, T.L. (2003). Environmental security in Taiwan: Problems and solutions. *Experience and traditions of ethnic wildlife management, Ulan-Ude*. http://world.lib.ru/p/palej_t_l/ecolog.shtml

Pchelyanskaya, I. (2017). What is wrong with the air of Almaty? https://krisha.kz/content/articles/2017/chto-ne-tak-s-vozduhom-almaty

Radio Taiwan International (2011). News. Media Review. Taiwan. *Ecology*. http://english.rti.org.tw/

Ramazanova, U. (2017). Attention, disaster: What happens to the air of Almaty? www.the-village.kz/village/city/situation/827-zagryaznenie-vozduha/2017/12/12

Shen, L., Nieuwlaar, E., Worrell, E., & Patel, M.K. (2011). Life cycle energy and GHG emissions of PET recycling: Change-oriented effects. *International Journal of Life Cycle Assessment*, *16*, 522–536. DOI:10.1007/s11367-011-0296-4.

Shmelev, S.E. (2011). Dynamic sustainability assessment: The case of Russia in the period of transition (1985–2007). *Ecological Economics*, *70*(11), 2039–2049.

Shmelev, S.E. (2012). *Ecological Economics: Sustainability in Practice*. Springer.

Shmelev, S.E., (2017). Multidimensional sustainability assessment for megacities. In: Shmelev, S.E. (ed.) *Green Economy Reader: Lectures in Ecological Economics and Sustainability*, Springer, pp. 205–236.

Shmelev, S.E, & Rodriguez-Labajos, B. (2009). Dynamic multidimensional assessment of sustainability at the macro level: The case of Austria. *Ecological Economics*, *68*(10), 2560–2573.

Shmelev, S.E., & Shmeleva, I.A. (2009). Sustainable cities: Problems of integrated interdisciplinary research, Interscience. *International Journal of Sustainable Development*, *12*(1), 4–23.

State Fund of Ecological Information (2015). National report on the state of the environment and the use of natural resources of the Republic of Kazakhstan for 2011–2014. Astana. http://new.ecogosfond.kz/wp-content/uploads/2016/08/NDSOS_2011-2014.pdf

Taipei (2013). Taipei City Government. *Administrative Districts*. https://english.gov.taipei/Content_List.aspx?n=02D12F5BE6C0FC93

UN (2007). *Indicators of Sustainable Development: Guidelines and Methodologies*, 3rd ed. New York, www.un.org/esa/sustdev/natlinfo/indicators/guidelines.pdf

UN (2015). *Transforming Our World*. https://sustainabledevelopment.un.org/index.php?page=view&type=400&nr=2125&menu=1515

UN (2017). 'Progress towards the Sustainable Development Goals', Report of the Secretary-General, Economic and Social Council, E/2017/66. https://unstats.un.org/sdgs/files/report/2017/secretary-general-sdg-report-2017-EN.pdf

UNEP (2011). *Towards Green Economy: Pathways to Sustainable Development and Poverty Eradication*. http://drustage.unep.org/greeneconomy/sites/unep.org.greeneconomy/files/field/image/green_economyreport_final_dec2011.pdf

Ust-Kamennogorsk (2016). The development program of the territory of Ust-Kamenogorsk city for 2016–2020. N8/3-VI. http://oskemen.vko.gov.kz./ru/programm1.htm

WHO (2010). Health topics. Environment and health. Air quality. *Data and statistics*. www.euro.who.int/en/health-topics/environment-and-health/air-quality/data-and-statistics/2010.

Yu, W., Pagani, R., & Huang, L. (2012). CO_2 emission inventories for Chinese cities in highly urbanized areas compared with European cities. *Energy Policy*, *47*, 298–308.

Zhusupova, A. (2015). Dynamics of social inequality in Kazakhstan. Institute of World Economy and Politics (IWEP) under the Foundation of the First President of the Republic of Kazakhstan – the Leader of the Nation. Astana-Almaty. http://iwep.kz/files/attachments/article/2016-0921/dinamika_socialnogo_neravenstva_v_kazahstane.pdf

8 Multidimensional sustainability benchmarking of the cities of the Middle East and North Africa

Tobias Schnitzler and Stanislav E. Shmelev

1. Introduction

1.1 Background

In recent decades, the world has experienced *unprecedented*[1] urban growth. In 2015, close to 4 billion people – 54% of the world's population lived in cities – and that number is projected to reach 66% by 2050. Cities have become the primary human living space (ECOSOC, 2017; UNDESA, 2013; UNFPA, 2007; WGBU, 2016).

Most inhabitants of Middle Eastern and North African cities have an *ecological footprint*[2] many times higher than the earth can sustain, an issue becoming more critical as the world is facing challenges in providing people with resources (Global Footprint Network, 2018). Cities in these countries increasingly face numerous sustainable development challenges including shortages of water, housing, infrastructure, energy, health care and education, as well as transportation (Abubakar & Aina, 2016; Shmelev & Shmeleva, 2018). In order to stimulate a global transition to sustainability, it makes perfect sense to start in cities. Cities are responsible for the majority of our greenhouse gas (GHG) emissions and waste output, which are believed to be primary causes behind rapid climate change (ECOSOC, 2017; UN Habitat, 2011; WGBU, 2016). The current trends we observe in cities bring with them economic, social and environmental challenges (see Section 3.1) and this is the reason why cities are the focal point of experts' attention today (von Weizsäcker & Wijkman, 2018, pp. 30–32; WGBU, 2016).

When examining the global economy from a sustainability perspective, it is evident that society today is behaving in ways that are both socially and ecologically *unsustainable*. Our development since the Industrial Revolution has had significant impacts on the environment, and we are now well within the era of the Anthropocene, where the changes on earth can be largely attributed to destructive and widespread human impact (Steffen et al., 2011). The earth itself is a closed system to matter, but is open to energy, primarily in the form of solar energy (Victor, 1991; Common & Stagl, 2005).

Life on earth inhabits the biosphere, wherein living organisms exchange matter and energy with their ecosystems through natural cycles (Common & Stagl, 2005). Without the interference of human activity, these cycles oscillate through natural rhythms. However, today the growth and fatal actions of human society have resulted in negative impacts on these subsystems, and we are therefore facing a systematic sustainability challenge (Ny et al., 2006). Examples of such disruptions can be seen through increasing levels of pollutants and manmade chemicals in the natural world (Law & Stohl, 2007), increasing levels of GHG due to the burning of fossil fuels (Kennish, 2002).

Steffen et al. (2015) outline nine specific planetary boundaries: climate change, novel entities, stratospheric ozone depletion, atmospheric aerosol loading, ocean acidification, biogeochemical flows, freshwater use, land-system change and *biosphere integrity*[3] (Steffen et al., 2015). According to this view, humanity is approaching the planetary boundaries for land use change, global freshwater use, ocean acidification and the global phosphorus cycle. The analysis proposes that the boundaries for climate change, rate of biodiversity, loss and interference with the nitrogen cycle have already been overstepped (Steffen et al., 2015; Weizsäcker & Wijkman, 2018, p. 14).

Moreover, the structure of society functions within a system that no longer allows all individuals to meet their *basic human needs.*[4] This can be observed through social problems such as inequality and an erosion of trust within our social structure (Gustavsson & Jordahl, 2008). If we continue to behave in this way, the earth will lose its ability to provide us with the necessary resources and conditions to meet our human needs.

1.2 Problem statement and purpose

Cities are developing into centers of economic growth, and it is projected that by 2025, six hundred of the world's largest cities will produce more than 60% of the global GDP (UN Habitat, 2016). Cities produce 80% of global greenhouse gas emissions, making a significant contribution to anthropogenic climate change (ECOSOC, 2017). However, most Middle Eastern and North African cities do not currently have initiatives and policies in place that are sufficiently progressive to adapt to these challenges. Murray, Minevich and Abdoullaev (2011) identify current waves of human crisis experienced throughout the world as a clear indication that our old institutions are inconsistent with a complex and fast-changing world. Borja (2007) points out expensive effects, such as difficulties in waste and resource management, increased air pollution and other concerns such as traffic concentration, and that resource systems have been developed in isolation from each other (Shmelev & Shmeleva, 2018, 2019). Moreover, physical and technical challenges such as deteriorating and outdated infrastructures within cities have been identified by Borja (2007) and Grimm et al. (2008).

However, the characteristics of cities make them an excellent platform to experiment and test future sustainability policies. Cities hold the potential to be sustainable because they are self-organizing learning systems, which allow people to be creative and work with each other (Innes & Booher, 2000). Evident issues such as high living density and a dependence on shared resources place Middle Eastern and North African cities in the position as being platforms for sustainable development, since they have characteristics under which sustainability can be modeled (World Bank, 2014).

The different *stakeholders*[5] have goals, interests and needs within planetary boundaries which make it almost impossible to find solutions that are sustainable. This chapter will thus show a case analysis of Dubai, Doha, Abu Dhabi, Rabat, Tunis, Jeddah, Algiers, Casablanca, Alexandria, Riyadh and Cairo. The use of outranking MCDA methods will help planners and urban designers to understand the many aspects of sustainability, their interactions and how planning processes and policies could support Middle Eastern and North African cities achieving sustainable development. So far there has been limited research on the extent to which various policies could be effective in promoting sustainable urbanization in the Middle East.

The analysis of the different indicator-based international assessment frameworks shows a difference in focus with regard to economic, environmental and social issues (Shmelev & Shmeleva, 2018, p. 2). The sustainable development goals (SDGs) indicators are more focused on the problems of developing countries (Shmelev & Shmeleva, 2018, p. 2; UN, 2016). The United for Smart Sustainable Cities program establishes the criteria to evaluate ICT's contributions in making cities smarter and more sustainable, and to provide cities with the means for self-assessments in order to achieve the sustainable development goals. Moreover, the ISO 37120 'Sustainable Development of Communities' standard offers more clarity in defining specific indicators. The European City Statistics project has a great range of classifications and typologies.

We will use the methodological framework presented in Figure 1.1. in Chapter 1 of this book as a theoretical foundation for further analysis in the present chapter.

Building sustainable cities entails integration and coordination among sectors. For instance, a spatial plan would need to accommodate industry, residential housing and green areas, to be integrated with adequate space for access to public transportation. Some win-win solutions could emerge since investment in green infrastructure can imply a reduction of CO_2 emissions, while protection of green areas can improve management of groundwater sources (UNDESA, 2013, p. 73).

We would like to define the indicators which will be used in our chapter as a product of detailed comparison of the indicator sets presented earlier (Shen et al., 2011; Shmelev, 2017). We chose the 16-indicator set because of the simplicity of matrix: each group or pillar of sustainable development is represented in this set. Generally speaking, a large set of criteria was assessed, including economic indicators (gross regional product at Purchasing Power Parity (PPP), the number of Fortune 500 companies headquartered in the cities, the disposable income per head at PPP, consumer price inflation rate and unemployment rate), smart indicators (number of patents per 1,000 inhabitants, average broadband Internet speed, number of underground stations per 1,000,000 inhabitants and creative industries employment), socio-cultural indicators (life expectancy at birth, share of population aged 24–65 with a higher education and the Gini index of income inequality), and environmental indicators (CO_2 emissions per person, NO_2 emissions, SO_2

emissions, PM_{10} concentration, domestic water consumption, municipal solid waste, recycling rate, green space and renewable energy). In sum, the 16 indicators for cities chosen are based on a *sustainable cities framework* (Shmelev & Shmeleva, 2009; Shmelev, 2017).

2. Case description and data

2.1 Background information on the cities in our study

Table 8.1 presents the data for the above-mentioned 16 indicators. Economically, Doha, Dubai and Abu Dhabi have a high per capita income (at PPP in 2010 prices). Inflation is low at 0.4% in Rabat and Casablanca. Furthermore, unemployment has been highest in Tunis (15%), followed by Alexandria (13.3%) and Cairo (13.3%). From an innovation point of view, Dubai, Doha and Abu Dhabi have exhibited significantly higher rates of innovation with 0.06 patents registered per 1,000 inhabitants, whereas all other cities were characterized by a figure of 0.02 or below. The average broadband Internet speed is highest (22 Mb/c) in Dubai. The number of underground stations per 1,000,000 inhabitants is highest (38.7) in Doha. From a social point of view, life expectancy at birth is highest in Doha with 77.88 years, followed by Dubai (76.95) and Abu Dhabi (76.67); and the lowest in Alexandria with (71.12). The level of higher education attained in Doha is 27.10 % for all of those residents aged 24–65 years. This is higher than in Jeddah or in Riyadh. Algiers and Cairo exhibit low levels of education, with only 8% of residents achieving tertiary education. Income differentiation in Rabat and Casablanca are high, illustrated by a Gini index of income inequality of 39.50%. In the environmental dimension, Doha is characterized by 45.47 tonnes of CO_2 emissions per person per year. The PM_{10} average annual concentrations are highest in Riyadh with 368 in mg/m^3. Jeddah (959.1 m^3) and Cairo (937 m^3) have the highest domestic water consumption per person per year. For comparison, Algiers (196 m^3) or Tunis (296.2 m^3) perform better. In the field of circular economy, Alexandria and Cairo generate significant amount of municipal solid waste of 438 and 456.9 kg/person per year, 12.5% of which is recycled. In comparison, Tunis generates less (295.65 kg). It is important to mention that Abu Dhabi has the highest recycling rate with 17% and Casablanca the lowest with 4%. The percentage of renewable sources in the energy mix has been highest in Rabat and Casablanca (33.63%) and lowest in Riyadh (0.05%).

In the following sections, the history and background of eleven North African and Middle Eastern cities will be described. A special focus will be placed on policies and action plans.

2.1.1 Abu Dhabi

Abu Dhabi is the largest emirate in the United Arab Emirates, occupying 84% of the national landmass territory. Abu Dhabi city in the emirate is the federal capital of the UAE. The estimated population of the emirate around mid-2014

Table 8.1 Middle Eastern and North African cities[6, 7]

City	*Gross Regional Product, PPP, $ billion*	*Disp. income per head, PPP, 2010 USD*	*Cons. price inflation rate, %*	*Unem., %*	*No. of patents per 1,000 inhabi-tants*	*Average broadband Internet speed, Mb/c*	*Number of undergr. stations per 1,000,000 inhabitants*	*Life expect. at birth, years*	*Share of population aged 24–65 with a higher education*	*Gini Index of income inequality, %*	*CO_2 emissions per person per year, t*	*PM_{10} average annual concentrations, mg /m^3*	*Domestic water consumption, m^3 per person per year*	*Municipal solid waste kg per person per year*	*Recycling rate, %*	*% renew. sources in the energy mix*
Dubai	82.87	34.14	2.30	0.30	0.06	22.0	15.62	76.95	17.95	31.00	23.30	91.00	916.10	605.90	11.50	0.46
Doha	102.20	66.89	3.40	0.20	0.06	5.30	38.70	77.88	27.10	39.00	45.42	168.00	390.20	485.45	12.50	0.40
Abu Ahabi	360.11	34.14	2.50	2.28	0.06	7.10	0.00	76.67	17.95	31.00	37.44	132.00	889.20	605.90	17.00	7.00
Rabat	41.18	2.91	0.40	9.90	0.01	9.30	0.00	75.31	11.00	39.50	1.74	31.00	411.00	532.90	4.00	33.63
Tunis	20.50	4.15	4.90	15.00	0.02	3.10	0.00	75.34	16.32	35.80	2.58	90.00	296.20	295.65	5.00	7.37
Jeddah	104.65	18.47	2.20	5.70	0.02	3.16	0.00	74.23	22.30	32.00	19.53	161.00	959.10	602.25	11.00	0.05
Algiers	45.00	4.47	2.90	10.60	0.00	3.70	5.59	75.64	8.03	27.60	3.72	65.00	196.00	441.65	8.00	4.10
Casablanca	37.90	2.91	0.40	9.90	0.01	8.00	0.00	73.61	11.00	39.50	1.74	61.00	427.20	474.40	4.00	33.63
Alexandria	32.40	2.39	10.10	13.40	0.01	6.00	0.00	71.12	8.70	31.80	2.20	120.00	937.00	438.00	12.50	11.48
Riyadh	163.48	18.47	2.20	5.70	0.02	14.70	0.00	74.08	22.00	32.00	19.53	368.00	959.10	602.25	11.90	0.05
Cairo	102.17	2.39	10.40	13.40	0.01	10.10	6.57	71.53	8.00	31.80	2.19	179.00	937.00	456.90	12.50	11.48

was 2.65 million (comprising 1,766,140 males and 890,308 females). This makes Abu Dhabi a medium-sized urban center. There were 507,479 UAE nationals and 2,148,969 expatriates. The average annual population growth rate for the years from 2005 to 2014 was 7.6 per cent. The Emirate of Abu Dhabi lies on the coast of the Arabian Gulf and is bordered by Sultanate of Oman to the east, the Kingdom of Saudi Arabia to the south and the west and the Emirate of Dubai to the north-east. Over the recent decades, the Emirate of Abu Dhabi has undergone a significant transformation in economic growth. Due to the vision of Abu Dhabi's leaders and the abundance of oil and natural gas reserves, the emirate has grown to become a dynamic hub and major competitor on the global stage. While recognizing the success of the oil sector, Abu Dhabi is working hard to reduce its reliance on hydrocarbons and broaden the emirate's economy. Investment in infrastructure, tourism, transport, health and education is continuing, in line with the government's 2030 economic plan (Official Portal of the UAE Government, 2018). The expectation was the creation of a long-term road map for economic progress through the establishment of a common framework aligning all plans and policies. Abu Dhabi's Policy Agenda is built around nine pillars that are meant to shape the emirate's future: a large empowered private sector; a sustainable knowledge-based economy; an optimal, transparent regulatory environment; a continuation of strong and diverse international relationships; the optimisation of the Emirate's resources; premium education, healthcare and infrastructure assets; complete international and domestic security; maintaining Abu Dhabi's values, culture and heritage, and a significant and ongoing contribution to the federation of the UAE. The Government of the UAE has committed itself to direct public policy to strengthen them, i.e. infrastructure development and environmental sustainability as key priorities. That means creating a professionally designed and well-managed urban environment in the cities, complete with world-class traffic and transport systems. Therefore, a comprehensive 2030 urban structure framework plan for the capital was developed and published (Abu Dhabi Council for Economic Development, 2008, pp. 5–7).

2.1.2 Alexandria

Alexandria is the second largest city in Egypt and a major economic center, extending about 32 km (20 mi) along the coast of the Mediterranean Sea in the north central part of the country. The estimated population of Alexandria is 5.172 million people and it is a large urban center. The city's low elevation on the Nile Delta makes it highly vulnerable to rising sea levels. Alexandria was originally a small, ancient Egyptian town founded c. 332 BC by Alexander the Great (Government of Alexandria, 2018). Today, Alexandria is both an important industrial center, because of its natural gas and oil pipelines from Suez, and a popular tourist destination. However, the city is suffering from economic and social problems: a high unemployment rate, inadequate housing conditions and extensive pollution. The governorate of Alexandria tried to address these problems in a City Development Strategy (CDS). However, the goals in the CDS's first phase were inadequate. The new Alexandria CDS is based on three stakeholder pillars: (1) local economic development, (2) participatory urban upgrading, and (3) the environmental rehabilitation of Lake Marriout. The CDS also addresses key challenges such as the development of illegal settlements, lack of employment opportunities, and environmental degradation. The project includes the establishment of a CDS department for implementation and long-term development sustainability, and priority development projects and potential sources of project funding. CDS's vision and action plan forms an overarching umbrella to organize and leverage donor support using broad-based stakeholder participation. Alexandria represents an ideal location for a CDS project, because planning and working together to improve the city has relevance for many groups. The notion of 'resurrecting' Alexandria received enormous support from decisions to build the Bibliotheca Alexandrina and pilot CDS processes. The business community, in conjunction with the Alexandria governor's extraordinary efforts and support, has created synergy and opportunity for investment in public infrastructure and in developing a vision for Alexandria (AFD, 2018; Metropolis, 2018).

2.1.3 Algiers

Algiers, with a total population of 2.9 million inhabitants, is the largest and most important city in Algeria. It is the country's capital city and a medium-sized urban center. Nicknamed '*Alger La Blanche*' or 'Algiers the White' because of its glistening white buildings that can be seen sloping up from the sea, it is located on the Mediterranean Sea, in the country's central region. A number of surrounding clusters of islets have been turned into a part of the port. Algiers is Algeria's main administrative center and also the country's main harbor, serving as both a shipping center and a principal Mediterranean refueling station. Algiers' main industries are oil refining, metal works, food production and petrochemicals. The main export products are iron, grain, citrus fruit, wine, phosphates, oil and vegetables (Algeria, 2018). Moreover, a review of the Master Plan of Urban Planning (PDAU) was initiated in 2007, and Algiers has undertaken the implementation of its new Master Plan, which was carried out and completed in 2014 by the engineering office 'Park Expo' on behalf of *la Wilaya* of Algiers (province of Algiers). The new Master Plan offers a global and integrated vision. This view seems to focus on environmental aspects, with the proposal of several agri-parks on the city's periphery to de-densify and slow down urban sprawl on the fertile lands of Mitidja. Parallel to the Master Plan, a Strategic Plan was developed; a plan that proposes concrete projects in the city of Algiers, playing the role formerly assigned to the Land Use Plan (POS). The Strategic Plan proposes concrete projects. The contracting authority is the county, but the municipality implements the projects – its role is also not clarified in the approach. The Strategic Plan of Algiers was thought out for the long term – specifically, for 2030. It includes six thematic plans, which are in detail: (1) the white plan, which is aimed at restructuring the city, especially with the creation of new facilities and a broad housing program, (2) the green plan, to restore the ecological balance, (3) the blue plan, primarily concerned with rational use of drinking water and recycling waste water, rather than dumping it in the sea, (4) the mobility plan, which will regulate and promote public and collective transport, (5) the conservation plan, to energize job creation and encourage initiative, and (6) the social cohesion plan, based on a policy of proximity and equipment in existing neighborhoods (Chabou-Othmani, 2015). Its implementation was planned in four phases: Phase 1: 2012–2016: Embellishment of the city; Phase 2: 2017–2021: The landscape of the Bay of Algiers; Phase 3: 2022–2026: The requalification of the periphery, and Phase 4: 2027–2031: Consolidation of the city (Algeria, 2018; Chabou-Othmani, 2015).

2.1.4 Cairo

Cairo is the capital of Egypt. The city's metropolitan area is one of the largest in Africa, the largest in the Middle East and the Arab world, and the 15th largest in the world. The city is associated with ancient Egypt, as the famous Giza Pyramid complex and the ancient city of Memphis are located in its geographical area. Cairo has long been a centre of the region's political and cultural life, and is titled 'the city of a thousand minarets' for its preponderance of Islamic architecture. With a population of over 9 million spread over 3,085 square kilometers, Cairo is by far the largest city in Egypt, with a large urban center, and an additional 9.5 million inhabitants living in close proximity to the city. Cairo, like many other megacities, suffers from high levels of pollution and traffic. Cairo's metro, one of two in Africa (the other being in Algiers, Algeria), ranks among the 15 busiest in the world, with over 1 billion annual passenger rides. Cairo is an expanding city, which has led to many environmental challenges. The city's air pollution is a matter of serious concern, with Greater Cairo's volatile aromatic hydrocarbon levels higher than many other similar cities. Air-quality measurements in Cairo have also been recording dangerous levels of lead, carbon dioxide, sulphur dioxide, and suspended particulate matter concentrations due to decades of unregulated vehicle emissions, urban industrial operations, and chaff and trash burning. There are over 4,500,000 cars on the streets of Cairo, 60% of which are over ten years old, and therefore lacking modern emission-cutting features such as catalytic converters. Cairo has a very poor dispersion factor due to sparse rainfall and the city's layout of tall buildings and narrow streets, which create a bowl effect (Khoder, 2007). Furthermore, in the Greater Cairo (GC) Urban Development Strategy, the following evident aspects were found: (1) achieving good living conditions for all sections of residents, (2) achieving traffic fluidity and public transportation suitable for all income brackets, (3) achieving traffic fluidity and public transportation suitable for all income brackets, and providing services, (4) providing services, housing and job opportunities for all sections of residents, (5) providing safe, clean, unpolluted and uncrowded environment with green areas; and (6) achieving prosperity of cultural and tourist activities (Ministry of Housing, Utilities, and Urban Communities, 2012).

2.1.5 Casablanca

Casablanca is the biggest city in Morocco; located on the Atlantic Ocean, the city was modeled on the city of Marseilles by the French in 1920s. Casablanca has 3.36 million inhabitants and is a medium-sized urban center. It is the capital of the Greater Casablanca region and is the principal economic capital, and Morocco's primary naval base. It is, however, not the administrative or political capital of the country. Casablanca is situated in the country's most fertile zone, the heart of Morocco, as well as being a major source of mineral wealth. One of the busiest ports in the Maghreb, the city has seen rapid development from the time of the French protectorate; it has grown much bigger than Marseilles, and seems like a European city. The Berbers were the city's first settlers in the tenth century BC. It then became a major fort for the Phoenicians as well as the Romans. Casablanca's rulers were from the kingdom then known as Anfa, until the year 1068, when it was conquered by the Almoravids. Under the rule of the Merinids in the fourteenth century it became a major port and by the fifteenth century it had become an independent state again. It gained notoriety for being a safe harbor for pirates and privateers, which led to its destruction by the Portuguese in 1468, who then built a military fortress there in 1515 which came to be called 'Casabranca' – 'white house' in Portuguese. In 1580, it was taken over by Spain until it was later made a part of Portugal in 1640. The 1755 earthquake led them to abandon the city which had been nearly completely destroyed. Sultan Mohammed ben Abdallah, grandson of Moulay Ismail, reconstructed the city in 1756–1790. At that time it was called 'ad-Dar al Bayda' ('white house' in Arabic) and 'Casa Blanca' in Spanish. From here onwards the city expanded with the growth of the British textile industry, as Morocco became an importer of wool beginning in the nineteenth century. In 1956, Morocco regained its independence from France (Marokko Info, 2018). Nowadays, the authorities' renewed attention led to the September 2014 launch of the Greater Casablanca Development Plan (Plan de Développement du Grand Casablanca, PDGC) – a US$3.4 billion initiative over six years (2015–2021) aimed at increasing policy and investment coordination between central and local government in the Casablanca agglomeration. Moreover, the Programme de Développement des Infrastructures Routières (PDIR 2015–2020) was also launched. In detail, this integrated strategic vision is divided into four strategic poles namely: (1) The living conditions pole, aiming to reconcile the city with its inhabitants; (2) the platform pole, aiming at optimizing the mobility of inhabitants and visitors in terms of time, cost and quality; (3) the economic excellence pole, aimed at maintaining Casablanca's leadership as Morocco's economic powerhouse, and (4) the animation pole aimed at developing a differentiated entertainment platform (Hachimi Alaoui, 2017).

2.1.6 Doha

Doha is the capital and largest city of Qatar. Situated almost midway down the east coast of the Qatari Peninsula, Doha is the country's center of administration, finance, culture, transportation and social services. The modern city grew from the fishing and pearling port of al-Bida, which at the end of the nineteenth century had around 12,000 inhabitants. The town's economy depended to a large extent on pearling, and the busy port had some three hundred pearling ships in 1939, just before the industry collapsed. After oil revenues began enriching the emirate in the 1960s, the city grew rapidly. Its simple one- and two-story stone, mud, coral block and timber dwellings were replaced by high-rise apartments and offices, palatial villas, and tree-lined subdivisions supported by modern infrastructure. The city's waterfront is lined by a gracefully curving roadway and landscaped walkway, or corniche. Although the oil and gas industry dominates the local economy, fishing and trade also bring activity to the port town. According to the 2016 census, the city had 1.3 million inhabitants. Because most of the city's residents are non-Qataris, the character of the city resembles others in the Persian Gulf such as Abu Dhabi, Dubai and Dhahran, where there are large numbers of Iranians, Indians, Pakistanis, Filipinos and Bangladeshis who influence the types of restaurants and the items sold in the markets. Doha is a medium-sized urban center (*Encyclopedia of the Modern Middle East and North Africa*, 2014). The Qatar National Master Plan (QNMP) is the spatial representation of the Qatar National Vision 2030. It provides comprehensive spatial planning for the state of Qatar at the national, municipal and local levels. It ensures the full integration of all planning efforts along with their effective implementation. The vision is to 'create a role model for sustainable urban living and the most livable towns and cities in the 21st century.' The Qatar National Master Plan, which was announced on January 1, 2018, will include the development of three key central districts along Doha's coastline, with a City Centre precinct at Doha's heart, based on six guiding principles. These are: (1) quality of life for all, (2) identity, (3) environmental values, (4) economic growth and diversification, (5) connectivity of people, and (6) ownership of planning. The QNMP has 17 strategic planning objectives, 61 policies, and 200 policy actions ranging from the short to long term (Ministry of Municipality and Environment, 2018a).

2.1.7 Dubai

The Emirate of Dubai is the second largest of the seven United Arab Emirates but has the biggest population at over 2.1 million inhabitants. It is a medium-sized urban center. Size has been synonymous with Dubai as it continues to build the first, largest and biggest constructions in the world. Dubai's dynamics are always transient and ever-changing, with its constant urge to construct something better and bigger than the previous. If there was a Palm Island, Nakheel thought of the World Island. Burj Al Arab seemed too timid when Burj Khalifa cropped up, distancing itself to being a loner in the crowd. Dubai constantly dwells in a suppressive competition with itself trying to magnetize tourists to a dreamy world of attractions and unheard-of luxuries. The emirate's scoring points lie in its entrepreneurial abilities to create the inconceivable, found in its tourist attractions, landmarks, shopping centers, parks, nightlife and hotels. Although it strictly safeguards its traditional practices, it allows space for other religions to breathe, a rare quality amongst the conservative Arab world. Thankfully, it has been successful in shielding itself from extremism, much-needed for it to survive. Today, Dubai has emerged as a cosmopolitan metropolis that has grown steadily to become a global city and a business and cultural hub for the Middle East and the Persian Gulf region (Government of Dubai, 2018). Moreover, the Dubai 2020 Plan provides a flexible strategy to guide Dubai's urban development to 2020 and beyond. It is based on background analysis and synthesis relevant to prevailing environmental aspects and urbanization contexts including the historic thresholds of urban development, and also responds to the impacts of the global economic downturn since 2008. The strategic directions of the Dubai Urban Masterplan 2020 concerned with various aspects including: (1) protect and facilitate economic opportunities, (2) provide flexibility for growth, (3) consolidate development projects, (4) conserve natural systems, (5) facilitate social needs, (6) optimize infrastructure and services, (7) broaden connectivity, accessibility and movement, and (8) promote a sustainable open space system (Dubai Municipality, 2018). Furthermore, it is also important to mention that the World Expo in Dubai in 2020 will be the first to be held in the Middle East and North Africa (NEMA) and South Asia region (Government of Dubai, 2018).

2.1.8 Jeddah

Jeddah is a city located in Saudi Arabia and the Red Sea coast is an important urban center in western Saudi Arabia. It is the largest city in Makkah Province, the largest seaport on the Red Sea, an important commercial hub and second largest city in Saudi Arabia after the capital, Riyadh. The city population currently stands at over 4.08 million inhabitants and it is rapidly increasing, and the more it continues to increase the more waste will be generated. It is a medium-sized urban center (Abubakar & Aina, 2016). Jaddah is the Arabic word for 'grandmother.' According to eastern folk belief, the tomb of Eve, considered the grandmother of humanity, is located in Jeddah and was mentioned by Sir Richard Burton in his translation of the *Book of a Thousand Nights and a Night*. In 1975, the tomb was sealed with concrete by the religious authorities to prevent Muslims from praying there. Jeddah is the principal gateway to Mecca, Islam's holy city, which Muslims are required to visit at least once in their lifetime. Jeddah is the most cosmopolitan and tolerant of all Saudi Arabia's cities, hosting expatriates from all over the world. It's also the second largest commercial center in the Middle East, after Dubai (Saudi Network, 2018). However, the existing land-use pattern and transport planning practices cannot keep pace with the city's rapid urban growth and the resulting sustainability challenges (Abubakar & Aina, 2016). With the aim of revitalizing the city, the Jeddah Plans was developed, a framework for sustainable and balanced future growth. It will redefine the city's future, creating a series of well-connected centers and a network of quality open spaces, with improved access to amenities and public transportation. Operating on a huge scale, the project involves a record number of stakeholders to consult, critique and reach consensus on a new path forward for Jeddah. The Jeddah Plans invest in ensuring the prime location of metro stations, encouraging residents to decrease their dependence on cars and giving them improved access to amenities and jobs. The creation of a people-friendly public realm is also critical to the plans' success. It includes the following aspects: (1) strategic planning, (2) governance, (3) spatial and urban planning services, (4) landscape and environmental sustainability, and (5) infrastructure services (Samer Sani, 2012).

2.1.9 Rabat

Around a million people reside in the capital city of Morocco, Rabat, which is one of the imperial cities of Morocco and boasts many enticing historical monuments and extraordinary places of interest. Rabat is between a small and medium-sized urban center. The history of Rabat begins in the third century BC with a settlement called Chellah on the banks of the river Bou Regreg. The Romans conquered Chellah in 40 AD and transformed it into their colony of Sala Colonia. Rome retained the settlement until 250 AD until it was taken over by the Berber monarchs. In 1170, because of its military significance, Rabat obtained the name 'Ribatu l-Fath,' which means 'stronghold of triumph.' Yaqub al-Mansur, an Almohad Caliph, moved the capital of his kingdom to Rabat. In fact, he constructed the city walls of Rabat and the Kasbah of the Udayas (known as the world's biggest mosque). But Yaqub died before the completion of the Kasbah and the remains of the incomplete mosque still stand at present. In 1912, the French occupied Morocco and established the territory (Marokko Info, 2018). Nowadays, the city's recent urban regeneration projects, focused on construction, tourism, industry and technology, are a clear sign that something is changing in Rabat. The new master plan aims to reorganize the seven strategic sectors of the city, granting them greater independence through the creation of new infrastructures. The sectors are in detail: (1) agriculture and fisheries, (2) industry, (3) energy, (4) mining, (5) building and public works, (6) merchant services, as well as (7) non-merchant services.

2.1.10 Riyadh

Riyadh has a large urban center and a population of 6.5 million, 64% of whom are Saudi citizens. Riyadh has the highest urban growth average, as the population has been expanding at a rate of 8.1% per annum (Abubakar & Aina, 2016). Furthermore, the greatest number of expatriates living and working in the city are from Bangladesh, India, Pakistan, Indonesia and the Philippines. Riyadh has become one of the most cosmopolitan cities in the Arab world. It is very much a working capital, in which the business of government for a country of 29 million people is interlaced with the activities of commerce, industry, banking and finance, diplomacy and academia. Unfortunately, Riyadh is not immune from acute and growing traffic congestion throughout much of the day. Indeed, there are times, particularly during rush hour, when traffic becomes gridlocked. Drivers stuck in their vehicles may not always appreciate it, but their misfortune is a testament to the rapid success of one of the world's fastest-growing cities. In Riyadh, for example, roads take up almost 40% of the city's construction costs (Abubakar & Aina, 2016). A new public transport system, consisting of a metro of six interconnecting lines, integrated with a bus rapid transit network, which will use dedicated highway lanes, will revolutionize the way in which the people of Riyadh and its hundreds of thousands of visitors who come each year, will travel around the capital. By 2020, when the public transport system will have been fully bedded in and citizens become used to its immense advantages, the capital's roads are going to be markedly less congested. The minority, who still feel the need to travel by car, will be experiencing faster journeys, generally free of the frustrations of traffic jams. Moreover, 40% of Riyadh's water requirements are met from local artesian wells, while the remainder is pumped to the capital from desalination plants on the Gulf coast at Jubail. Despite the extra expense, water remains relatively cheap, though charges are now being raised incrementally, to encourage better conservation of this precious asset. Water is piped to 96% of the capital's population and the system continues to expand, as new outlying areas of the city are developed, for both domestic and business use (Arriyadh City Website, 2016). Moreover, the most important project undertaken is the MEDSTAR project (Metropolitan Development Strategy for Arriyadh). MEDSTAR proposes a gradual spatial change of the city's organization into a multi-center system which locates new concentration in the new development areas. In detail: (1) five new metropolitan sub-centers with higher-order services for people in the outward growing residential developments, located 15–20 km away from the city centers each in different directions, (2) activity spines to link the metropolitan sub-centers with the historical city, (3) the historical city as a center of such unique services as the central mosque, the seat of the governor, the metropolitan Municipality, the High Court and headquarters of banks and international companies, (4) the central area, defined by the historic city, the area of the Royal Court and Diplomatic Quarter and the future central park at the old airport (these nodes are to provide the base for an imposing image of the capital), (5) new suburban cities each will cater for 1 million inhabitants, and (6) an effective public transportation system as the backbone of the structure plan with major transport corridors defined and relevant technologies proposed. Moreover, the MEDSTAR project specifies the distribution of main land uses, activities, business centers, transport system, public facility network, environmental requirements and open areas (Arriyadh Development Authority, 2013).

2.1.11 Tunis

Tunis, capital of Tunisia, is situated at the southern end of the Gulf of Tunis, which opens out to the eastern part of the Mediterranean Sea; the initial site was established by a narrow isthmus between the *sebkha* Essijoumi (old saltworks) in the south-west and El-bouheïra (lagoon) in the east. Tunis's climate is soft and enchanting, the average temperature being 11°C in winter and 26°C in summer. The city of Tunis has a medium-sized urban center, with a population of 900,000, and a tourist visitor rate of a million annually. An always warm reception and a serene peace of mind are strengthened by a confirmed tolerance over the centuries. With Islam as the main religion, the city embraces the Jemâa (mosque) Ezzitouna, the Cathedral of St Vincent de Paul and the Great Synagogue of Tunis, the three main places of worship for the Muslim, Christian and Jewish communities (Commune Tunis, 2018). In late 2017, the Tunisian government began the development of the 'National Transport Master Plan for 2040.' This is likely to prioritize necessary expansion of transport networks for the future, which may make it easier to mobilize private and domestic investments for the sector. The plan ensures the integration and overall effectiveness of the strategy with other sectors such as urban planning, energy and environment. In terms of sustainable development, some issues have already been identified, such as the reduction of road accidents and energy consumption (Oxford Business Group, 2018).

2.2 Linear aggregation

We carry out a linear aggregation with different weights representing various policy priorities, e.g. economic, smart, social and environmental (Shmelev & Shmeleva, 2018, p. 9). In our opinion, such a linear aggregation could be easier to interpret than complex outranking MCDA results for a decision maker. The procedure could be outlined as follows. Adding up to 1, each group of indicators receives a weight of 0.25 under the equal priorities and 0.5 when a policy preference is placed on this group. The remaining groups receive a weight of 0.16 respectively.

Since the number of indicators in various indicator groups differ, the weights of individual indicators are calculated accordingly.

The results for Dubai, Doha, Abu Dhabi, Rabat, Tunis, Jeddah, Algiers, Casablanca, Alexandria, Riyadh and Cairo are presented in Table 8.2.

In a linear aggregation with preference towards environmental issues, Doha has the best score (0.61) and Alexandria the worst (0.299). The research shows that, along with serious concern about the environment, Doha exhibits extremely strong economic performance (0.69), followed by Abu Dhabi, Dubai and Riyadh. Alexandria shows the weakest economic performance (0.16). From a social point of view, Doha has still the best performance (0.67), and Cairo (0.27) as well as Alexandria have the worst (0.22). Doha leads the ranking in this region under smart policy priorities as well (0.69). For comparison, in the equal weights scenario, the outcomes are the same. Doha has the best score (0.66), and Alexandria the worst (0.22).

Figure 8.1 shows the linear aggregation and the preferred groups. It is important to point out that Doha, Abu Dhabi and Dubai perform best and Casablanca, Cairo and Alexandria worst.

2.3 MCDA tools

Recent years have seen rapid development in research activities in the field of multi-criteria decision aids (Roy, 1996; Cinelli et al., 2014). They have resulted in various streams of thought and methodological formulations for the assessment of complex decision challenges (Meyer and Bigaret, 2012, p. 1). MCDA

Table 8.2 Ranking of cities

City	*Equal*	*City*	*Environmental*	*City*	*Economic*	*City*	*Social*	*City*	*Smart*
Doha	0.66440548	Doha	0.60978435	Doha	0.6879849	Doha	0.66982774	Doha	0.69002492
Abu Ahabi	0.61295234	Abu Ahabi	0.5809586	Abu Ahabi	0.67044385	Abu Ahabi	0.63710534	Abu Ahabi	0.56330156
Dubai	0.56730376	Dubai	0.5389857	Dubai	0.58479747	Dubai	0.61127519	Dubai	0.53415667
Riyadh	0.46558782	Rabat	0.48545046	Riyadh	0.48694897	Riyadh	0.5105148	Riyadh	0.45874694
Jeddah	0.40369541	Algiers	0.46567632	Jeddah	0.43125276	Jeddah	0.47346387	Rabat	0.32669498
Algiers	0.38830301	Casablanca	0.45611123	Rabat	0.37186325	Algiers	0.44444753	Jeddah	0.31490755
Rabat	0.38109896	Tunis	0.44851097	Algiers	0.35484309	Tunis	0.38576935	Casablanca	0.29192519
Tunis	0.35018956	Riyadh	0.40614059	Casablanca	0.34784324	Rabat	0.34038714	Algiers	0.28824511
Casablanca	0.3462776	Jeddah	0.39515746	Tunis	0.28156694	Casablanca	0.28923076	Tunis	0.28491098
Cairo	0.26738301	Cairo	0.32123964	Cairo	0.20730367	Cairo	0.25688975	Cairo	0.28409896
Alexandria	0.2202254	Alexandria	0.29951614	Alexandria	0.16124648	Alexandria	0.22278449	Alexandria	0.19735449

Source: authors' calculation.

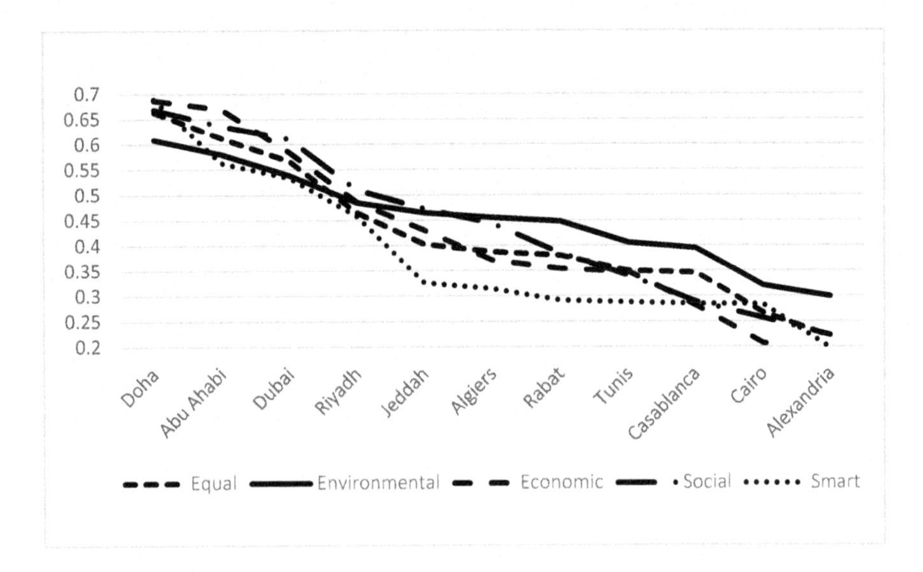

Figure 8.1 Linear aggregation and preferences

Source: Own composition

has been used for urban sustainability assessments and decision support due to the opportunities it offers for finding solutions between conflicting goals and priorities (Roy, 1996).

Taking the complexity of sustainability evaluation and the need to offer a path towards the achievement of a sustainable development future into consideration, decisions should be made in a reliable, structured, and transparent way (Cinelli et al., 2014, p. 140). MCDA could contribute to such a challenge and consists of a group of approaches which have the possibility to account for multiple criteria, in order to support individuals or groups to rank, select, or compare different policies or technologies (Cinelli et al., 2014, p. 140; Munda, 2008). Furthermore, MCDA gives the methodological and mathematical tools to operationalize *incommensurability*.[8] It is possible at both the macro and micro levels of analysis as well as in project evaluation (Martinez-Alier et al., 1998, p. 277).

Martinez-Alier at al. (1998, p. 278) mention that in their view ecological economics rests on a foundation of weak comparability of values. MCDA gives the opportunity to take into account economy–environment interactions, because it is multidimensional in nature and in this way weak or strong sustainability could be operationalized (Martinez-Alier et al., 1998, p. 281). Munda (1997) add that an evident consequence of *noncompensability*[9] is that it is possible to operationalize the idea of strong sustainability. It is based on the assumption that certain sorts of natural capital are defined as critical, and it is not substitutable by man-made capital (Bartelmus, 2013, p. 26; Ekins et al., 2003)

In conclusion, since MCDA techniques allow the researcher to take into account conflictual, multidimensional, incommensurable and uncertain effects of decisions, they build a promising assessment framework for sustainable cities (Martinez-Alier et al., 1998, p. 285; Roy, 1996; Shmelev & Shmeleva, 2018, p. 2). Hence, the area of MCDA is focused on providing solutions to decision issues where many criteria and preferences must be taken into account.

2.3.1 Outranking tools

We chose outranking MCDA tools because sustainability issues are more and more managed for multiple uses. Decision makers have not only economic challenges, but also those of social justice and non-market values of recreation and nature conservation, for example. Generally speaking, economic, ecological and socio-cultural sustainability are required. It is important to mention that multifunctionality calls for multi-objective sustainability management planning and decision support.

To assess sustainability of performance measured by an indicator, we would like to discuss the threshold values. These values should consider (1) the goal of applying the assessment tool set by the decision makers, (2) legal threshold values (e.g. air pollution threshold), or (3) other reference values (Wiek & Binder, 2005, p. 598).

We can effectively capture strong sustainability through technical parameters of MCDA tools. One of the most relevant groups of MCDA tools for sustainability assessment are the outranking approaches that could capture the essence of stronger sustainability. A robust example of this approach is the ELECTRE family of methods designed by Professor Bernard Roy.

2.3.2 ELECTRE III

ELECTRE are preference aggregation-based methods. They work with pair-wise comparisons of the alternatives (Cinelli et al., 2014, pp. 140–142). Moreover, the method can consider imprecise data or indifference and preference thresholds. The use of thresholds is an important and useful feature of the MCDA approach. Imprecise information could be used and the level of compensation among the criteria can be limited (Cinelli et al., 2014, p. 146).

Furthermore, Callon et al. (2001) recommend that any weight sustainable city indicator takes public participation in decision making into account. It is important to mention that the concept of threshold refers to a scientifically demonstrated reference value. It applies mainly to environmental challenges (e.g. air pollution). The critical value should stay in line with a recognized, generally arbitrary, reference value from standards (Callon et al., 2001).

According to the set of 16 criteria selected for the evaluation, ELECTRE III, under the assumption that a 5% difference in the value of each criterion is possible for preference or domination (q) and less than 5% presents an indifference (p) coupled with the assumption of equal weights, reveals the following results. Doha is the most sustainable city; followed by Abu Dhabi.

2.3.3 DIVIZ

The guidelines of DIVIZ give step-by-step instructions on how to (1) define alternatives, (2) define criteria and weights, (3) set up preference thresholds (a special attention to these should be paid), (4) generate a resulting outranking relation, and (5) visualize the result. DIVIZ enables the ability to combine programs implementing MCDA. DIVIZ is an easy software tool used for creating and using complex workflows of MCDA algorithms. Such workflows are often called 'methods,' for example, in the ELECTRE method by Roy (1996). Furthermore, the foundations for such multi-criteria sustainability assessment for cities was outlined in our previous work (Shmelev, 2017).

3. Case analysis

3.1 Worst practices

The results for Cairo show that the city's citizens drive private vehicles and might not immediately adopt public transportation that might help in reducing energy consumption and air pollution. In addition, Cairo does not have the adequate infrastructure that facilitates electric car use and emissions reduction. Unfortunately, renewables are hardly used in Cairo and Egypt in general, and thus urban performance on environmental indicators is relatively low in Cairo. Potential for improvement resides with the Solar Plan, which includes adding 3.5 GW (2.8 GW CSP and 700 MW PV) of solar energy by 2027. The Ministry of Electricity and Renewable Energy has signed seven Memoranda of Understanding worth US$500 million for solar and wind projects in Egypt. According to the Wind and Solar Atlas, opportunities exist for a feed-in tariff (FIT) and for wind and solar plants in the East and West Nile areas that would produce around 31,150 MW from wind and 52,300 MW from solar (Ministry of Housing, Utilities, and Urban Communities, 2012).

Tunis has a very low overall economic performance, and is likely to prioritize necessary expansion of the building and transport networks for the future, which may make it easier to mobilize private and domestic investments for the sector. The 'National Transport Master Plan for 2040' ensures the integration and overall effectiveness of the strategy with other sectors such as urban planning, energy and environment. In terms of sustainable development, some issues have already been identified, such as the reduction of road accidents and of energy consumptions (Oxford Business Group, 2018).

Alexandria has very poor performance on all indicators. Key challenges, such as the development of illegal settlements, lack of employment opportunities and environmental degradation, are not being tackled. In sum, Alexandria suffers from economic and social problems: a high unemployment rate, inadequate housing conditions and extensive pollution. Taking these into consideration, many questions emerge regarding the future challenges facing Alexandria to achieve sustainability, from providing basic services and upgrading and integrating settlements, to working to give the 'green' agenda the same priority as other pressing necessities. The research acknowledges that addressing the green agenda will be the crucial task in the years to come.

3.2 Best practices

The best performing cities in our database are Doha, Abu Dhabi and Dubai. Opportunities for sustainable development are rich in Doha. This Arab state is preparing for life after oil. Furthermore, Doha is performing stronger on profit factors than on social and environmental issues due to its strong economic development and energy efficiency. Cost and ease of doing business are other areas where Doha performs best; however, the city scores lower in terms of transport infrastructure. Taking this sector in Doha into consideration, the Minister of Energy and Industry said that Qatar Rail, one of the largest and most sustainable infrastructure projects, has set strict environmental requirements for the design, construction and operation of all stations. This would save at least 30% of carbon emissions compared to normal rates. The initiative aims at reducing about 4% of Qatar's carbon emissions within the overall goal of reducing 17% of total emissions by 2022. It also aims to increase the proportion of electric and hybrid vehicles in the country by 10% by 2030 and develop the infrastructure needed to support this trend (Ministry of Municipality and Environment, 2018b).

The government of Abu Dhabi has committed itself to direct public policy to strengthen infrastructure and have development and environmental sustainability as a key priority. That means a professionally designed and well-managed urban environment in the cities, complete with world-class traffic and transport systems. Therefore, a comprehensive 2030 urban structure framework plan for the capital was developed and published (Abu Dhabi Council for Economic Development, 2008, pp. 5–7).

Local businesses in Dubai are not just a simple service, for tourists and local communities, but also a significant element of local identity and a part of the cultural heritage of the place. Whereas pedestrianization has been widely adopted as a means to revitalization and as a key factor in promoting tourism, evidence has revealed that the buy-in of local businesses (as one of the main beneficiaries) is vital to ensure the success of pedestrianization schemes and to get support for future schemes. In addition, MBR, in the desert at Seih Al-Dahal nearly 30 miles south of Dubai, is now a massive array of solar panels that spreads for miles and continues to grow. It is the largest solar park in the world and in 2016 achieved the lowest prices for solar-powered electricity world-wide. As such, it seems a fitting facility for Dubai, the small emirate that thrives on notability and spectacle. The huge solar park, when complete, will provide a good foundation for Dubai to reach its renewable energy goals. The Dubai Integrated Energy Strategy expects solar to make up 25% of the emirate's energy mix by 2030, with the proportion of natural gas declining to 61% (the rest being nuclear and 'clean coal'). Looking further out, the Dubai Clean Energy Strategy 2050 intends that 75% of Dubai's total power output will come from clean energy by 2050. As the UAE has little in the way of wind resources (it's too far south for much wind), most of this clean energy will come from the sun (Government of Dubai, 2018).

4. Discussion and conclusion

4.1 Discussion

Here we will focus on policies and actions (i.e. electric cars and renewables). Whilst it is not really possible to write of the existence of a unique model of 'Middle Eastern and North African cities,' the cities in this region share many common traits and similarities. As the cradle of civilization, Middle Eastern and North African cities are marked by a common geographic framework around this valuable coastal area and by a long history, which they all share. The region is marked by the paths of many years of cultural and economic exchange. Within the various urban structures, interdependence and successive domination can still be seen today. Once compact, the cities are now spreading, often along the coasts. They are invading the outskirts, swallowing up hitherto independent villages and outlying farming land. Although walking still predominates as a way of getting from A to B in the cities to the south and east, increasing use of the private car encouraged by public policy aimed at mass motorization as well as the deterioration in the quality of service provided by public transport has triggered uncontrolled urban sprawl and increased car dependency in most cities.

The cities in Saudi Arabia, with the country's vast oil resources, would seem an unlikely example for renewables. It is important to mention that Saudi Arabia's location and climate mean it has plenty of promising sites for solar and wind farms. This is also valuable for other cities in the Middle East and North Africa.

With a view on pedestrianization, the al-Azhar Street axis was a major traffic route through and into Old Cairo. Now that a tunnel has been constructed to take through traffic, it has been proposed that the street should be pedestrianized along with adjacent squares. The pedestrianization scheme is assessed in terms of the views of various stakeholder groups including residents, merchants and planners. An alternative, less drastic pedestrianization scheme is suggested, one that would favor continued commercial and residential functions rather than encouraging gentrification trends. The rehabilitation of Old Cairo remains a delicate and controversial issue (Ministry of Housing, Utilities, and Urban Communities, 2012).

4.2 Conclusion

Sustainable development indicators for Middle Eastern and North African cities are expected to contribute towards calculating if the cities are showing progress towards achieving strong sustainability. Taking the aspect of strong sustainability into consideration, to enhance the links between indicators and policy process, diverse formulations of strategies were used to promote the instrumental and conceptual use of indicators.

Sustainable urbanization has become one of the most significant development challenges. It is apparent that there is a need for cautious implementation of some of the policies. In our opinion, the governments have to act quickly to address problems with air quality and waste management. Furthermore, the gap between the expected growth of transport and the political targets for stabilization and reduction of CO_2 creates a tremendous challenge for urban areas in Middle Eastern and North African cities. It is often difficult for people living in rural areas to reduce car use, and the implication is, therefore, that cities must provide a greater share of the reduction of CO_2 and other pollutants than their share of population.

In addition, technology has had tangible impacts on urban planning. Technology has changed the form of world cities, especially with regard to transportation and communication, as one can see in Doha, Abu Dhabi and Dubai. Technologies such as autonomous, self-driving cars and car-sharing schemes may help us rethink how cities could be developed in the future. Such technologies could have the potential to transform sizable areas dedicated to cars and transportation vehicles, such as the conversion of parking lots for commercial or recreational use. It is also important to mention that the use of car-sharing schemes can encourage people to walk and use public transportation systems

Notes

1 Never before known or experienced.
2 A way to define ecological footprint would be to call it the impact of human activities measured in terms of the area of biologically productive land and water required to produce the goods consumed and to assimilate the wastes generated.
3 Biological integrity is associated with how "pristine" an environment is and its function relative to the potential or original state of an ecosystem before human alterations were imposed. Biological integrity is built on the assumption that a decline in the values of an ecosystem's functions are primarily caused by human activity or alterations.
4 A traditional list of immediate basic needs is food (including water), shelter and clothing. Many modern lists emphasize the minimum level of consumption of basic needs of not just food, water, clothing and shelter, but also sanitation, education, and healthcare.
5 Definition: A person, group or organization that has interest or concern in a process. Stakeholders can affect or be affected by the organization's actions, objectives and policies.
6 Baseline is the year 2014. Data was not available for Indicator 2 (Companies – Number of Fortune 500 companies headquartered), Indicator 9 (Creative industries employment, in %), Indicator 14 (NO_2 emissions average annual concentrations, in mg/m3), Indicator 15 (SO_2 emissions average annual concentrations, in mg/m3) and Indicator 20 (Green space as % of total urban space).
7 Sources: UN World Urbanization Prospects, 2011; REN21, 2016; World Bank, 2012, 2014.
8 Incommensurability means that it has no common basis, measure, or standard of comparison.
9 Noncompensable refers awards of service-connection which VA determines do not warrant the award of monetary compensation.

References

Abubakar, I.R. and Aina, Y.A. (2016): "Achieving sustainable cities in Saudi Arabia: Juggling the competing urbanization challenges". In U.G. Benna and S.B. Garba (eds), *Population Growth and Rapid Urbanization in the Developing World* (Chapter 3). Hershey, PA: IGI Global.

Abu Dhabi Council for Economic Development (2008): *The Abu Dhabi Economic Vision 2030*. www.abudhabi.ae/cs/groups/public/documents/publication/mtmx/nju0/~edisp/adegp_nd_131654_en.pdf (accessed 20 September 2018).

AFD (Agence française de développement) (2018): *Alexandria: Regenerating the City. A Contribution Based on AFD Experiences*. Paris: AFD.

Algeria (2018): *Algiers. Explore Algeria's Dynamic Capital City*. www.algeria.com/central/algiers/ (accessed 17 September 2018).

Arriyadh City Website (2016): "Riyadh in year 2016". www.arriyadh.com/Eng/Ab-Arriyad/Content/getdocument.aspx?f=/openshare/Eng/Ab-Arriyad/Content/Riyadh-in-year-2013.doc_cvt.htm (accessed 17 September 2018).

Arriyadh Development Authority (2013): *Riyadh Metropolitan Development Strategy: Executive Summary*. Riyadh: Saudi Arabia.

Bartelmus, Peter (2013): *Sustainability Economics: An Introduction*. New York: Routledge.

Borja, Jordi (2007): "Counterpoint: Intelligent cities and innovative cities". Universitat Oberta de Catalunya (UOC) Papers: E-Journal on the Knowledge Society, 5. www.uoc.edu/uocpapers/5/dt/eng/mitchell.pdf (accessed 11 August 2018).

Callon, M., Lascoumes, P., and Barthe, Y. (2001): *Agir dans un monde incertain. Essai sur la démocratie technique*. Paris: Éditions Seuil.

Chabou-Othmani, M. (2015): "Sustainable urban requalification in Algiers as a way to recover deteriorated areas", *Transactions on Ecology and The Environment*, 193, 15–25.

Cinelli, Marco, Coles, Stuart R., and Kirwan, Kerry (2014): "Analysis of the potentials of multi criteria decision analysis methods to conduct sustainability assessment", *Ecological Indicators*, 46, 138–148.

Common, Michael and Stagl, Sigrid (2005): *Ecological Economics: An Introduction*. Cambridge: Cambridge University Press.

Commune Tunis (2018): *Visiting Tunis*. Presentation. www.commune-tunis.gov.tn/publish/content/article.asp?id=19010 (accessed 1 September 2018).

Dubai Municipality (2018): *Municipality Business. Planning and Construction. Dubai 2020 Urban Masterplan*. Dubai.

ECOSOC (Economic and Social Council) (2017): *Progress towards the Sustainable Development Goals. Report of the Secretary-General*. www.un.org/ga/search/view_doc.asp?symbol=E/2017/66&Lang=E (accessed 7 September 2018).

Ekins, P., Simon, S., Deutsch, L., Folke, C., and De Groot, R. (2003): "A framework for the practical application of the concepts of critical natural capital and strong sustainability", *Ecological Economics*, 44, 165–185.

Encyclopedia of the Modern Middle East and North Africa (2014): "Places. Asia. Arabian Peninsula Political Geography. Doha". www.encyclopedia.com/places/asia/arabian-peninsula-political-geography/doha (accessed 17 September 2018).

Global Footprint Network (2018): *Mediterranean Ecological Footprint Trends*. www.footprintnetwork.org/content/images/article_uploads/Mediterranean_report_FINAL.pdf (accessed 26 September 2018).

Government of Alexandria (2018): *Why Alexandria?* www.alexandria.gov.eg/Alex/english/index.aspx (accessed 19 September 2018).

Government of Dubai (2018): *About Dubai*. www.dubai.ae/en/aboutdubai/Pages/default.aspx (accessed 19 September 2018).
Grimm, N., Faeth, S.H., Golubiewski, N.E., Redman, C.L., Wu, J., Bai, X., and Briggs, J.M. (2008): "Global change and the ecology of cities", *Science*, 319(5864), 756–760.
Gustavsson, Magnus and Jordahl, Hendrik (2008): "Inequality and trust in Sweden: Some inequalities are more harmful than others", *Journal of Public Economics*, 67(3), 348–365.
Hachimi Alaoui, Nadia (2017): "A 'Time' to act: The 2015–20 Development Plan for Greater Casablanca in development as a battlefield", *International Development Policy series*, 8, 189–219.
Kennish, Michael J. (2002): "Environmental threats and environmental future of estuaries", *Environmental Conservation*, 29, 78–107.
Khoder, M.I. (2007): "Ambient levels of volatile organic compounds in the atmosphere of Greater Cairo", Atmospheric Environment, Air Pollution Research Department, National Research Centre, Dokki, Giza. *Atmospheric Environment*, 41(3), 554–566.
Law, Kathy and Stohl, Andreas (2007): "Arctic air pollution: Origins and impacts", *Science*, 315, 1537–1540.
Marokko Info (2018): *Rabat: The Resplendent City of Rabat*. www.marokko-info.nl/rabat-city-morocco/ (accessed 18 September 2018).
Martinez-Alier, J., Munda, G., and O'Neill, J. (1998): "Weak comparability of values as a foundation for ecological economics", *Ecological Economics*, 26, 277–286.
Metropolis (2018): *City Development Strategy for Sustainable Development: Phase II, Alexandria, Egypt*. https://policytransfer.metropolis.org/case-studies/egypt-alexandria-sustainable-city-development (accessed 7 October 2018).
Meyer, P. and Bigaret, S. (2012): "Diviz: A software for modeling, processing and sharing algorithmic workflows in MCDA", *Intelligent Decision Technologies*, 6(4), 283–296.
Ministry of Housing, Utilities, and Urban Communities (2012): *Greater Cairo Urban Development Strategy. Part I: Future Vision and Strategic Directions*. Cairo: Egypt.
Ministry of Municipality and Environment (2018a): *Doha Municipality Vision and Development Strategy*. Doha: Qatar.
Ministry of Municipality and Environment (2018b): *Qatar National Master Plan*. Doha: Qatar.
Munda, G. (2008): "The issue of consistency: Basic discrete multi-criteria 'Methods'", in *Social Multi-Criteria Evaluation for a Sustainable Economy*, Heidelberg: Springer. www.springer.com/gp/book/9783540737025
Munda, G. (1997): "Environmental economics, ecological economics and the concept of sustainable development", *Values*, 6(2), 213–233.
Murray, Art, Minevich, Mark, and Abdoullaev, Azamat (2011): "Being smart and smart cities", *The Future of the Future*, October 2011, 20–23.
Ny, Henrik, MacDonald, Jamie, Broman, Göran, Yamamoto, Ryoichi, and Robèrt, Karl-Henrik (2006): "Sustainability constraints as system boundaries: An approach to making life-cycle management strategic", *Journal of Industrial Ecology*, 10(1–2), 61–77.
Official Portal of the UAE Government (2018): *About the UAE. The Seven Emirates. Abu Dhabi*. www.government.ae/en/about-the-uae/the-seven-emirates/abu-dhabi (accessed 17 September 2018).
Oxford Business Group (2018): *The Report: Tunisia 2018*. London, UK.
REN21 (2016): *Renewables 2016: Global Status Report*. Paris: REN21 Secretariat.
Roy, B. (1996): "Classement et choix en présence de points de vue multiples (la méthode ELECTRE)", *Revue française d'informatique et de recherché operationelle*, 2, 57–75.
Samer Sani, O.B. (2012): "Towards more effective urban planning in Jeddah, Saudi Arabia". PhD thesis. RMIT University, Melbourne: Australia.

Saudi Network (2018): *Jeddah.* www.the-saudi.net/saudi-arabia/jeddah/ (accessed 18 September 2018).

Shen, L.Y., Ochoa, J.J., Shah, M.N., and Zhang, X. (2011): "The application of urban sustainability indicators: A comparison between various practices", *Habitat International*, 35, 17–29.

Shmelev, S.E. (2017): "Multidimensional sustainability assessment for megacities". In S. Shmelev (ed.), *Green Economy Reader: Lectures in Ecological Economics and Sustainability* (pp. 205–236). Dordrecht, NY: Springer.

Shmelev, S.E. and Shmeleva, I.A. (2009): "Sustainable cities: Problems of integrated interdisciplinary research", *International Journal of Sustainable Development*, 12(1), 4–23.

Shmelev, S.E. and Shmeleva, I.A. (2018): "Global urban sustainability assessment: A multidimensional approach", *Sustainable Development*, 1–17.

Steffen, Will, Grinevald, Jaques, Critzen, Paul, and McNeill, John (2011): "The Anthropocene: Conceptual and historical perspective", *Philosophical Transactions of the Royal Society*, 369, 842–867.

Steffen, W., Richardson, K., Rockström, J., Cornell, S.E., Fetzer, I., Bennett, E.M., Biggs, R., Carpenter, S.R., de Vries, W., de Wit, C.A., Folke, C., Gerten, D., Heinke, J., Mace, G.M., Persson, L.M., Ramanathan, V., Reyers, B., Sorlin, S. (2015): "Planetary boundaries: Guiding human development on a changing planet", *Science*, 347(6223): 1259855. doi:10.1126/science.1259855 (accessed 18 August 2018).

UN (2016): *Sustainable Development Goals. New Urban Agenda. Key Commitments.* www.un.org/sustainabledevelopment/blog/2016/10/newurbanagenda/ (accessed 2 October 2018).

UNDESA (2013): *World Economic and Social Survey 2013, Towards Sustainable Cities.* www.un.org/en/development/desa/policy/wess/wess_current/wess2013/Chapter3.pdf (accessed 1 October 2018).

UNFPA (2007): *Linking Population, Poverty and Development: Urbanization; A Majority in Cities.* www.unfpa.org/pds/urbanization.htm (accessed 11 August 2018).

UN Habitat (2011): "Hot cities: Battle-ground for climate change". http://mirror.unhabitat.org/downloads/docs/E_Hot_Cities.pdf (accessed 14 July 2019).

UN Habitat (2016): "New Urban Agenda adopted at Habitat III, UN Habitat". https://unhabitat.org/new-urban-agenda-adopted-at-habitat-iii/ (accessed 24 September 2018).

UN World Urbanization Prospects (2011): *World Urbanization Prospects: The 2011 Revision.* New York: United Nations.

Victor, Peter A. (1991): "Capital theory", *Ecological Economics*, 4, 191–213.

von Weizsäcker, E.-U. and Wijkman, A. (2018): *Come On! Capitalism, Short-termism, Population and the Destruction of the Planet – A Report to the Club of Rome.* New York: Springer Science+Business Media LLC.

WGBU (Wissenschaftlicher Beirat der Bundesregierung Globale Umweltveränderungen) (2016): *Der Umzug der Menschheit. Die transformative Kraft der Städte.* Berlin: AZ Druck und Datentechnik Berlin.

Wiek, A. and Binder, C. (2005): "Solution spaces for decision-making: A sustainability assessment tool for city-regions", *Environmental Impact Assessment Review*, 25, 589–608.

World Bank (2012): "Who needs smart cities for sustainable development?" http://web.worldbank.org/WBSITE/EXTERNAL/TOPICS/EXTSDNET/0,,contentMDK:23146568~menuPK:64885113~pagePK:7278667~piPK:64911824~theSitePK:5929282,00.html (accessed 24 August 2018).

World Bank (2014): "Inflation, consumer prices (annual %)", https://data.worldbank.org/indicator/FP.CPI.TOTL.ZG?end=2014&start=2010 (accessed 4 March 2019).

9 City Poverty Indexes

Participatory approaches to 'Leave No One Behind'

John W. Taylor and Mohammad Kamruzzaman Palash

Introduction

The City Poverty Index is both a tool and a process that allows local governments and stakeholders to visualize concentrations of poverty in cities, and provide them with information to reveal the often invisible conditions of the urban poor. In many cities across the developing world, very little information is available to local government planners and decision makers; this makes it challenging for government officials to use evidence to formulate policies and put in place effective interventions that address the causes of poverty. There is also very little participation by the poor in efforts to identify and qualify urban poverty. These two issues create an opaque picture of urban poverty conditions in cities in the developing world, and this leads to urban poverty either remaining invisible, or the formulation of poverty-reduction policies and programs that are either incomplete or insufficient. To address this gap, a city-scale participatory approach has been developed in Bangladesh through the National Urban Poverty Reduction Program (UNDP Bangladesh, 2018). It fosters the collaboration of residents of poor communities and local ward officials, and gathers and organizes essential information to create a comprehensive city-level poverty database and series of maps: the City Poverty Index. This index has been used to inform city-wide poverty reduction strategies in several cities across Bangladesh, providing local governments with a means to prioritize pro-poor interventions and plans. It has also helped bring about consensus-based and targeted poverty-reduction approaches that can have an important impact on more equitable budgeting, infrastructure investments, and improved public services for the poor; all contributing to reducing urban poverty in target cities.

The New Urban Agenda puts forth the challenge to policy makers, local governments, and civil society to '*leave no one behind*' but little guidance is provided on what methods and approaches can ensure the transformation of that goal into a reality. As the trends of urban economic growth and urbanization continue to bring prosperity and growth to countries across the developing world, increasingly large numbers of the poor are left out. Instead of economic and social conditions improving for all urban residents, the benefits are uneven, with poverty becoming the unfortunate yet stark reality of cities across the world. Inequality threatens to leave out many at the bottom, impeding gains in health and security, and putting at risk the achievement of the Sustainable Development Goals. Uneven development also poses a threat to political stability, environmental and social sustainability, and puts those living and working in precarious and vulnerable conditions at risk. Therefore finding ways to address urban poverty is a critical challenge for cities, a challenge that presents both opportunities, in terms of increased productivity, diversity and vitality of urban economies, but also threatens major problems, such as social unrest, public health epidemics and economic stagnation, if efforts fail.

One of the paradoxes in addressing urban poverty is that in most cities of the developing world in daily life the poor are visible and ever-present, yet in official terms very little is known about them. It doesn't take long walking around a city such as Dhaka, Bangladesh, to see that poverty is everywhere. The bustling streets are filled with informal vendors, low-income dense settlements sprawl across large swaths of the city, and deteriorated infrastructure seems to be a rule not an exception. All these visual clues make visible the presence of large numbers of poor people who live on the margins of rapidly growing megacities such as Dhaka. However, in terms of available data, about where slums are located, their conditions, the profile of poor households, or any maps or data about them, it can seem like poverty is not an issue at all. This lack of visibility results in low levels of awareness about issues that drive poverty in cities, and contributes to their absence in public discourse and public policies. As a result few functional policies are in place to address urban poverty issues at scale. The lack of information and public debate also lead to less political pressure to address it as a problem, and therefore national and local governments are not pushed to develop pro-poor policies that could bring about wide-scale change.

A lack of information about urban poverty

One of the factors that makes responding to urban poverty so difficult is an absence of information that can help to provide a sense of scale, locate the issue, and identify the critical factors that make the poor so vulnerable. Without any reliable information, planners, decision makers and pro-poor advocates have very little evidence with which to inform discussion and the development of policies. Individual surveys of individual low-income communities are often undertaken by NGOs or by local governments for pilot projects and this does contribute to an understanding of specific areas of the city. However, such surveys are focused on specific communities and fail to provide an overall picture of poverty at the city-scale. Thus perceiving conditions at a more comprehensive scale, that encompasses the overall dimensions of slums and the numbers of poor, are not possible with accuracy (Boonyabancha, 2009; Archer et al., 2012). The picture available is not proportionate to the size of the problem. In many cities in Bangladesh, for example, there are over a thousand slums spread across their entire extent of the city, with the population of these slums sometimes outnumbering residents of non-slum areas (UNDP, 2018). Surveying just a handful of slums in a city cannot reveal the staggering extent of these slums; it just represents the tip of the iceberg.

In addition, the information that *is* often available to local governments about urban poverty is unreliable and incomplete. Demographic data about urban populations usually only counts those residents who are officially registered, and many of the poor fail to register themselves, or are unable to do so, given that they have migrated from other areas. This means that the estimates of urban poor residents, and hence their housing and basic service needs is underestimated. The location of slums is also not fully recognized because mapping slum areas across cities is uncommon, and rarely carried out in coordination with slum communities themselves. This activity in and of itself is challenging because city officials may feel that slums are not their priority and may be unfamiliar with identifying or visiting them. Slums can be hidden away and difficult to access, and due to their mutative and transient nature, they can emerge or change quickly. Without comprehensive poverty maps, city governments are unable to really visualize the extent to which slums are present in their cities, or begin to spatially analyze them, for example, to identify gaps in public services and how to address physical and social vulnerability issues. There is therefore an apparent need for urban data to quantify and measure urban poverty; however, poor people's invisibility is not only a technical issue in need of a solution. It is often deliberate and convenient, and in some cases the poor are unwilling to be recognized at all.

The urban poor lack a voice and recognition

At its root, the invisibility of the poor is a governance issue, one of legitimate recognition and representation. If the poor are not officially recognized, they are not able to raise their voices or legitimately push local governments for improvements to public services or larger budget allocations. This situation results in part due to rural migrants entering cities without the official papers that recognize them as belonging to the urban areas they have moved to. While they may have committed to living in cities many years ago, they remain registered in other districts, and therefore are not yet being recognized as legitimate citizens of the communities they now inhabit. This is compounded by the fact that many of them may also occupy lands that are not legally their own. Without official residency and often without legal land tenure status, they are not eligible to receive services and security. While the poor come to cities to find jobs and access economic opportunities, and contribute significantly to local economies with their labor and productivity, they are not fully recognized as belonging.

Curiously, this ambiguous and uncertain situation may benefit both local governments and urban migrants. Invisibility might benefit the poor, who otherwise would be forced away from convenient yet vulnerable housing arrangements and to pay taxes for public services. In the case of local governments, they don't feel responsible for making long-term and sometimes large-scale investments in parts of the city, thereby saving scarce resources. One consequence of their unofficial status is that the poor's lack of access to basic services is easily overlooked; they are forced to use informal mechanisms, often through the black market, and pay much higher rates for services, such as water and electricity, compared to those who access them legally. Instead of investing in poor settlements where many residents are not officially registered as voters, local governments may favor investing in efforts that respond to needs of more politically enfranchised voters, such as infrastructure projects that are more visible and whose impact could be shared with other income groups.

Lacking tools and resources

Another reason local governments struggle to respond to urban poverty is because they are ill equipped to do so; they lack tools, financial resources, and human resources that are skilled in managing complex poverty conditions. While many mayors do prioritize improving the conditions in low-income settlements and the lives of the poor, responses are often piecemeal and reactive. They fail to perceive root causes of complex issues and prioritize actions accordingly. One of the reasons is due to financial constraints; in Bangladesh, transfers from national government largely support the budget of local governments, and local revenues are still inadequate. This hinders ambitious plans to develop more ambitious and large-scale policies to bring about systemic changes, for example, extending water supply infrastructure or improving drainage to informal settlements. Local government policies have also traditionally been driven by central government and technical expertise has limited experience of designing sensitive and progressive pro-poor policies, as they had not previously been required to do so. Government officials also often lack the experience of interacting with poor communities and bringing about collaborations that could result in wider impacts. Without tools that can support their efforts, and without sufficient human and financial resources, these limitations further constrain local governments from the impact that they could potentially make to reduce poverty in cities.

The City Poverty Index: informing decision making and mobilizing the poor

The City Poverty Index is both a participatory process and an instrument to inform decision making and pro-poor planning. It results in a set of city-level maps that give stakeholders information about the range of conditions of poor settlements across the city, locate hotspots of poverty and identify public services gaps. Importantly, it is a participatory process that involves the poor, and in doing so recognizes their contributions and importance in the city as citizens. It also involves lower levels of local government – the ward level – by actively engaging with ward councilors, the lowest level of elected official. This section describes: (1) the collective community mapping process that brings together community-level data to create maps and expand knowledge of low-income settlements across the city, (2) the validation of this information by local stakeholders to authenticate its accuracy, and (3) the aggregation and analysis of this information to form city-level datasets and the City Poverty Index. Ultimately, the outcome of the process is comprehensive view of urban poverty that can be shared with all stakeholders to inform the development of pro-poor actions and policies.

Step 1: the city-wide participatory mapping process

The city-wide participatory mapping process collects information about low-income settlements, and in doing so provides various opportunities to involve citizens and community members. Participation can occur in several different stages of this process – through data collection, validation, prioritization of local issues, and finally, by sharing information. Participation in the process encourages residents to contribute to decision making, such as through the identification of their planning needs, participatory budgeting and problem-solving community-level problems. The mapping process harnesses the day-to-day knowledge of low-income settlements by those that know them best – the residents. The result is a map of the boundary of each settlement and a database of indicators about their conditions.

The process described in this chapter took place in eleven cities in Bangladesh between 2017 and 2019 enacted by the United Nations Development Programme (UNDP), as part of the National Urban Poverty Reduction Programme (NUPRP). NUPRP is funded by the government of the United Kingdom and is implemented in an institutional partnership with the Bangladesh government's Ministry of Local Government, Rural Development and Co-operatives. The mapping process brought together the active involvement of community leaders from poor communities across each city, ward councilors, mayors and residents. Given the positive response and utility of the City Poverty Indexes, this mapping approach will be repeated in these same cities every two years, and the initiative will be scaled-up to another twenty or more cities across the country beyond 2019.

The data collection process starts with a city-level consultation meeting with the local government council, city officials and community leaders as a first step to recognizing and responding to the needs of the urban poor in each city. During the meeting, stakeholders are informed about the objectives and the process of the poor settlement mapping and how different groups are expected to contribute. Shortly afterward, the second stage occurs at the ward level in which the ward councilor convenes a meeting in the community center with residents of the poor communities within the ward. During the meeting, an orientation is given about how mapping can benefit local communities in the ward, and what community surveyors have to do. Aerial images of the ward are shared to explain the mapping tools and survey methods. Following the orientation, the surveyors take the aerial map (printed on durable paper) to the field to visit each of the poor settlements.

At this stage, location and boundaries of the poor settlements are identified by conducting a transect walk and walking with residents around them. The surveyors use the map to draw the boundary lines and also bring together residents in focus-group discussion to discuss and fill in a scorecard which lists 16 poverty indicators. The survey provides a rapid assessment of the settlement's conditions, it is split into three sectors – infrastructure, social well-being, and housing – with each having a set of indicators. For infrastructure, the seven indicators are road access, drainage, solid waste, sanitation, electricity, water supply and street lighting. For social well-being, the four indicators are related employment, income, education (school attendance of the children) and social problems. For housing, the five indicators are housing conditions, land tenure, eviction, land ownership and occupancy. The survey asks the respondents from each slum to score on a scale of 1 to 4 what they think best represents the conditions in their community. In each case, 1 represents the worst conditions (for example, no drainage or no access to piped water) and 4 represents optimal conditions (for example, drainage infrastructure is present and functional, or the entire community has good access to piped water). In addition to these 16 indicators, additional information is also collected such as an estimate of the number of poor households, the age of the settlement, and the name of the slum. In delineating the slum's boundaries on the map, a measure of the settlement's size can be created, which is useful in determining population density estimates.

The fourth and final step at the ward level is a public meeting where the ward-level settlement map, the data about the conditions of each slum, are shared with the ward councilor and local residents for review. The ward councilor checks the data's validity to ensure that it is accurate, and if so, endorses it publically. This is an important step and should be carried out in front of community representatives because it signifies that the ward-level information is officially approved at the level closest to the community level. This allows it to be then utilized as part of the city-level dataset and aggregated with the information from other wards. Once all the ward-level information has been endorsed, all the maps and data are compiled into a city-level database to be analyzed and presented back to stakeholders from throughout the city during a city-level results-sharing workshop.

Step 2: aggregating and organizing data at the city level

Once the maps and data indicating poor settlements and poverty conditions have been collected and validated, the data is organized and aggregated into a city-level dataset. This can be a significant task because data entry is involved, and linking the database to maps requires knowledge of Geographic Information Systems (GIS). This stage of work was thus undertaken by the UNDP team. In the city of Narayanganj, for example, there are over 1,500 individual slums, spread out over 27 wards (UNDP, 2018); each dataset contains a great deal of information and a lot of data entry was required. The linkage of the database with the geographic information allows for it to be analyzed into a series of thematic maps. Each one presents varying layers of poverty conditions for the whole city, such as access to piped water or access to sanitation. The information is organized and presented in a way that allows for participants to make use of it, even if they are not planners or familiar with data analysis. For example, the use of bar charts makes comparative analysis possible, and the combination of population and area to create population density maps helps to normalize data and make it more useful to compare other indicators, for example, the intensity of use of certain public facilities. Finally, the division of data into quartiles can simplify viewing by presenting a range of data into a range of four colors (on chloropleth maps) so that relative intensity can be judged. This is particularly useful when it is important to identify which areas of cities have greatest deprivation and thus greatest need. While the database management and GIS mapping functions were conducted by technical staff from the UNDP, efforts have been made to train local officials to be able to do so themselves. The section below summarizes some of the important ways in which the urban data about the conditions in poor settlements is organized and aggregated.

a) ***Data is organized to allow comparative analysis***

Data about individual settlements is aggregated to the ward level, and this means that it is possible to make comparisons between wards. The ability to compare wards is significant because it allows for spatial comparisons to take place at the city level, so that relative levels of, for example, access to piped water, can be visualized comparing all the wards. It is evident from thematic maps that access to different services varies widely depending on what part of the city slums are located. These maps have generally shown that in urban peripheries, basic services are comparatively much less accessible than in city centers and more established settlements.

Another way in which comparisons can be made is through the presentation of information using bar charts, which provide a simple visual reference. The information on each map is accompanied by a bar chart that indicate where levels are high or low, and even how wards compare to a city's average for that indicator. When pointing out which wards might require special attention, the bar chart is helpful because it provides more of a quantitative reference to supplement spatial information.

A further data presentation technique is the creation of population density maps that offer another tool to support comparative analysis. In order to understand the conditions in slums, it is useful to consider population density because it sheds light on how intensively resources are shared, how many households occupy limited spaces, and the multiplier effects of problems or solutions could have. For example, when considering the negative impact of the lack of access to sanitation, areas of high population density would be those that would be worst affected by a public health epidemic. Likewise they would be the areas that would benefit the most by investments in this sector. Thus it is a useful map to have available during public discussions, since it could help accompany other maps and serve as a reference, indicating comparatively in which areas of cities would investments make the biggest impact.

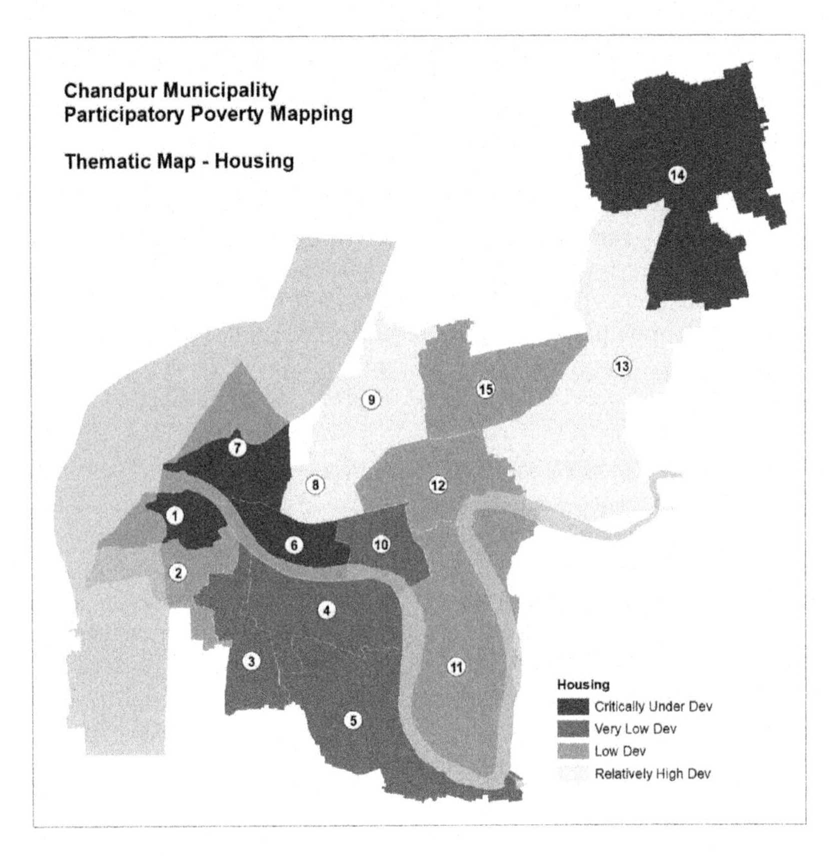

Figure 9.1 Conditions of housing in the poor settlements by wards of Chandpur Municipality

b) ***Simplifying data into ranges for easy viewing***

Thematic information can be separated out visually into different categories to enable easy interpretation and facilitate decision making. For thematic maps and the overall City Poverty Index, the data has been divided into quartiles, four categories, which reveal different levels. For simplicity's sake, these categories indicate: (Q1) wards with settlements with the lowest average levels of access which concentrate the worst conditions, which are critically underdeveloped; (Q2) wards with settlements with levels of access to services are slightly higher but still very low; (Q3) wards with settlements with comparatively higher access, are low, and (Q4) the fourth quartile with relatively better-off conditions. The presentation of information in this way helps point the attention of users to those wards in the Q1 and Q2 categories which have most urgent need, compared to the other wards.

Infrastructure Aggregate Score

Sum of the Scores of Drainage, Electricity, Roads, Sanitation, Street Lighting, Waste, and Water per Settlement

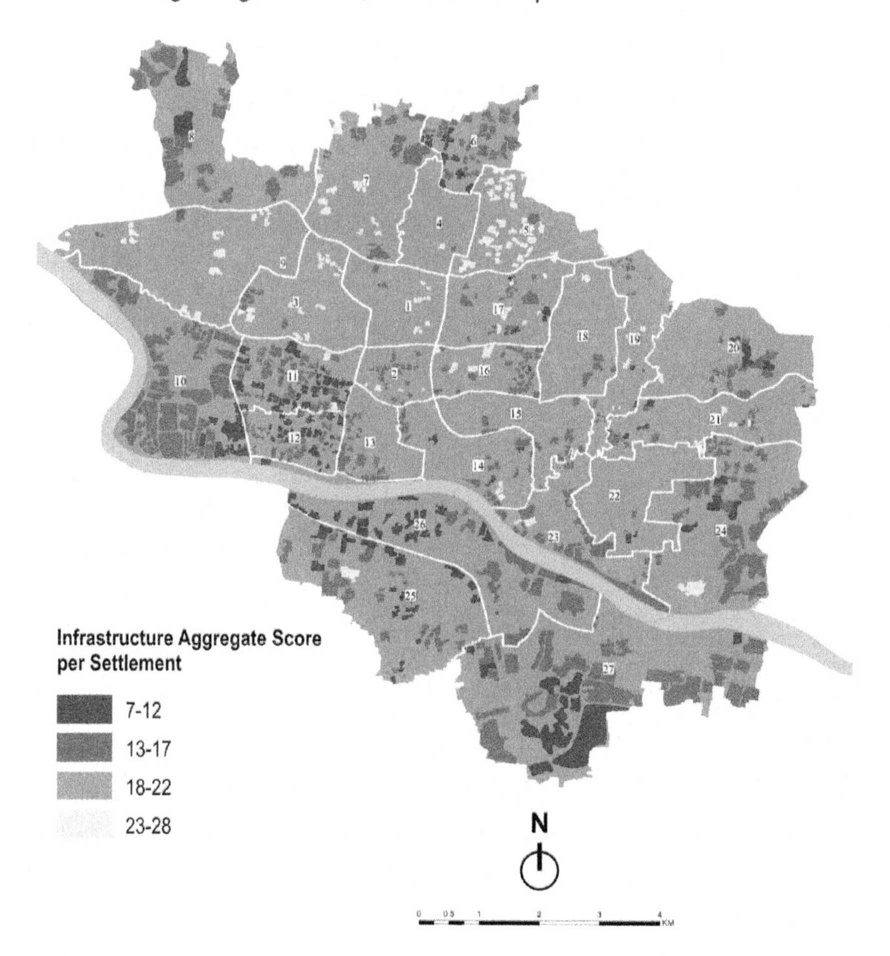

Figure 9.2 Settlement-level Infrastructure Index, made up of seven infrastructure indicators of the poor settlements for the city of Sylhet

c) ***Similar indicators are aggregated into thematic indexes***

In addition to the individual maps of each indicator, it is useful to evaluate the condition of the city based on different themes, for example, in terms of basic services and infrastructure, well-being and livelihoods, and housing. To do so, the data from several indicators can be aggregated thematically, combining for instance the condition of access roads, sanitation, access to water

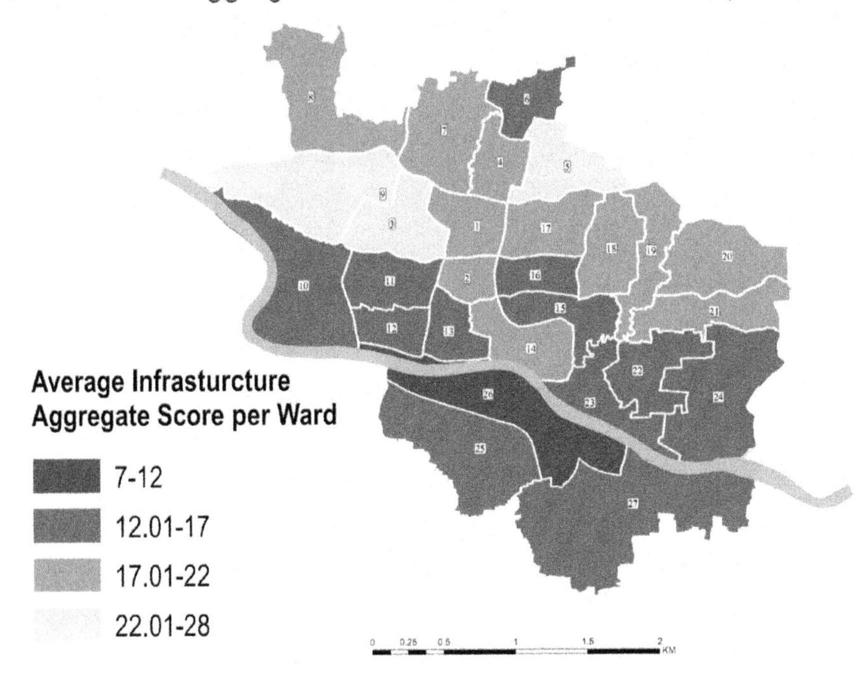

Figure 9.3 Ward-level Infrastructure Index, made up of an aggregate of seven infrastructure indicators from poor settlements for the city of Sylhet

and to trash collection, into one single 'infrastructure index.' This allows the workshop participants to engage in a discussion about infrastructure conditions in general, and not have to focus specifically about any one of the different indicators separately. This is helpful for facilitating wider discussions and not allowing the discussion become too fragmented by the list of individual indicators.

Step 3: presenting the maps to involve stakeholders in defining poverty and prioritizing indicators

Once the maps of poverty conditions have been created, they are presented during a city-level workshop to stakeholders. The workshop seeks to explain the findings, build stakeholders' understanding of the issues, and help interpret the root causes of poverty and priorities for the city. Stakeholders are convened by the mayor's office to ensure that not only is there a good turn-out, but also that a range of people are represented. These include leaders of urban poor communities, ward councilors, civil society representatives and city officials. For many, it is the first time that they have seen mapped information about slums in their city and it can be an eye-opening experience for them. The workshop is called the 'City Context Workshop' and it adopts a pedagogic approach, in which the maps are hung on a wall in a gallery format and a facilitator indicates what each one represents. Workshop participants rotate around the room, from map to map, and then discuss in working groups what each map might mean or what it might miss.

Providing time and facilitating discussion allows stakeholders to reflect on the maps and also to directly note down on them additional information that might be necessary to better understand the context. Workshop facilitators encourage participants to use markers on maps to draw or write down information and gather relevant missing information. Splitting into smaller discussion groups that review thematic issues brings people from different walks of life together to share their perspectives. These stakeholders include city officials, poor community residents and civil society representatives, working shoulder-to-shoulder and exchanging ideas and observations. At the discussion's conclusion, the different groups share back with the plenary what their groups had discussed and what additional observations are necessary to complete the picture.

The final session of the City Context Workshop provides the last remaining information needed to complete the City Poverty Index: defining the most relevant indicators to measure poverty for *their* city. City stakeholders are requested to identify which of the 16 indicators are most important, or most relevant, to be considered in defining poverty. In cities, for example, where flooding is a main concern, they may choose to highlight the indicator concerning drainage and road access, since mobility and waterlogging may be priority concerns for the local government and the poor. Participants select the highest priority indicators for their city, while relegating others to secondary or tertiary importance. The indicators are then given a *weighting* using a scale of 1 to 3. Thus in cities where flooding is an urgent concern, the indicator for 'access to drainage' might be prioritized by giving it a weighting of 3. Weighting indicators helps to recognize the differences in poverty conditions between cities, where the factors that contribute to it are not necessarily of equal value or relevance.

This process of prioritizing poverty indicators, involving the poor communities and other stakeholders, helps to better define poverty in the city – it is participatory. This can be done through voting and tallying up the scores from the participants engaging stakeholders to define what poverty means for them. Each city's poverty index will be different and be adapted to each city's realities and concerns. Once the indicators have been prioritized, they are then ready to be analyzed and the City Poverty Index determined.

Step 4: creating and presenting the City Poverty Index

Once the maps and data has been endorsed by stakeholders, and the weightings assigned to the indicators, the City Poverty Index can be calculated. To create a fair comparison of the data as an index score, it is important to normalize it.

Normalization can be done in a number of ways, but in this case the poor population, poor population density, were used to provide a way to better compare the index scores. This allows wards where there are relatively low numbers of poor residents, to be compared fairly with those with higher numbers. Population density is another factor that is useful to introduce because the poor settlements with high population density are appeared to be worse in physical and socio-economic conditions, in comparison with the poor settlements with low density. Using the poor population density in calculating the index score, helps determining the wards that need most attention for the maximum impact on poverty reduction in the city.

When the City Poverty Index has been completed, it is presented and explained to city-level stakeholders in a public forum. This workshop completes the process of collecting and analyzing the urban poor data, but should serve as the beginning of a strategic planning process or lead to other pro-poor interventions. The workshop brings together those who contributed to and were instrumental in bringing the information together, including the ward councilors, community leaders and the mayor, but also includes other relevant groups such as local NGOs, public service providers and groups advocating for the poor. The workshop begins by sharing the whole process, citing how at each step the data had been reviewed and validated by contributors and elected officials. Other contributing information, such as how the city's poverty indicators were prioritized, should also be explained, so that it becomes clear how the Index came about. The resulting maps are then displayed and provide reference for local government to inform their pro-poor strategy.

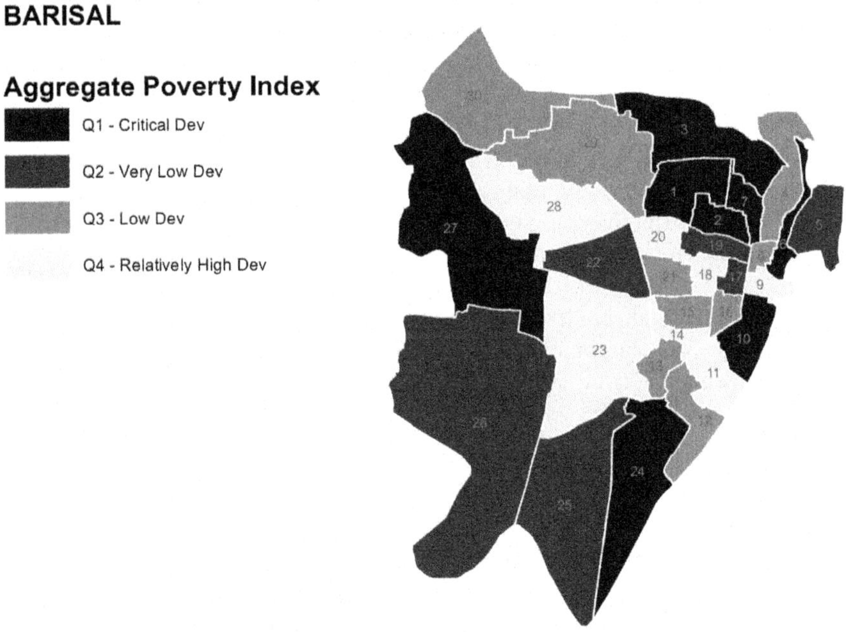

Figure 9.4 The City Poverty Index for the city of Barisal

The collective review and validation of the process and the resulting data is critical to the acceptance, use and overall success of the City Poverty Index. Since data collection and analysis occurs at different levels (the community, ward and city levels), it is important that at each stage there is sufficient review and endorsement so as to give credibility and legitimacy to the final result. At the community level, the data has to be collected by local communities and reviewed with other residents. At the ward level, the ward councilors should convene a public meeting to review and certify the data, asking for any further comments and for agreement, before endorsing it. Finally, at a city-level meeting, the local government convenes a meeting inviting city-level stakeholders, including the mayor (as well as all the ward councilors and community leaders) to review the process and endorse the results. This concludes the process of gathering and verifying the information and allows the maps and information to be credible for official use by local government.

Using the City Poverty Index in Bangladesh

The National Urban Poverty Reduction Program (NUPRP) in Bangladesh is using the City Poverty Index to support local governments to better engage stakeholders in city-wide dialogues about poverty, and to take decisions about resources' allocation to address poverty conditions. The NUPRP project seeks to persuade local governments to adopt a proactive attitude to urban poverty, recognizing the existence and needs of poor communities, and working with them to systematically address them (Alkire, 2015). But local governments that don't have information about slums, or any way of sorting which ones have greater needs than others, are at a loss for where to start. This is where the City Poverty Index serves not only as a way to collect information and understand conditions where no official information existed before, but also to initiate dialogue between local governments, ward councilors, and the residents of poor communities, about what actions are needed to take (Shmelev, 2017).

It helps that the City Poverty Index presents a single picture of the status of the conditions in a city's poor settlements in the form of a map, because a diverse group of stakeholders, including representatives of poor communities and ward councilors, are unfamiliar with the complexities of poverty data. The Poverty Index thus serves to simplify and clarify the issue of how poverty is distributed by showing how poverty conditions vary across the city. This is a point of departure for discussions about the root causes and what measures are needed to address these conditions. Identifying underdeveloped hotspots where poverty conditions are concentrated is a way that doesn't overcomplicate the great range of different indicators or poverty-related sectors, collapsing them into a single index score. Such over-complication could lead to exclusion, rather than greater inclusion, of stakeholders in the discussion.

As a starting point, the presentation of the City Poverty Index is an opportunity to convene and engage stakeholders, allowing them to see poverty as a collective issue that stretches across the whole city. As mentioned above, the city-stakeholder workshop presents an objective view of poverty conditions, demonstrating that the maps and data have been processed through a comprehensive, rigorous and transparent analysis. Through recapping the whole process, which of course includes consultation about the selection of indicators, and giving weighting to specific indicators on issues of greatest importance to the city, the workshop participants are reminded that the Index is a result of a collective process. This helps to depoliticize and qualify the results which the maps reveal – in the Index, participants see that some wards of the city concentrate poverty at much higher levels than others; even though people may believe conditions in wards across the city to be equally bad.

Refocusing the attention of workshop participants at the city level forces them to consider poverty in new ways. This is significant because ward councilors and poor community residents are more accustomed to thinking in terms of their own communities, or advocating for resources for the problems they see immediately before them, without necessarily considering the bigger picture. By giving them a means to compare conditions fairly reveals a more complex picture in which some city areas are seen to be much worse off than others. It also may show patterns of exclusion that would otherwise not be self-evident.

For example, it was common that stakeholders remarked how poverty conditions were far worse in urban peripheries, which they hadn't originally realized previously. Also the Index score for the condition of settlements in central areas, which is what many might associate with impoverished conditions, is often higher than those in the periphery. This indicates that central areas' basic services are more accessible and conditions are comparably more favorable compared to those in the periphery. The slums in the central areas of cities, while very densely populated, perhaps with trash piling up or with deteriorating sanitary conditions, often have far better access to roads, electricity, water and sanitation, as well as to jobs and other goods found in central areas. Giving participants the evidence to observe this is an important first step in having them think of poverty beyond their own communities' self-interests.

NUPRP uses the City Poverty Index to then put forward a strategic approach to poverty reduction, the targeting of limited project resources towards those poor communities in the wards with highest concentrations of poverty. This constitutes an initial idea of a 'city poverty reduction strategy' because it is one that takes into account the needs of the whole city, but is selective in its focus. By revealing which wards concentrate the worst conditions, it puts forward a rationale that public investments should go to the areas where they can leverage the most impact on improving the lives of the poor.

NUPRP offers this as a starting point for deciding where the project's poverty-reduction budget should be allocated, suggesting that 50% of funds (including livelihoods and education grants, as well as infrastructure funds) be directed to those wards with the highest levels of poverty, while 25% go to those that fall into the next highest index score. There are four categories of ward (most critically underdeveloped, very low development, medium, well-developed) and so a ratio of how to spend the resources is suggested as 50:25:15:10 to each of them. In some cases, this ratio is accepted, while in other cases discussion might alter the proportions, allocating less to communities in the wards with the highest poverty concentration, and more to the others. In those cities where this has worked, what is significant is that city stakeholders have agreed upon a rationale for the strategic allocation of pro-poor resources; for why those resources should go to some areas instead of others, and that by targeting those resources on areas of highest deprivation, they will make a far bigger impact than communities where poverty is not quite as acute.

The City Poverty Index provides mayors with an important tool to support them in warding off *elite capture*, namely convincing councilors or influential individuals or groups that resources should go to where they are needed most (Dasgupta & Beard, 2007). Without objective evidence to support the claim that resources are more needed in one area over another, it is possible for them to be misused, or to not be optimized. Mayors might use resources to bolster communities that have supported them in the past while ignoring those where they are less popular, or councilors may push the mayor for grants for their own communities in exchange for votes – considerations that do not take into account prioritizing needs based on poverty conditions. The Index score thus helps to qualify discussions in a way that makes need most evident, and eliminates the issue of not having enough information to know what those needs are (Shmelev & Shmeleva, 2018). The result is that investments are targeted towards wards with high concentrations of poverty, and they are scaled back in those wards with far lower poverty levels.

Challenges in using the City Poverty Index

Introducing tools such as the City Poverty Index can be challenging: the Index presents a new and unfamiliar way of addressing issues. This requires stakeholders to be willing and able to adapt, and also requires sufficient support to accompany what can be a difficult transition. In the cities where the NUPRP project has been active, there have been mixed results but three main challenges emerged: (1) the problem of local governments adapting to different ways of working, which required that they change how they think about budget and planning; (2) adopting a new scale of action to take on the issue of poverty, now seeing it as a city-wide one and thereby becoming a more complex and far-reaching issue, and (3) overcoming inherent prejudices and reservations of stakeholders to work together with the poor. The above challenges are briefly described below.

a) **Introducing institutional change to budgeting and planning**

Undertaking change is never easy and this is certainly the case when that change requires transforming local government processes. This is challenging because government officials are trained in predetermined methods and, given the prevalence of top-down processes, there is little scope for new perspectives or changes to occur to those processes. The City Poverty Index presents a very different way of thinking about budgeting and planning, one that for example specifically focuses on the poor (as opposed to focusing on the city at large), and that also involves so many stakeholders in its creation and decisions about its use. In many cases, local governments are reluctant to introduce radical changes, and they may also feel they lack the technical support to do so on their own (or when support from development projects is removed). Therefore mayors and other officials may feel willing to adopt the new methods with the technical support offered by a development project (such as the UNDP's NUPRP) and the funds that it provides, but consequently feel unwilling to invest their own resources and efforts in following up this process in successive years because such change is difficult to maintain continuity.

b) **Shifting the scale of reference is unfamiliar**

Another change which the City Poverty Index introduces is shifting people's viewpoint from that of specific communities to thinking at the scale of the whole city. For many who are directly involved in pro-poor interventions, those at the frontlines of working with slums and advocating for the needs of the poor, they think of specific community needs. Community leaders and ward councilors are thus much more familiar with, and incentivized, to defend the interests of, and actively push for, the development of their own communities. The City Poverty Index presents a city-level picture and in doing so it indicates certain wards as being more needy than others; implying that some community needs are more pressing than others. If stakeholders bring a perspective of local interests, then it is difficult to shift the discussion to a bigger picture conversation about poverty. This has been a constant challenge faced during the NUPRP and one that has prevented comprehensive dialogues from emerging, as some stakeholders feel that such discussion draws attention away from their constituents and local interests.

c) **Generating buy-in and accepting the participation of the poor**

The participation of the poor in the City Poverty Index process is new and unfamiliar for many stakeholders, Local governments have not traditionally been open and inclusive in engaging poor communities to address their needs. This is of course because many residents of poor communities are not recognized officially, and many communities are not legal, and so government officials adopt a more traditional approach, some questioning the legitimacy of poor residents' participation. As a result, their views are disqualified or marginalized for being uninformed (Arnstein, 1969). Another problem is that by making transparent a data-drive budget allocation rationale, over an approach that draws from the discretionary power of local governments and elites, can threaten those interests. Making the process more inclusive and transparent threatens elite capture that is associated with local budgeting processes. Conducting a dialogue between different stakeholders also requires facilitation support to ensure that different voices are heard and alternative viewpoints can be expressed (Sanoff, 1990). Thus a final challenge is that of being able to support the multi-stakeholder workshops with sufficient facilitation support to ensure that they are sufficiently inclusive and participatory.

The impacts and benefits of the City Poverty Index

The introduction of the City Poverty Index to cities in Bangladesh seeking to address urban poverty is far from complete or widespread, but some important observations can be made about its impacts after the first few years of its use. The following three positive impacts show increased understanding of poverty issues, greater visibility and participation of the poor in city-level forums and planning workshops, and the use of the Index to develop plans and budgets; these impacts indicate how the Index is altering the ways in which the needs of the poor are traditionally addressed.

a) **Local governments and stakeholders have a greater understanding of poverty conditions**

When the NUPRP project began in 2016, very little was known about the extent of poverty conditions in cities; this hindered the development of pro-poor policies at a large enough scale to make a difference. Some cities were operating in the dark, not knowing how many slums there were, where they were, or really having an understanding of the needs of the urban poor. The participatory community mapping process was a first step to bringing the poor into a broader conversation about the conditions in slums; it was able to put information on the table in front of stakeholders that started people thinking and problem solving.

The City Poverty Index provided mayors and government officials a more nuanced understanding of the needs of poor communities, indicating how they are different in different parts of the city. In Sylhet, in the north of Bangladesh, the mayor adopted an approach that focused on investing resources on drainage, as flooding and waterlogging were revealed to be critical in areas of the south-east of the city, while in other areas his approach focused on job creation, where public services were already at much higher levels. The ability to differentiate conditions and needs is instrumental in allowing local governments to adopt more targeted approaches adapted to the diversity of conditions experienced by the poor. In Mymensingh, another city in the north-west of the country, the Index demonstrated that poor communities in the expanding periphery required a strong emphasis on infrastructure investments, such as roads, drainage and water, while the city-center slums required connecting with *existing* services. These communities already had access to roads, water and electricity, but for the most part they were illegally connected because the slums were not officially recognized. The mayor started to consider the possibility of recognizing slums in the city center and connecting them with existing services such as trash collection and the city's official water supply, and reconsidering how their tenure status could be made official.

The urban data and information provides evidence with which informed decisions can be taken. In many of the cities where the NUPRP operated, mayors struggle to adopt a targeted approach because focusing on some wards as opposed to others always implied wards which would receive more public investments than others. Mayors drew upon the City Poverty Index as evidence that conditions were not the same throughout the city and that a targeted approach was necessary. For example, in Sylhet the maps and Index gave the mayor confidence in his ability to direct actions and select those areas where they were most needed, providing evidence to contend with ward councilors who advocated for their communities' needs, but without seeing the bigger picture of the city's conditions overall.

b) **More participation and representation of the poor**

The use of the City Poverty Index has led to much broader participation of the poor in terms of pro-poor policy making and planning, both at the ward level, and in city-level processes. One of the reasons is because the participatory process to map settlements and collect data served to mobilize them, and raise awareness of the role they can play when they engaged with local government officials. Meeting and collaborating with their ward councilors, at first in the mapping exercise, and later in discussions about conditions and priorities, brought poor residents into much closer contact than before. In some cases, this opened the door to establishing improved working relationships with their elected officials, bridging a symbolic divide that exists between poor citizens and those higher up who seem inaccessible or disinterested. The bottom-up mapping process, which drew upon the organization and efforts of low-income community leaders, showed ward councilors and other government officials the impact of their contributions and their willingness to support city-wide efforts.

By highlighting the conditions of the poor, they have also benefitted from more opportunities to participate in planning and budgeting discussions. Shining a light on the issue has led to their inclusion in a number of additional governance processes that they were never a part of before. In the cities of Chitaggong, Khulna and Narayanganj, the city-level Community Development Committee Federations, which represents the interests of poor communities, have been invited to form part of City Level Consultation Committees. These are formally recognized bodies that play a consultative role in discussing and deciding upon urban policy. These federations have also been invited to participate in discussions regarding pro-poor budgeting, planning for climate resilience, and pro-poor economic development planning. While participation alone is not enough to shift government policy, it is a meaningful beginning to become a more visible and vocal stakeholder in the city, representing the interests of a traditionally marginalized group. Instead of being overlooked and left out, their participation symbolizes their gaining recognition. Thus the sharing of the City Poverty Index played an important role in opening up other spaces of participation.

c) **Providing guidance on how to tackle urban poverty**

Beyond bringing attention to poverty and including the poor, the City Poverty Index provides guidance to local governments about what to do next. By displaying hotspots of concentrated poverty, the Index can be used as a road map on how investments in infrastructure, improved basic services and social programming can be used to reduce poverty where it is most acute. Instead of presenting a picture that confuses or leads mayors and local government officials to feel uncertain how to proceed, by indicating degrees of priority clearly shows them where to begin. The mayors of Sylhet and Mymensingh are using the maps to indicate how public service gaps can be filled, so that water and sanitation services can be expanded to the poor. In fact, in the eleven cities where the NUPRP has been active from 2017, urban poverty reduction strategies have been developed that directly draw on the Index to prioritize poor communities and allocate resources accordingly. Thus it serves as the starting point for the definition of pro-poor strategies, budgets and plans.

Conclusion

Tools such as the City Poverty Index, that map and provide analysis of complex and often invisible issues, such as poverty conditions, can help more and more stakeholders better understand and put forward ways to make their cities better. Like many challenging problems, urban poverty does not simply require a technical fix; there are deeply rooted societal causes that prevent sustainable solutions, and thus the participation of the poor in discussing and formulating alternatives is essential. Once more stakeholders have awareness of the root causes of poverty conditions and their extent, further policy measures can be taken. It can thus be seen as a first important step to better addressing the problem (Shmelev, 2017). As we have seen, in Bangladesh the Index can lead to pro-poor budgeting, initiatives to fill public service delivery gaps, even empowering the public at large to become more involved in their cities and address neighborhood-level issues through community initiatives. Identifying hotspots of poverty or defining the location of slums might also open the door for development banks or other organizations to progressively upgrade slums, starting from those with the worst conditions first. Thus the future implications for such a tool are promising because it can lead to many subsequent policies and actions. It does so in part by shifting interest away from just pockets of poverty to the scale of reference that is more expansive and inclusive – the scale of the city.

The New Urban Agenda challenges mayors, policy makers, activists, governments and community leaders across the world to '*leave no one behind*' (UN Habitat, 2017). This means finding ways to include the marginalized, and help them be a part of the upward trajectory and growing prosperity of cities. As cities across the developing world continue to grow, the hardships and problems faced by the urban poor are becoming more acute and urgent, underlining how important it is to understand and address their needs. But doing so requires more than infrastructure and technical solutions alone, it requires *leaving no one out* of plans to develop cities, listening to the urban poor and giving them a voice, and ensuring that the services and opportunities needed to reach the lowest rung of the ladder are there so that they can access those opportunities that come with growing and developing cities. The use of data and maps to analyze and solve problems cannot only be for cities of the developed world and wealthier countries; its use can and should be harnessed for the developing world too. Putting tools such as the Urban Poverty Index in the hands of mayors, government officials and poor communities can build their capacity, improve governance by encouraging the poor to collaborate with local governments, and lead to improved conditions and services. It starts with an important first step: to understand who the poor are, recognize them, and include them in building a better city.

References

Alkire, S. et al. (2015) *Multidimensional Poverty Measurement and Analysis*. Oxford: Oxford University Press.

Archer, Diane, Chawanad Luansang, and Supawut Boonmahathanakorn (2012) 'Facilitating community mapping and planning for citywide upgrading: the role of community architects', *Environment and Urbanization*, 24 (1): 115–29.

Arnstein, Sherry R. (1969) 'A ladder of citizen participation', *Journal of the American Institute of Planners*, 35 (4): 216–24.

Boonyabancha, Somsook (2009) 'Land for housing the poor – by the poor: Experiences from the Baan Mankong nationwide slum upgrading programme in Thailand', *Environment and Urbanization*, 21 (2): 309–29.

Dasgupta, Aniruddha, and Victoria A. Beard (2007) 'Community driven development, collective action and elite capture in Indonesia', *Development and Change*, 38 (2): 229–49.

Sanoff, Henry (1990) *Participatory Design: Theory and Techniques*. Raleigh, NC: H. Sanoff.

Shmelev, S. (ed.) (2017) *Green Economy Reader: Lectures in Ecological Economics and Sustainability* (Studies in Ecological Economics). New York and Dortmund: Springer.

Shmelev, S.E. and Shmeleva, I.A. (2018) 'Global urban sustainability assessment: A multidimensional approach', *Sustainable Development*, 26 (6): 904–20.

UN Habitat (2017) *The New Urban Agenda*. http://habitat3.org/wp-content/uploads/NUA-English.pdf

UNDP (2018) *Multidimensional Poverty Index (MPI)*. hdr.undp.org/en/2018-MPI

UNDP Bangladesh (2018), 'Urban poverty profile of Narayanganj City, National Urban Poverty Reduction Program'.

10 Cities and renewable energy

David Elliott

Introduction: the need for change

Urban areas are major energy users, with demand for heating, cooling, lighting and transport currently being met in most countries by the combustion of fossil fuels. That contributes to air pollution problems and to climate change, making life in cities, and elsewhere, harder. The health impacts of air pollution in some newly expanding cities in Asia have brought home the need to change the way in which energy is generated and used, as also have longer-term worries about climate change, not least since that may well lead to flooding of coastal cities as sea levels rise and storm intensities and frequencies increase (IPCC, 2014). The obvious response, apart from remedial adaptation efforts to try to deal with these impacts, is to stop burning so much fossil fuel. In that around 80% of the energy used globally at present comes from these sources, that will not be easy. However, it seems possible that it can be done, over time, chiefly by switching to the use of renewable energy sources and using energy more efficiently (Elliott, 2015). This chapter looks at the options available, and at the part that cities can play in making the transition. Their role is pivotal, as over 70% of global carbon emissions come from energy use in cities (Appleby, 2018).

Urban energy options

Most energy production activities are, at present, carried out at sites outside of cities, although cities, with their high populations, actually consume the bulk of energy used in most countries – around 75% of global primary energy (UN, 2016). Most of this energy is imported into cities. That is clear in terms of vehicle fuel and natural gas, but it is also the case for electricity: although there are sometimes power plants inside cities, most of it is usually made outside of cities. Conventional fossil-fired power plants have usually been large and sited near to fuel access points, for example near coal fields, major ports, or rail access. They also need access to large amounts of cooling water. Remote location, aided by tall-chimney flue-gas dispersal, also reduces some of the local impacts of noxious and acid emissions. New cleaner-burn technologies can reduce some of these emissions further, as can the 'end of pipe' flue gas clean-up approach. However, there is no way to continue to use dirty fossil fuels, such as coal and oil, without producing carbon emissions, and capturing and then storing these emissions in geological strata is likely to be expensive, with uncertain long-term implications: their sudden release is always a risk (Elliott, 2019).

Nuclear plants are also usually sited remotely, in part to get access to cooling water (from lakes, rivers, or the sea), but also to reduce risks to the population from radiation exposure in the case of major plant integrity disasters. In theory, smaller nuclear plants might be sited near or even in cities, but so far this option is undeveloped, and might face strong local opposition on safety grounds, with security issues being an added concern.

By contrast, some of the renewable energy technologies are well suited to cities, the most obvious being solar energy. Solar heat collectors can be installed on rooftops to provide space and water heating, and roof-mounted photovoltaic (PV) solar cells can provide electricity direct to consumers. Most cities have extensive roof space, some of it conveniently orientated to collect solar energy, and installing these emission-free and non-invasive systems is relatively easy and quick.

Clearly, some cities will be better placed geographically than others to use solar energy, although the incident solar energy at the equator is only around twice that in far northern or southern latitudes, and in the latter areas, any solar heat energy available has much higher value and the heating season is longer. Electricity-producing photovoltaic solar cells do not need hot sun to work, just bright sunlight, which can be common, at times, in far northern/southern winters. That said, clearly, cities in the sunbelt zone will have the advantage of overall sunnier weather and clearer skies.

Globally, by the end of 2017, there was over 470 gigawatts (GW) of solar heat collector capacity in place, much of it in China, mostly used for domestic water heating, and also, globally, around 400 GW of PV solar electricity-generating capacity, with perhaps about a third of that being city-based domestic or institutional projects (REN21, 2018). To put that in perspective, there is around 400 GW of nuclear capacity in place globally.

However, although the potential for solar is large, there are spatial limits to how much solar energy can be collected in cities. This may be seen from a simple example. It is possible for well-insulated and designed individual houses to collect most of the energy they need (power and heat) through rooftop solar systems, assuming some storage and/or top-ups from other sources to meet night-time and winter demand. But that is not the case when dwellings are piled on top of each other in high-rise apartment blocks. Then the roof area stays the same, but the number of home (or other) units, and the total energy needed from the roof, is multiplied. It may be possible to compensate for this by having solar energy collectors on other larger roof spaces (e.g. on warehouses), or on suitable walls (e.g. of office buildings), but those non-domestic buildings may need the energy themselves. The use of open areas for ground-mounted solar collectors is another option, but open spaces are precious in urban environments, as are garden areas.

It may be possible to redesign cities so that they present larger areas to the sun for solar heat collection and PV generation. New types of translucent PV material are also being developed which can turn glazed areas into solar energy collectors while still acting as windows. However, the energy conversion efficiency of such solar window systems tends to be low. Architects and designers have produced some ambitious proposals for urban layouts with solar PV canopies covering much of the upper regions, and, although some might find that oppressive and confining, it might be one way cities could go (Zed, 2018). However, for the moment, the surface areas available for PV deployment are limited to rooftops, and, less effectively, walls, with shadows from nearby buildings then being an issue.

Even so, solar can play an important role in cities, as elsewhere, with, in addition to solar heat applications, many domestic consumers across the EU and the US and elsewhere having already installed solar PV units. They are often linked to net-metering or Feed-in-Tariff (FiT) schemes, which allow consumers to export any excess electricity produced to the grid and import power from the grid to meet shortfalls, e.g. at night and in the depths of winter. Some consumers are also installing battery storage systems to allow themselves more flexibility. There are some limits to that. It is expensive to store power for any length of time, even overnight, so as to provide power when there is no solar input (see Box 10.1).

Box 10.1 Battery storage limits

Tesla's 13.5 kWh rated Lithium Ion Powerwall domestic battery unit currently retails in the UK at nearly £6,000, including inverter (Tesla, 2018). Assuming you could charge that fully during the day from a typical 2–5 kW PV array on your roof, it would power a standard 2 kW heater overnight (i.e. in this case, for 6.7 hours). If you wanted more power than that at night (e.g. to charge your electric vehicle), you would need more batteries. Moreover, since on some days there would not be much output from the PV arrays, you would need many more batteries and probably more PV arrays, multiple units, to cover, say, a week of dull skies. Clearly that is unviable for home applications and most domestic users rely on grid imports to give full power cover. More advanced flow battery systems are emerging, but they only really make sense for use by utilities at the larger scale, to provide short-term grid balancing (Elliott, 2017).

However, playing to its strengths, solar PV is well suited to meeting daytime electricity demand in office buildings and retail outlets, with good seasonal matching to summer daytime air-conditioning loads. As climate change impacts more, the latter will grow and meeting this demand with PV avoids adding the carbon emissions that would be associated with using conventional mains electricity, much of which will still be fossil derived, at least for a while. With costs falling, PV seems certain to spread across many applications, and, in addition to the booming domestic sector, it has been taken up for use at commercial and industrial sites, many of them in urban areas (see Box 10.2).

Box 10.2 Large-scale commercial/ industrial solar PV

Commercial and industrial buildings with large roof areas offer an ideal site for PV solar – a roof that earns its keep. Some have been installed on retail company warehouse rooftops, like the 1.2 MW array on German discount store Lidl's logistics centre, and industrial applications are expanding, e.g. General Motors has an 11.8 MW PV array at its plant at Zaragoza in Spain, Toyota's parts centre in Belgium has a 1.8 MW array, Goodyear Dunlop's logistics centre at Philippsburg in Germany has a 7.4 MW, while Volkswagen has a 2.4 MW array at Wolfsburg in Germany. In the US, there is the FedEx Ground Woodbridge distribution hub in New Jersey (2.42 MW) and Google's 1.6 MW array and the Coca-Cola Plant (329 kW) in Los Angeles (Wikipedia, 2019).

PV solar has also found applications in some energy-intensive industrial plants, for example with plans for the Tata Steel plant in Scunthorpe in the north-east of England to be supplied with power from the nearby 38 MW Raventhorpe Solar Park, meeting around 7% p.a. of Tata's power needs. The advent of new, more energy-efficient production technologies for high-temperature industrial processes, such as electric arc furnaces, induction furnaces and dielectric heaters, could make it possible for solar energy to play more of an industrial role, although this will probably still be limited within cities.

Although solar PV is now expanding, wind energy has been the big success story for renewable generally, with nearly 540 GW installed globally by the end of 2017 (REN21, 2018). However almost all of this is in rural, or, to a lesser extent, offshore areas. There have been some attempts to use wind energy systems in cities. Small devices can be mounted on roofs, but typical wind speeds in cities are low (and often erratic/turbulent, due to the interference from tall buildings) and the efficiency of small devices is also low. Wind device output is proportionate to the *square* of the rotor radius, so 1,000 small 1 kW turbines yield much less energy proportionately than a single large 1,000 kW turbine, and the latter can be sited in multiple units in windier up-hill rural areas, that being very advantageous, since with the output is proportionate to the *cube* of the wind speed. It is possible to put larger devices on the top of major buildings, or in open spaces, where wind speeds might be a little higher, but basically, in most cases, apart from 'brown-field', industrial, or dockland sites, wind power generation is not well suited to urban environments (see Box 10.3).

Box 10.3 Urban wind power

Ford Dagenham has installed three large 1.8 MW wind turbines at its diesel engine plant in Dagenham, in the north-east of London, supplied by green energy company Ecotricity, who have also installed three more at

Avonmouth Docks near Bristol (Ecotricity, 2016). Several other large wind projects, serving office and warehouse loads, have been developed, and some urban 'brown-field' wasteland sites may offer more opportunities. However, smaller domestic-scale urban wind projects have usually not fared so well. An early study of micro-wind projects by the UK Building Research Establishment concluded that 'in many urban areas they are unlikely to pay back either their carbon emissions or the home owner's costs for installation and maintenance' (BRE, 2007). It is possible that new designs of more efficient 'building-integrated' micro or mini wind turbines will emerge, and small-scale wind can play a role in some locations, but, in general, unlike PV solar, not in most urban contexts (Elliott and Taylor, 2000).

What cities do have, and produce continually, is large amounts of bio-wastes, mostly the result of human processing of food imported from outside the city. Some of this, along with other material wastes, can be used for energy generation. Sewage gas is one of the cheapest renewable energy sources, and using solid post-consumer wastes (which can have a biological content) for energy production is also arguably a better idea than sending them to landfill sites outside the city. However, the amount of energy that can be produced in this way is limited. The volumes of waste are usually not sufficient to sustain the energy needs of urban dwellers, since, to put it simply, human beings are not efficient converters of food into energy and even less efficient at producing high energy-value waste. Its nutrient value is however another issue: burning bio-waste, and so destroying valuable organic material (useful for farming), may be a very poor idea in ecological terms.

That is not to say that some bio-wastes cannot provide a significant energy input in cities, although there are environmental and locational issues. Sewage treatment plants are not the best of neighbours and tend to be located in urban fringes or outside the city, as are landfill plants, even if some can be used for methane gas collection. Despite tight emission regulations in many countries, waste combustion plants can also be perceived as problematic, with there being fears about toxic emissions, e.g. from burning plastics. Pyrolysis (decomposition brought about by high temperatures) and improved gasification techniques may reduce some of these concerns, but it can be argued that, rather than being burnt, waste should be minimised at source or be recycled, with any useful materials being reclaimed.

Energy recovery from wastes certainly raises some complex issues. Simple mass-burn of unsorted waste has many environmental problems and, arguably, is not suited to cities. However, in addition to pyrolysis and gasification, there are options for less aggressive forms of waste processing and energy recovery. For example, there are some interesting projects for collecting urban food wastes for conversion to biogas via anaerobic digestion (AD). Over 1.6 million tonnes of food are being recycled as energy through AD now in the UK, compared to 0.3–0.4 m tonnes in 2010 – a small but useful resource (DEFRA, 2017). There is also the intriguing idea of supplying a forge in Sheffield with methane from an anaerobic digester fed with food and other waste from a nearby waste recycling centre (REM, 2017).

There are some other renewable options that may be relevant within some cities, including small-scale hydro and geothermal energy. Some cities already make use of local small-scale hydro and geothermal heat from aquifers. Deeper 'enhanced geothermal' projects may also prove viable for electricity production, with some projects being looked at in the UK, for Newcastle and Manchester, but, as with hydro (and wind), the big resource is outside of cities.

The need for imports

The simple message from the brief review of some of the key urban renewable energy options above is that cities cannot meet all their energy needs from urban renewable sources. This can be presented in terms of the relative levels of energy demand and energy generation per unit area: urban densities are just too high for their energy needs to be met from renewable sources in urban areas. See Box 10.4 for some numerical data on energy densities.

Box 10.4 Energy densities

According to a review by the International Energy Agency, the spatial density of energy demand within city urban areas typically ranges between 10 and 100 W/m^2 of land area. The IEA says that the energy demand density 'is influenced by population density and average income such that a wealthy suburb of large detached houses with gardens may have a similar energy density (around 10–50 W/m^2) to a poorer suburb of apartment buildings that accommodate more people per km^2 but who use less energy for heating and cooling and own fewer appliances. High-rise building areas might reach energy densities between 500 and 1 000 W/m^2, industry between 200 and 700 W/m^2, with energy-intensive industries such as steel mills and refineries being above 1 000 W/m^2.'

By contrast, the IEA puts the energy density range for solar at 10–30 W/m^2, depending on location and assuming uninterrupted deployment in open land. The energy densities for wind and biomass are put even lower (0.5 and 1 W/m^2 respectively), and the IEA concludes that, overall, 'the high energy densities of some urban areas would require large areas of "catchment" land outside of the city boundary if renewable energy is to be used to increasingly supply the energy needs within the boundaries of a large city' (IEA, 2009).

That conclusion is strengthened if the energy needed for transport is included. In theory, some transport requirements in cities are less than in rural areas, given the shorter distances between likely travel nodes, and many of these transport requirements can be met by energy-efficient mass-transit systems. However, given the large, usually very mobile, city populations, and increasing urban sprawl, the energy needed for transport in cities is not insignificant (it can reach

30% or more of total energy use in some cities), with most of it at present being based on imported fossil fuels or electricity.

As with other types of energy use, more efficient, less energy-intensive technologies may be introduced (electric taxis, biogas-powered trucks and vans) along with improved public transport. Policies supporting walking and cycling can also help to avoid fuel use. Certainly energy-efficiency gains are possible in all sectors, in buildings especially, via better insulation and improved design. However, there are limits to how much energy can be saved by technical improvements, as well as limits to how much behavioural change might be expected. The EU aims to improve energy efficiency overall by 32.5% by 2030, and France and Germany both have targets of reducing their total national energy use by 50% by 2050. These are ambitious targets, with reduced urban energy use clearly being an important element. Nevertheless, even if these energy-saving efforts are successful, there will still be a need for energy, and cities seem likely to remain relatively high energy users. They can and should do as much as possible to avoid energy waste, and to supply energy from low carbon sources. However, as should be clear from the analysis above, they will have to continue to import energy.

The need for urban energy imports is heightened by the fact that renewables such as wind and solar are variable. Even with battery back-up, as noted earlier (see Box 10.1), it would be very hard for solar projects in a city to meet urban energy needs reliably around the clock and throughout the year from local energy inputs, especially in the case of larger commercial and industrial energy users. There can be long lulls in renewable supply. So there will be a need for extra energy inputs to balance the variations in solar (and wind) availability.

While local batteries can provide short-term back-up (for a few hours) for power supplies, larger stores will be needed for longer-term balancing (days and occasionally weeks) and most of these would have to be sited outside cities. There may be some exceptions (see Box 10.5), but, in general, long-term bulk storage is space consuming and most of the systems available, such pumped hydro reservoirs and underground cavern compressed-air stores or hydrogen gas stores, by their nature cannot be in cities.

Box 10.5 Biogas storage: the return of the gasometer?

There is an urban option for storage of biogas, as was done in the past in the UK for coal-derived 'town gas', in giant 'gasometers'. They acted as buffer stores to balance out variations in demand from consumers and supply from urban gas-generation plants (the so-called 'gas works'). That approach might be reinstated, with biogas produced from urban bio-waste sources and used for heating or perhaps even for power generation. However, as was noted earlier, the urban bio-waste resource is relatively limited, so biogas would have to be imported from outside. In which case the value of storing it inside cities would be reduced.

Storage is not the only, or necessarily the best, option for balancing variable renewables. It is expensive. It is usually easier and cheaper to generate more power to meet any lulls. Small back-up generators, perhaps using stored biogas, are an option for the short term and can be located in cities, but there are limits to that (see Box 10.5). A more radical approach is to widen the energy collection footprint so as to make use of better, more geographically dispersed locations. Renewable inputs vary significantly by location, so linking up to resources elsewhere, using extended power grid networks, can help with balancing. Moreover, the wider the geographical spread, the better the balancing since it will encompass areas with different weather patterns, with links to other countries making it even better.

So we have a geographic issue here, and some interesting trade-off choices. Transmitting power over any distance involves energy losses, so, on that basis, you would want the external sources to be as near as possible to the users in cities. However, the sources are mostly *not* near cities, e.g. for wind they are in hilly areas or out to sea, while hydro is in usually remote mountainous areas. Moreover, as just noted, sources *far* away may actually be *better* for balancing. Fortunately, long-distance transmission using High Voltage Direct Current (HVDC) supergrids involves much less energy loss than would be the case with the use of conventional AC power grids. They can lose up to 10% of the energy per 1000 km. With HVDC, by contrast, it is only around 2%/1000 km.

The upshot of this is that cities that happen to have good renewable resources reasonably near to them may benefit economically to some extent, but the benefit of balancing by using power imported from far away may offset that advantage. Clearly, in practice, it will depend on the location. In some cases, there will be large sources nearby. For example, some coastal cities can have large offshore wind farms near them (although 'near' may mean 20–30 miles away). Other cities may be able to rely on large rural wind farms and solar farms nearby, but some will also have to make use of power from large offshore wind, wave and tidal projects, most of which are likely to be much further away. Similarly for hydro and geothermal projects, all usually being remote from cities, with some of the balancing power maybe having to come via imports from even further away, including from overseas.

So the simple message is that, whereas in the past some cities have had large power plants inside them, or nearby, dedicated to supplying energy to them, this will be far less viable with renewables. Cities will be partly reliant on imports of electricity via long-distance power grids, as well as possibly imports of energy in the form of biomass, or more likely biogas, from local rural sources, but also possibly from abroad. Of course, cities always were reliant on fuel imports for power plant inside or near them, but now the type of imports will change.

These new arrangements and requirements will have a range of implications, locally, nationally and internationally, as will the overall shift to using renewable energy. A report from the International Renewable Energy Agency 'A New World – The Geopolitics of the Energy Transformation', noted that, as IRENA's director general put it, the renewable transition from fossil fuels was 'a move away from the politics of scarcity and conflict to abundance and peace with new opportunities for many countries' (IRENA, 2019). Certainly, the geographic concentration of fossil fuels in just a few countries has had a significant impact on the wealth and security of many nations. An energy transformation driven by renewables could bring radical changes, and possibly enable the democratisation of access to energy sources. However, that will not be automatic. There may also be conflicts over access to renewable resources, since they too are not distributed equally around the world, with some areas and some cities being better placed than others.

Local energy, global energy and scale issues

While cities can produce significant amounts of energy from within their own boundaries, as we have seen, most are also likely to need to draw energy from external renewable energy resources including, crucially, for balancing. Fortunately, the overall global renewable resource is very large, enough to supply all of humanity's energy needs if the conversion and distribution technologies can be developed fully (Elliott, 2013).

However, as already indicated, not all of this is located conveniently for cities.

In most places around the world, the renewable resource base is large, but its extent varies by location. Generally speaking, countries in and around the equatorial 'sun belt' have better solar resources than those in the north and south (World Bank, 2016), but the latter have better wind regimes (DTU, 2018). However, in practice that pattern is modified by local features and conditions: wind speeds are higher in elevated areas and near (and off) the coasts, and obviously cloud cover can reduce local ground-level solar intensity. Some of the other sources are defined by unique local geological and geographical features: mountains and rivers for hydro, geothermally active sites for geothermal plants. Large-scale concentrated solar thermal power plants, with mirror or lens focusing, must be located in areas with direct non-diffuse sunlight, such as deserts. That is also true of large-scale focused-solar PV projects. Offshore wind, wave and tidal stream projects, obviously enough, are only possible at sea or in large estuaries (IRENA, 2018).

Many countries are now promoting significant expansion of renewable energy use at the national and local level (Elliott, 2013; Elliott, 2015). Box 10.6 provides a snapshot of some of the key national and regional targets. Cities in these countries and areas are likely to be well placed to be able to expand their use of renewables, since they can draw on these external resources. That will give them an incentive to support ambitious national renewable energy programmes and possibly also to invest in these projects, as well as developing their own.

Box 10.6 Global renewable energy targets: some national ambitions

The European Union, covering 28 countries (as of July 2019), aims to derive 20% of all of its energy (not just electricity) from renewables by 2020 and 32% by 2030. Some member states have already achieved that or more, with, for example, in 2016, Austria obtaining 34% of its energy from renewables, Latvia 37%, Finland 39%, and Sweden 54%, while hydro-rich Norway was at 70% and Iceland 73% (Eurostat, 2018). Longer-term ambitions include Germany's target of getting 80% of its electricity from renewables, mostly wind and solar, by 2050, and Denmark's plan to be totally carbon free by 2050.

Renewables have been expanding very rapidly in China – wind especially – with around 300 GW of wind capacity installed. Although coal still dominates, the aim is to expand non-fossil fuel energy so that it supplies 30% of electricity by 2030. The US obtains around 15% of its electricity from renewables at present. In his 2011 State of the Union speech, President Obama called for 80% of US electricity to come from 'clean' sources by 2035, though that included nuclear and possibly gas and so-called 'clean coal', options which Donald Trump is now much more likely to support and indeed has. Even so, renewables are doing well in the US, mainly since they are now economically attractive.

Renewables are also doing well in many newly developing countries. For example, Argentina already obtains over 40% of its power from renewables, Peru 48%, Venezuela 53%, and El Salvador over 54%, all benefiting from large hydro inputs, as do some countries in Africa, such as Angola and Zambia. Indeed, the availability of large hydro inputs means that over sixty countries globally already obtain the bulk of their electricity from a renewable source, including nearly all the electricity in Norway, most of it in Iceland, and up to 60% in Austria, Canada, New Zealand and Sweden. The further expansion of large hydro may be limited by environmental concerns, but small- to medium-scale hydro can be less invasive, and wind and solar are expanding very rapidly around the world. Geothermal and biomass sources are also likely to expand. Geothermal energy is already used to meet heat and power needs of several cities near suitable resources, notably in Iceland. There may be land-use and environmental constraints on some types of biomass expansion, but, as we have seen, there are significant agricultural bio-waste options.

The current global renewable electricity figure is around 25% (REN21, 2018), with many scenarios suggesting that, as renewable costs continue to fall and expansion accelerates, it could reach at least 50% by 2050 and possibly much more. There are scenarios looking to 80–90% or more of global electricity coming from renewables by 2050, as well as high contributions to heat and transport needs (Elliott, 2015; LUT/EWG, 2017; Jacobson et al., 2017).

While cities will have to remain parasitic on energy supplied from outside their borders to some extent, that is not to say there are not major options for supplying and using green energy efficiently within cities, and improved technology may increase the percentage. For example, more efficient solar PV cells are emerging, so that more energy can be obtained from smaller areas of cell. However, for the type of non-focusing systems likely to be used in cities (as opposed to large utility projects in desert areas), we are talking about an energy conversion efficiency increase from around 10–15% to perhaps 20%, depending on location. Incremental technical improvements like that, although welcome, will not change the basic spatial constraints within cities significantly. So attention needs to be paid to using technologies like this in the best way, so as to get the most effective utilisation. For example, hybrid PV and solar thermal heat-collector systems are emerging ('PVT'), which open up some possible new lines of optimal deployment, avoiding conflicts for roof space between PV solar and solar heat collectors.

There are some other optimal deployment issues. For example, there is the issue of scale. In the urban context, with high population densities, it makes sense to have *collective* energy provision. Rather than having solar heat collectors (or PVT cells) on individual rooftops and installing heat storage facilities in each house, it is more efficient to have large shared solar rooftop arrays and large shared community heat stores. These can average out the variations in heat use across a large number of houses. Large heat stores also have lower energy losses than small stores, since the surface-area-to-volume ratio is lower.

Solar inputs are only one option for heat. Community heating networks, supplied with heat from other low carbon sources, such as biogas and wastes, can also play a role and make sense in urban areas where there is high residential density. Other community-scaled low-carbon options include using large heat pumps to extract ambient heat energy for use in district heating networks, and the use of waste heat from conventional power plants, operated in combined heat and power (CHP) mode, feeding into district heating networks (see Box 10.7).

Box 10.7 District heating

District heating networks provide a way to supply heat efficiently to consumers and businesses in cities from a variety of sources, including renewable sources, and are increasingly seen as a key option for sustainable urban heat supply (UNEP, 2015). They are already in quite wide-scale use. For example, around 60% of Denmark's domestic and commercial heat is supplied by district heating networks. Some use heat from fossil fuel-fired plants, some of them run in combined heat and power mode, with some now fired by biomass, but by 2050 Denmark wants 40% of this district heating load to be solar fed, including projects with *inter-seasonal* heat stores, allowing summer heat to be used in winter. Some already exist (SDH, 2018;

(continued)

(continued)

Planenergie, 2018). District heating networks can also be fed by large heat pumps using ambient energy sources. About 60% of the total energy input for Stockholm's Central Heat Network is provided by six 180 MW heat pumps, total heat supply capacity 420 MW (thermal), drawing heat from the sea. Helsinki in Finland has a 90 MW heat pump plant, feeding its district heating network. That supplies over 93% of Helsinki's heat, drawing heat from waste water outflows and the sea (JRC, 2012).

A review by the UK Energy Technologies Institute noted that 'in European cities, heat networks are a major way of providing heat into buildings. For example, in Germany every town with a population of over 80,000 has at least one heat network.' In the UK, only around 2% of heating was provided by heat networks, but the ETI estimated that 'nearly half of heat demand could be met by heat networks' (ETI, 2018).

The discussion so far has rather arbitrarily made a simple distinction between urban and rural areas, whereas in fact there are gradations. District heating may be viable in some suburban areas and in medium-sized towns, with the power plant perhaps being on the outskirts, using biomass collected from the hinterland. Some towns, or indeed cities, may also have wind and solar farm projects, or biogas waste-digestion plants, on their rural borders. That is the case for some of the large number of community-based green energy projects that have emerged in towns and villages in Germany in recent years.

There may also be possibilities for local co-ownership by city dwellers of larger projects further away, including solar farms and wind farms. That offers a possible solution to what could be a growing problem of urban and rural conflicts, as cities import more power from rural areas.

Rural and urban conflicts

The fact that cities will need to import power should not be surprising. Cities have always had to rely on rural areas for food and also, in many cases, for water, as well as for energy. That will not change. In fact, a switch to sustainable energy may require cities to be *more* reliant on rural areas, for example to house large-scale wind and solar farms, possibly along with biofuel and biomass plantations. This may of course lead to conflicts. Rural communities may resent the imposition of what they may see as intrusive projects which, given that there are generally many more people in cities than in rural areas, will mainly benefit urban dwellers. There has certainly been local opposition to some wind farms in the UK, as well as to some solar farms. If cities need to import more energy from rural areas, problems like this may increase, unless steps are taken to limit potential conflicts of interest.

Rural communities have always serviced cities, most obviously by producing food, and earned a living from doing so. When that has not been the case, there have sometimes been significant conflicts, as for example when rural valleys are inundated to create reservoirs primarily to provide water for cities to use, or major hydro projects for large-scale energy production, with power fed to the national grid. There may be compensation offered for loss of farm land, or for having to move home, but there have been occasions when rural communities have objected strongly. The most obvious examples are in developing countries, where some large hydro projects have been opposed strongly by local communities, who stood to get few of the benefits but incur most of the impacts. Battles like this are still occurring around the world. Perhaps less well known, there have also been conflicts over water reservoir projects, including, historically, in the UK. For example, in 1956, residents from a Welsh village, threatened with inundation to make a reservoir to supply water to Liverpool, marched through the streets of Liverpool in protest. A transformer on the site was firebombed.

Conflicts like that have yet to occur in relation to UK wind and solar farms, although in the case of wind farms, resentments are sometimes expressed about 'greedy outsiders making money by despoiling the landscape', or, more prosaically, about undermining local house prices (Elliott, 2003). It has been similar for solar farms in recent years. Perversely, what we may also see is opposition to these rural projects by city dwellers, who may want the countryside to remain 'unspoilt', as a recreational resource, and, in some cases, to protect views from their second homes. More generally, rural communities may object to the way city dwellers use and perceive the countryside. Clearly, there may be some conflicts between urban and rural value systems and economic interests.

As reliance on renewables expands, these conflicts will have to be dealt with. Girardet has talked of the need for a 'regenerative city' approach, seeking to address not just urban renewal on sustainable lines, but also improvements in 'the relationship between cities and their hinterland, and beyond that with the more distant territories that supply them with water, food, timber and other vital resources' (Girardet, 2010). That should clearly include energy resources.

Local ownership and local power

As mentioned earlier, one approach to dealing with this issue might be through increased local ownership. It has certainly helped to avoid opposition at the local level to wind farms, as is evident from Denmark, where most wind projects are locally owned and usually welcomed, and indeed sought after. As the Danes say, 'Your own pigs don't smell.' Local ownership provides an economic incentive and the opportunity for direct local control.

There are of course a range of factors shaping how easy it is to move to local ownership: including the availability of suitable support schemes and local orientations (Willis, 2016). Local ownership certainly opens up a wide range of technical, social and political issues (Koirala et al., 2016) as well as opportunities and constraints (Brummer, 2018).

However, what matters here is whether this idea can be spread to cities and cover a wider range of projects, including larger ones further away, allowing urban populations to get involved, and bridge the urban–rural divide. Although it is hard to see how that could work in the case of large-scale hydro projects, smaller projects may be less of a problem. City dwellers might buy shares in rural wind farm projects and even in offshore projects, as has happened with the large, partly municipally owned wind farm just off the coast from Copenhagen. The same might be possible for solar farms in rural areas. Although care will have to be taken to avoid exploitative relationships, economic and social links can thus perhaps be established, even if the technology is outside the city and its power is imported.

However, there are limits. Solar farms obviously need to be in open rural areas, which are usually well away from cities. Some offshore wind farms are now being sited 30, 40 or even more miles offshore (Hornsea 3 is over a hundred miles west of Hull in Humberside), with floating deep sea units enabling projects to be sited even further out.

That may stretch any sense of 'local' to the limit. Indeed, as already noted, by their nature, most large renewable energy resources and balancing projects will be remote from cities – some possibly even in other countries. While in technical terms, that may not matter, since power from them can be transmitted long distances with low-energy losses via HVDC supergrids, clearly it would be easier to build a sense of connection with projects near cities.

There may be some opportunities for that. For example, although the most productive onshore wind farms tend to be in hilly areas remote from cities, some offshore wind farms can be nearer to coastal cities. In the UK context, for example, there is one project off the south-east coast near Brighton. Somewhat further out there are sites off the north-west coast from Liverpool/Merseyside and off the east coast of Scotland, about fifty miles or so from Edinburgh. The US Eastern Seaboard offers similar possible tie-ups for New York, Boston, and so on. There is also a good resource off the west coast, so Los Angeles, San Francisco and San Diego could benefit, and similarly for some of the large coastal cities in Japan and China.

So, although these projects may still be some distance away, 20–30 miles or more in some cases, the idea of interlocking patterns of co-ownership of projects, possibly by some type of municipal arrangement, may not be a non-starter. If nothing else, it might be possible for cities to, as it were, 'adopt' an offshore wind farm, even if it was some way away.

Local ownership and control of energy projects does, however, imply a more intimate involvement, and in that regard, solar inputs are much more available in many places and solar technology is well suited to direct local ownership. There are many consumer-owned projects now in existence, many of them in urban and suburban areas. However not everyone can participate in these. As noted earlier, residents in apartment blocks will not all have access to a roof area, so they cannot all actually have PV arrays themselves. Nevertheless, collective ownership of projects nearby on suitable surfaces is an option. A 2015 US National Renewable Energy Laboratory (NREL) report said that 49% of US households and 48% of businesses were at that point unable to 'host a PV system of adequate size or virtually net meter an entire system themselves'. In which case, one way in which

PV might expand could be by widening access, by sharing the output from PV on rooftops of multi-dwelling buildings, or in nearby land supplied to buildings where it was hard to fit PV. The NREL said

> off-site shared solar and arrays on multi-unit buildings can enable rapid, widespread market growth by increasing access to renewables on readily available sites, potentially lowering costs via economies of scale, pooling customer demand, and fostering business model and technical innovations. Shared solar could represent 32–49% of the distributed PV market in 2020, growing cumulative PV deployment in 2015–2020 by 5.5–11.0 GW.
>
> (Feldman et al., 2015)

It is some way off from collective/shared ownership like this, within cities, bridging the social and economic gap between those in flats and those in houses, to bridging urban–rural divides, but the principle might be extended, with those in areas with few renewable resources participating in projects in areas where there are more. The rural–urban divide may in fact become blurred, as projects spread across cities, suburban areas and rural areas. It is interesting that nearly 40% of German renewable capacity is now locally owned, some by household domestic PV 'prosumers', many of whom are in urban and suburban areas, but some also by energy co-ops in towns and rural areas. The latter have followed on from rural wind co-ops that had spread across Denmark, with community-based green energy co-ops now spreading across Germany, using wind, solar and/or biomass.

While most of these community projects are in rural areas, in towns and villages, there are some within the urban context. At present they, and private prosumers, usually feed excess power to the grid, after meeting their own needs. But there are moves to take control of local distribution as well as generation, in urban areas especially, in some cases via municipal schemes. Interestingly, there are some examples of surplus local PV solar generation in parts of some German cites, being exported to the rest of the city (Mittal, 2016). It might be possible to spread this idea, and community-controlled distribution networks as well, much wider, to also engage with people in rural areas.

Clearly, the energy system is changing. Local self-generation initiatives are changing the nature of the energy market in Germany and elsewhere, challenging the dominance of the conventional energy utilities, generators and distributors. Some have seen this as a major social and political challenge to the status quo (Debor, 2014), although it is still relatively fragile, and for example, may be constrained by changes in the support system for renewables and the demise of the feed-in-tariff schemes. Nevertheless, even given the new more market-led approach, with new ideas emerging about local generation and the use of local short-term battery storage, there are some interesting new systems being developed for peer-to-peer power trading and managing the new smart grid systems, which may help community-level initiatives (Sonnen, 2018).

The longer-term future for 'local power' in Germany, as elsewhere, is unclear, but a report from Greenpeace, looking to 2050, argues that perhaps 70% of energy should be generated from local decentralised projects, with only 30% being imported from larger centralised, presumably more remote schemes, so as to enhance local control and avoid extra grid links (Greenpeace, 2015). That degree of local autarchy may have social and environmental attractions, but, depending on exactly how 'local' is defined, it would make balancing locally variable supplies and demand harder.

The Greenpeace study is essentially concerned with possible overall global patterns running up to 2050, not with specific national or regional situations. They point out that their suggested 70%/30% split 'is not based on a detailed technical or economic optimization; in each location, the optimum mix is specific to local conditions. Further detailed studies on regional levels will be needed to better quantify the split between distributed and large-scale renewable generation better'. Nevertheless, the basic principle of choosing an appropriate balance between small/local and large/centralised does have some relevance, including to the urban–rural split and to opportunities for local ownership and control (see Box 10.8).

Box 10.8 The Greenpeace large/small 70%/30% split

Greenpeace obviously favours local-level generation. However, the organisation says that 'while a large proportion of global energy in 2050 will be produced by decentralised energy sources, large-scale renewable energy will still be needed for an energy revolution', pointing to the higher efficiency of large projects like offshore wind farms and even 'large concentrating solar power (CSP) plants in the Sunbelt regions of the world'. So that means long-distance grid links.

It admits that opting mainly for smaller-scale local systems would come at a price, since it would mean installing more local generation capacity than would otherwise be needed, so as to cope with local renewable supply lulls: 'Making local plants bigger (over-sized) is less economical than installing large-scale renewable energy plants at a regional scale and integrating them into the power system via extended transmission lines'. Nevertheless, a little optimistically, Greenpeace says,

> in principal, over-sizing local generation locally would reduce the need for large-scale renewable generation elsewhere as well as [for] upgrading the transmission network. In this case the local power system will evolve into a hybrid system that can operate without any outside support.
>
> (Greenpeace, 2015)

That may be possible for some locations, if total independence is a priority, but does not seem likely to be possible for many cities.

It is hard to avoid the conclusion that some longer-distance imports will be needed from a range of more remote sources, including perhaps from other countries, via supergrid networks (Elliott, 2016). That could help balance local variations in supply and demand across wider areas, linking to areas with better, more reliable resources. So a 70%/30% small local/large remote split may not be suited in every case and certainly it seems unlikely to be possible for most cities to generate 70% of their power, even given strenuous efforts to reduce demand. A 50%/50% internal/external source split may be a good challenging target in most cases.

Whether that would make it easier to avoid urban–rural conflicts is unclear. It rather depends on how 'local' is defined. In the fully 70%/30% Greenpeace model, presumably there would be many local projects in rural areas, but also in cities, and little interaction between them, although as Greenpeace admits, they would need larger amounts of local generation and back-up capacity. In a 50%/50% split, there would be much more traded power, with an increased potential for conflicts, since more of this would be coming from rural areas to cities.

However, this level of rather abstract analysis may be of little value. In reality, there is likely to be a mixed pattern, with locally owned community projects in some rural areas trading with other such communities and projects, but also with cities, and possibly with community energy projects within them. In addition, and more likely to generate conflicts, there may be some larger commercial projects in rural areas feeding power to all of them, on a national or regional basis, possibly also with international power grid links. In which case, local ownership would only be part of the pattern.

Attitudinal change and the energy transition

The prescription of local cooperative/community ownership, as at least part of an equitable strategy for developing sustainable energy supplies for cities and more generally, assumes there would be demand for this form of involvement. That may not be the case. Most people may be happy just to buy whatever power is offered. Certainly, few people, wherever they lived, have in the past shown much interest in where their energy came from. Coal mines and power stations are not exactly tourist venues and for most people what mattered was the cost of what came out of the socket in the wall or the pipe to their house. Issues relating to the cost, and health and environmental impacts of power generation have of course led to political pressures for better regulation and in the past to nationalisation – an idea that may yet resurface. Certainly the UK Labour Party wants the energy sector to be taken back into public ownership, although via a range of ownership formats, including municipal enterprises and local community co-operatives (Corbyn, 2016).

Arguably, something like that could support sustainable approaches to energy generation. Certainly, as rural renewables begin to take the strain in supplying cities with energy, a new relationship between urban energy users and rural communities may have to develop and local co-operative ownership might be part of the way forward. If the proportion of the population living in urban areas continues to grow, as seems inevitable, maintaining good rural–urban relationships may become increasingly hard, but also more necessary. Being optimistic, that may not need major attitudinal changes. Initiatives related to ethical and sustainable organic food sourcing suggest that at least some city dwellers can develop an interest in what they consume and where it comes from. It is true that energy is more intangible, but wind farms do draw many visitors and it is conceivable that new forms of social interaction and even empathy could emerge.

All that said, for good or ill, more likely is a change in the economic relationship between rural and urban areas and populations. For the moment, that will only effect a limited number of people in rural areas. Only a small minority of people living in rural areas now work on or are connected with farms, but the expansion of wind and solar farms, and possibly biomass plantations, will perhaps involve more, and will certainly impact on local rural residents more, even if they are just commuters.

However, if and when more people in rural locations are able to take over more control of their area's renewable resources, via local co-ops and the like, the economic base could change. It is already the case that rural areas in the UK, and elsewhere in the EU, are changing, in part due to the worsening economic returns from farming for food production. That may be lamentable, but certainly farmers increasingly look to renewable energy projects as an alternative or supplement. If other members of the rural community also get involved, in the same way as may happen with some local energy project in cities, that would open up a range of interesting new social and economic issues and potential interactions.

Clearly, the expansion of the use of renewable energy in cities, although possible if helped by imports from rural areas, should not be at the expense of rural communities, but it may be that social and structural changes, new patterns of ownership and new attitudes can avoid conflicts. Certainly, as more cities embark on ambitious programmes, with high targets for renewable energy use, these issues will have to be addressed and the necessary changes supported.

Progress on the way ahead

Having ranged around the issue quite widely, it is now useful to take stock and look at what has been happening on the ground. Around the world, cities are beginning to think in terms of meeting their energy needs from renewable sources, so as to limit air pollution and climate change problems. The case for this transition is strong and there have been many interesting initiatives launched around the world, often led by city governments (IEA, 2009). In addition to many early initiatives across the EU (including in Freiburg, Germany and Woking, England), more recent high-profile examples include projects in China (Tianjin Eco-City) and the UAE (Masdar City), and initiatives in the US, where cities such as San Diego and San Francisco, and many others, have committed to 100% clean energy (Van Horn, 2015). Many more cities have since announced similar plans. Over a hundred cities claimed to be obtaining at least 70% of their electricity from renewable sources in 2017, compared to 42 in 2015 (Hunt, 2018).

For the future, there is a wide ongoing debate on the possibilities for greening cities, ranging from the pragmatic to the utopian (Girardet, 2015; Riley, 2017; Inhabitat, 2017; SES, 2018), with some fascinating proposals and inspiring campaigns. However, rather than engaging in further speculation or polemic, it is perhaps helpful to look at what has been done so far in terms of renewables, both internally and imported.

A 2015 Carbon Disclosure Protocol (CDP) report noted that, by then, globally, 308 cities had participated in a CDP programme aiming to better manage their climate change strategies, with 109 cities setting renewable energy or electricity targets (CDP, 2015a). Some are shown in Box 10.9. CDP said that 35% of the 162 cities that had reported details were getting 75% of their electricity from non-fossil sources, with, in many areas, hydro being a major supplier, producing over 72% of the mains power used in 45 Latin American cities (CDP, 2015b).

Box 10.9 Some city renewable energy targets (CDP, 2015a)

Aspen	100%
Santa Monica	100%
San Francisco	100%
Stockholm	100%
Vancouver	93%
Canberra	90%
Austin	55%
Adelaide	50%
Los Angeles	33%
Sydney	30%

The implication of that is that this energy was not generated actually in the city, but was being imported from outside. That is also likely to be true of the smaller wind input, although that may not be the case for the solar input, and that part has been growing. CDP's 2018 update notes that the number of cities then reporting had risen to over 570 and that 'some 184 cities now have solar energy in their electricity mix, while 189 report that they source wind energy' (Appleby, 2018).

Interestingly, the CDP data also shows that, while the so-called 'new renewables', such as wind and solar, are beginning to make their mark, in a minority of cases, nuclear electricity was also used. Nuclear is not a renewable source, and, at least at present, the technology is not available for power generation within cities.

It is also controversial, with, quite apart from the safety and security risks, the cost of nuclear being high and rising, while those for most renewables are falling (WNISR, 2018).

Leaving that issue aside, a review of a programme involving 13 cities in the EU Energy Cities initiative produced some indication of what was being actually produced *inside* some cities. It found that, by 2013, the cities involved in its programme could meet 44% of their energy needs from locally produced energy. However, much of that still involved the use of fossil fuels, albeit more efficiently, in local projects. The *renewable* proportion (4.38 TWh) was only around 16% of the local generation total (26.77 TWh), or about 7% of total energy use (Energy Cities, 2016a). It should hopefully be possible to increase the 'local generation' total (some of which can be low carbon), and also, crucially, the renewable proportion achieved by initiatives like this: the cities looked at had, reportedly, actually tripled their renewables share since 2010. Moreover, with low-cost PV, it should now be possible to do much better than 7%, and even better later on.

An updating Energy Cities/CLER/RAC report did indeed suggests that, by 2050, Frankfurt could be producing 25% of its energy from *internal* renewable sources and could be getting 25% from renewables in the wider metropolitan area, while Malmö in Sweden was aiming for 50% from inside the city, both of these targets being backed up by a 50% cut in energy use. Even more dramatically, Frederikshavn in Denmark was aiming for 100% by 2050 from the greater Frederikshavn region (Energy Cities, 2016b).

These are ambitious, possibly optimistic, targets, with in most cases, some power still being imported. The report notes that, for cities aiming for this sort of goal:

> in most cases, a city or metropolitan area, due to its density, will not be able to cover its energy needs with 100% renewable energy produced on its territory, even after markedly reducing them. It will therefore have to outsource wood, electricity and biogas, as it already does for food.

In which case, the cities will need to build good relationships with their hinterland. The report warns that

> for rural communities, the growing appetite of cities for their resources may be perceived as a threat. Rural areas very often do not meet their own needs with locally produced energy; so how could they consider supplying other territories? And rural or suburban territories which have had to put up with city annoyances in the past (installation of an incinerator, landfill site or the spreading of sewage sludge) are unwilling to see their environment spoilt to satisfy the needs of city dwellers.

However, optimistically, it says

> here lies an opportunity to create a win-win relationship between urban and rural land area. Cities have an interest in providing financial or technical

> support for the development of renewable energy production in neighbouring rural areas, thereby cementing local economic development and resilience. Likewise, rural communities, some of which are already engaged in a positive energy community approach, may also develop relations with cities, offering them access to their resources whilst keeping control of the projects and benefiting from locally generated revenue.

These seem very reasonable as aims and aspirations, seeking out shared interests, but as I have indicated above, the devil will be in the detail. There will be many issues to resolve, not least the terms of trade, so as to maintain stable and reliable energy supplies to cities and cordial co-operative relationships – what the Energy Cities report calls a new social solidarity, rather than an exploitative relationship. It will be interesting to see how that goes, as more cities join in the drive to get most of their energy from renewables.

Elsewhere in this book, mention is made of some of them, and it is relatively straightforward to see which renewable sources are likely to be the most relevant for nearby or imported power production in each case. Many of the cities looked at already have high hydro inputs, for example São Paolo (100%), Montreal (95%), Bogota (84%), Vancouver (81%) and Zurich (66%). For wind, Copenhagen already gets 47% of its power from wind, Adelaide (31%), Austin (23%), Denver (19%), Madrid (17%), Casablanca (12%) and Toronto (11%) (Environment Europe, 2019).

As noted above, offshore wind, already used by Copenhagen, could also supply many other coastal cities, including in China and Japan, as well as on the east (and possibly west) coasts of the US. Wave and tidal stream projects could also have a local input in some of these locations. Many cities can also make use of solar, directly and imported, and some already do, notably Los Angeles (20%). Certainly the US solar potential in many locations is large (Solargis, 2018). Solar (and CSP) is also highly relevant for much of Africa (Solargis, 2017), including for some of the large cities in the Global South (Wu et al., 2017).

As has been argued above, having good renewable resources nearby will give cities a significant advantage over those less well geographically favoured, although, since renewable supplies vary, the cities in favoured locations, as well as the others, will also have to get access to balancing inputs. Some may be able to rely on local storage for short-term balancing, but, when there are long lulls in local wind and solar inputs, most will also have to import power from other regions where there are inputs available at that point in time. So the local resource is only part of the picture. Fortunately, however, the global resource is very large and varied, and trading over wide areas can help balance local variations. That is why the development of long-distance supergrid links is likely to become increasingly important.

Some of this will just involve linking up regional and sub-national grids, but national grids are also increasingly being linked. Moreover, there are studies of much wider international links, for example across Asia (Gobitec, 2015; Movellan, 2016), between North and South America (Aghahosseini et al.,

2019), and even between Asia and Europe (Morgan, 2018) and Europe and North America (Purvins et al., 2018).

The main focus of this chapter has been on local generation and that should be able to provide most of what is needed most of the time, depending on how 'local' is interpreted, with cities importing power from nearby, but they may also have to import power from further afield. As has been argued, there will be social and political implications for importing power from rural areas, but there will also be wider geopolitical implications from importing power from further away. Decentralisation clearly is valuable and much can be done on that basis, but the switch to renewables cannot escape the need for some larger-scale grid-linked systems.

There are of course some possible wider (and wilder) extremes. For example, on one hand, there are those who look to a fully decentralised, simpler, low-growth, more frugal, mainly rural future, even possibly to the complete demise of cities as ecologically unviable, with all energy being locally derived (Trainer, 1995, 2010). On the other hand, perhaps less palatably, some self-styled 'Eco-modernists' look to vast hyper-centralised high-tech cities, powered mostly by nuclear fission, or more likely fusion, with GM food grown hydroponically and the natural environment (outside the city) left to revert to the wild (Asafu-Adjaye et al., 2015). This chapter has tried to map out the implications of adopting a middle route, although one that is nearer to the first than the second extreme.

Conclusions

As this chapter has explained, cities can get some of their energy from renewable sources within their boundaries, e.g. from rooftops and waste. Electricity can be generated from PV solar and heat from solar collectors, while heat and power can be obtained from bio-wastes, and in some cases geothermal sources within cities. However, it seems unlikely that many cities can generate *all* the energy they need themselves. Cities can and should seek to supply as much of the energy they need as possible from local urban sources, including wastes. But as with water and food production, many urban areas will still be dependent on rural areas, and in some case marine locations, for some of their energy. They will also at times have to import some power from further afield.

This does not have to be seen as a failure. While self-sufficiency is a good goal, it can become a dogma if applied too single-mindedly to any specific context. Indeed, in some cases, importing energy may be a more efficient way to use renewable energy. Certainly, as we have seen, there are limits to how much energy can be produced entirely within cities, even those with sunny climates, for example given the likely rise in air-conditioning demand, as climate change impacts more. In addition, renewables are mostly variable and will need systems to allow for balancing, including supergrid links, so that local supply-and-demand variations can be balanced across wide areas. That implies a wider framework than just generation within cities, linking up to where the resources are best available, which could be some way away, even overseas.

Fortunately, as has been noted, there are plenty of renewable resources available around the world, enough to meet urban needs, with many studies now suggesting that, if properly developed, and combined with energy saving, renewables could supply the bulk of the electricity needed in most countries and most of their other energy needs by around 2050 (Jacobson et al., 2017). However, there is much work to be done to reach that state, including, along with ambitious national renewable energy development, major energy-saving policies. In the context of cities, that means the development of improved building design, urban planning and transport policies, tailored to local conditions, although that may first need a better understanding of energy use in cities and how to change it (Crutzeig et al., 2015).

While as indicted above, much progress has been made due to grass-roots and prosumer initiatives, there is also a crucial role to be played in cities by local government, via municipal projects and support programmes. Certainly many of the national 'low carbon' energy scenarios that have emerged in recent years for the UK assume major local-level inputs from such projects, with many of them being in urban areas, and involving local heat and power generation, as well as community and consumer projects (RTP Engine Room, 2015; FES, 2018).

Local city governments, along with community organisations, can play key roles in developing energy projects within cities, with an extra incentive, in addition to the energy and environmental aspects, being the potential for local job creation and economic renewal (ICLEI, 2012). This has become a major issue in Africa, where youth unemployment is widespread. There have of late been some ambitious proposals for targeting aid to local job creation via investment in local renewable energy projects, with some of those projects being in cities (Elliott & Cook, 2018).

However, most of the renewable energy projects are likely to be outside of cities. Although there may be exceptions, and some cities may be able to sustain independent initiatives, it follows from the analysis above that the cities that are likely to do best, in terms of meeting all their energy needs from renewable sources, will be those in countries which are promoting renewables strongly, at the overall national level. Given that, as noted above, national renewable expansion programmes are accelerating in most countries around the world, the prospects for expanding the use of renewables in cities therefore look quite good, assuming appropriate relations can be maintained with rural areas. In addition, they can also expand renewable generation within their boundaries, while also reducing their energy demand as far as possible.

There will no doubt be setbacks and problems, with transport issues (only partly touched on in this study) perhaps being the hardest to deal with. However, one way or another, given careful attention to energy efficiency, it seems that, from a technological perspective, most cities should be able to move toward obtaining most of their energy needs (including for transport) from renewables supplies within a few decades, if suitable development paths are followed (IRENA, 2016). What this chapter has sought to explore are some of the related technical choices, but also some of the social issues.

References

Aghahosseini, A., Bogdonov, D., Barbosa, L., and Breyer, C. (2019) 'Analysing the feasibility of powering the Americas with renewable energy and inter-regional grid interconnections by 2030', *Renewable and Sustainable Energy Reviews*, Vol. 105, pp. 187–205: www.sciencedirect.com/science/article/pii/S1364032119300504

Appleby, K. (2018) 'Cities are harnessing the power of renewable energy: here's how', CDP: www.cdp.net/en//articles/cities/cities-are-harnessing-the-power-of-renewable-energy

Asafu-Adjaye, J. et al. (2015) 'An Ecomodernist Manifesto', 18 authors, self-published: www.ecomodernism.org/

BRE (2007) 'Micro-wind turbines in the urban environment', Building Research Establishment, Garston.

Brummer, V. (2018) 'Community energy: benefits and barriers', *Renewable and Sustainable Energy Reviews*, Vol. 94, pp. 187–196: www.sciencedirect.com/science/article/pii/S1364032118304507

CDP (2015a) 'Global Cities Report 2015', CDP Infographic: www.cdp.net/en-US/Pages/events/2015/cities/infographic.aspx

CDP (2015b) 'Cities Electricity Mix Map', CDP: https://data.cdp.net/Cities/2015-Cities-Electricity-Mix-Map/kwjr-j78z

Corbyn, J. (2016) 'Why Labour is putting energy reform at the heart of its green agenda', *The Guardian*, September 7: www.theguardian.com/environment/blog/2016/sep/07/why-labour-is-putting-energy-reform-at-heart-of-its-green-agenda-jeremy-corbyn

Creutzig, F., Baiocchi, G., Bierkandtd, R., Pichler, P., and Seto, K. (2015) 'Global typology of urban energy use and potentials for an urbanization mitigation wedge', *PNAS*, May 19, Vol. 112, No. 20, pp. 6283–6288: www.pnas.org/content/112/20/6283.full.pdf

Debor, S. (2014) 'The socio-economic power of renewable energy production cooperatives in Germany: results of an empirical assessment', Wuppertal Institut: http://epub.wupperinst.org/frontdoor/index/index/docId/5364

DEFRA (2017) *UK Food Waste Policy*, Department for Environment, Food and Rural Affairs, London: https://publications.parliament.uk/pa/cm201617/cmselect/cmenvfru/429/429.pdf

DTU (2018) *Global Wind Atlas*, Technical University of Denmark: https://globalwindatlas.info/

Ecotricity (2016) 'On site wind energy', Ecotricity company site: www.ecotricity.co.uk/for-your-business/on-site-wind-energy

Elliott, D. with Taylor, D. (2000) 'Renewable energy for cities: dreams and realities', paper for the UK-ISES Conference on 'Building for Sustainable Development', May 26, RIBA, London.

Elliott, D. (2003) *Energy, Society and Environment*, Routledge, London.

Elliott, D. (2013) *Renewable Energy: A Review of Sustainable Energy Supply Options*, Institute of Physics Publications, Bristol [2019 updated second edition forthcoming].

Elliott, D. (2015) *Green Energy Futures*, Palgrave Pivot, Basingstoke.

Elliott, D. (2016) *Balancing Green Energy*, Institute of Physics Publications, Bristol.

Elliott, D. (2017) *Energy Storage Systems*, Institute of Physics Publications, Bristol.

Elliott, D. (2019) 'Carbon capture and renewables', in Wood. G. et al. (eds) *The Long Goodbye to Fossil Fuels*. Palgrave, Basingstoke.

Elliott, D. and Cook, T. (2018) *Renewable Energy: From Europe to Africa*, Palgrave, Basingstoke.

Energy Cities (2016a) 'Energy Cities' members delivering on their Covenant of Mayors CO_2 reduction commitments', EU Energy Cities Network: www.energy-cities.eu/Energy-Cities-Members-delivering

Energy Cities (2016b) 'Cities heading towards 100% Renewable Energy', CLER, Energy Cities and Réseau Action Climat: www.energy-cities.eu/IMG/pdf/publi_100pourcent_final-web_en.pdf

Environment Europe (2019) *Environment Europe™ Sustainable Cities Database*: environmenteurope.org/

ETI (2018) 'District heat networks in the UK', Energy Technologies Institute: https://d2umxnkyjne36n.cloudfront.net/insightReports/District-Heat-Networks-in-the-UK-Final.pdf

Eurostat (2018) *Share of Energy from Renewable Sources*, Eurostat data: http://appsso.eurostat.ec.europa.eu/nui/show.do?dataset=nrg_ind_335a&lang=en

Feldman, D., Brockway, A., Ulrich, E., and Margolis, R. (2015) 'Shared solar: Current landscape, market potential, and the impact of Federal securities regulation', National Renewable Energy Lab, Golden: www.nrel.gov/docs/fy15osti/63892.pdf

FES (2018) 'Future energy scenarios', Consumer Renewable Scenario, National Grid, 2018 update: http://fes.nationalgrid.com/media/1363/fes-interactive-version-final.pdf

Girardet, H. (2010) 'Regenerative cities', World Future Council, Hamburg: www.worldfuturecouncil.org/fileadmin/user_upload/papers/WFC_Regenerative_Cities_web_final.pdf

Girardet, H. (2015) *Creating Regenerative Cities*, Routledge, London.

Gobitec (2015) Gobitec project website: https://gobitecdotorg.wordpress.com/

Greenpeace (2015) *Energy[r]evolution*, 5th edition, Greenpeace/Global Wind Energy Council: www.greenpeace.org/international/Global/international/publications/climate/2015/Energy-Revolution-2015-Full.pdf

Hunt, E. (2018) 'More than 100 cities now mostly powered by renewable energy, data shows', *The Guardian*, February 27: htttp://www.theguardian.com/cities/2018/feb/27/cities-powered-clean-energy-renewable

ICLEI (2012) 'Role of local governments in promoting renewable energy businesses: a contribution to a green urban economy', ICLEI Global Report, University of Amsterdam and ICLEI, Bonn: old.iclei.org/fileadmin/PUBLICATIONS/Global_Reports/ICLEI_Global_Report_2012_The_role_of_local_governments_in_promoting_renewable_energy_businesses_FINAL.pdf

IEA (2009) 'Cities, towns and renewable energy: Yes In My Front Yard', International Energy Agency, Paris: www.iea.org/publications/freepublications/publication/cities2009.pdf

Inhabitat (2017) 'China's vertical tower garden plan', April 2: https://inhabitat.com/chinas-first-vertical-forest-is-rising-in-nanjing/

IPCC (2014) *Fifth Assessment Report*, Intergovernmental Panel on Climate Change, Geneva: www.ipcc.ch/

IRENA (2016) 'Renewable energy in cities', International Renewable Energy Agency, Abu Dhabi: www.irena.org/-/media/Files/IRENA/Agency/Publication/2016/IRENA_Renewable_Energy_in_Cities_2016.pdf

IRENA (2018) *Global Atlas for Renewable Energy*, International Renewable Energy Agency: www.irena.org/GlobalAtlas

IRENA (2019) *A New World: The Geopolitics of the Energy Transformation*, International Renewable Energy Agency, Abu Dhabi: www.irena.org/publications/2019/Jan/A-New-World-The-Geopolitics-of-the-Energy-Transformation

Jacobson, M., Delucchi, M., Bauer, Z., Goodman, S., Chapman, W., Cameron, M., Bozonnat, C., Chobadi, L., Clonts, H., Enevoldsen, P., Erwin, J., Fobi, S., Goldstrom, O., Hennessy, E., Liu, J., Lo, J., Meyer, C., Morris, S. and Yachanin, A. (2017) '100% clean and renewable wind, water, and sunlight all-sector energy roadmaps for 139 countries of the world', *Joule*, Vol. 1, No. 1, pp. 108–121: www.sciencedirect.com/science/article/pii/S2542435117300120

JRC (2012) 'District heating and cooling', European Commission Joint Research Centre, Petten: http://setis.ec.europa.eu/system/files/JRCDistrictheatingandcooling.pdf"http://setis.ec.europa.eu/system/files/JRCDistrictheatingandcooling.pdf

Koirala, P., Koliou, E., Friege, J., Hakvoort, R.A., and Herder, P.M. (2016) 'Energetic communities for community energy: a review of key issues and trends shaping integrated community energy systems', *Renewable and Sustainable Energy Reviews*, www.sciencedirect.com/science/article/pii/S1364032115013477

LUT/EWG (2017) 'Global energy system based on 100% renewable energy', Lappeenranta University of Technology and Energy Watch Group, Berlin: http://energywatchgroup.org/wp-content/uploads/2017/11/Full-Study-100-Renewable-Energy-Worldwide-Power-Sector.pdf

Mittal, A. (2016) 'Revolutionary: Germany builds a Solar City that produces four times more energy than it consumes', *The Logical Indian*, April 14: http://thelogicalindian.com/environment/germanys-revolutionary-solar-city-that-produces-four-times-more-energy-than-it-consumes/

Morgan, S. (2018) 'EU study weighs linking power grid to China's', *Euractiv*, March 21: www.euractiv.com/section/eu-china/news/eu-looks-into-benefits-of-energy-silk-road-to-china/

Movellan, J. (2016) 'The Asia Super Grid: four countries join together to maximize renewable energy', *Renewable Energy World*, October 18: www.renewableenergyworld.com/articles/2016/10/the-asia-super-grid-countries-join-together-to-maximize-renewable-energy.html

Planenergie (2018) Planenergie website: http://planenergi.eu/activities/district-heating/seasonal-heat-storage/

Purvins, A., Sereno, L., Ardelean, M., Covrig, C.-F., Efthimiadis, T., and Minnebo, P. (2018) 'Submarine power cable between Europe and North America: A techno-economic analysis', *Journal of Cleaner Production*, Vol. 186, pp. 131–145: www.sciencedirect.com/science/article/pii/S0959652618307522

REM (2017) 'Sheffield forge could be the first in the UK to be powered by biomass and biogas', *Renewable Energy Magazine*, November 17: www.renewableenergymagazine.com/biomass/sheffield-forge-could-be-the-first-in-20171117

REN21 (2018) *2015 Global Status Report, Renewable Energy Network for the 21st Century*: www.ren21.net

Riley, T. (2017) 'What next for renewables in cities? The expert view', *The Guardian*, February 22: www.theguardian.com/sustainable-business/2017/feb/22/renewable-energy-cities-solar-wind-rio-delhi-nairobi-london-arup-vestas

RTP Engine Room (2015) 'Distributing power: A transition to a civic energy future', Report of the Realising Transition Pathways Research Consortium 'Engine Room': www.realisingtransitionpathways.org.uk/realisingtransitionpathways/news/distributing_power.html

SDH (2018) EU Solar District Heating Info hub: www.solar-district-heating.eu/

SES (2018) 'Singapore housing project', Science and Engineering Supporters video, Facebook, January 24: www.facebook.com/engineeringsciencee/videos/915204125307876/

Solargis (2017) Solar resource data obtained from the Global Solar Atlas, owned by the World Bank Group and provided by Solargis: https://globalsolaratlas.info/downloads/sub-saharan-africa

Solargis (2018) 'Solar resource map', Solargis: https://solargis.com/maps-and-gis-data/download/north-america

Sonnen (2018) Sonnen virtual community battery grid-link system: https://sonnenbatterie.de/en/sonnenCommunity

Tesla (2018) Powerwall battery details: www.tesla.com/en_GB/powerwall

Trainer, T. (1995) *The Conserver Society*, Zed Books, London

Trainer, T. (2010) *The Transition to a Sustainable and Just World*, Environbook, Annandale.

UN (2016) *Energy*, UN Habitat website: http://unhabitat.org/urban-themes/energy/

UNEP (2015) 'District energy in cities: unlocking the potential of energy efficiency and renewable energy', UN Environment Programme, Paris: www.unep.org/energy/portals/50177/DES_District_Energy_Report_full_02_d.pdf

Van Horn, J. (2015) 'San Diego passes strongest city-wide 100% clean energy law in America', *Ecowatch*, December 15: http://ecowatch.com/2015/12/15/san-diego-renewable-energy/

Wikipedia (2019) List of rooftop photovoltaic installations, Wikipedia entry: https://en.wikipedia.org/wiki/List_of_rooftop_photovoltaic_installations#In_operation

Willis, R. (2016) 'Cultural dimensions of community energy', Guest Blog, UK Energy Research Centre, London: www.ukerc.ac.uk/network/network-news/guest-blog-cultural-dimensions-of-community-energy.html

WNISR (2018) *World Nuclear Industry Status Report*, Independent annual review: www.worldnuclearreport.org/

World Bank (2016) *Global Solar Atlas*, World Bank/ESMAP/SOLARGIS: https://globalsolaratlas.info/

Wu, G., Deshmukh, R., Ndhlukula, K., Radojicic, T., Reilly-Moman, J., Phadke, A., Kammen, D. and Callaway, D. (2017) 'Strategic siting and regional grid interconnections key to low-carbon futures in African countries', *PNAS*, Vol. 114, No. 15: www.pnas.org/content/114/15/E3004.abstract

Zed (2018) Jingdezhen Ceramic Centre, Zed Factory City concepts: www.zedfactory.com/jingdezhen-ceramic-centre

11 How sustainable is smart and how smart is sustainable?

Irina A. Shmeleva and Stanislav E. Shmelev

Introduction

Ever since the introduction of the Smart for Sustainable Cities project by the UNECE and ITU, there is a growing interest in using smart solutions to promote sustainable urban performance. But is it so straightforward? Is sustainable by definition smart and smart effectively sustainable? We address these issues below.

The Rome Declaration adopted at the UN Forum on 'Shaping smarter and more sustainable cities: striving for sustainable development goals' in May 2016 declared that 'cities need to become smarter, with technological solutions deployed to address a wide range of common urban challenges' of sustainable development (UNECE & ITU, 2016). The EU's European Economic and Social committee considers smart sustainable cities to be a tremendous source of growth, productivity and employment. A smart sustainable city, according to UNECE, is an innovative city, that uses information and communication technologies (ICTs) and other means to improve quality of life, efficiency of urban operation and services, and competitiveness, while ensuring that it meets the needs of present and future generations with respect to economic, social, environmental as well as cultural aspects (UN ECOSOC, 2015).

The literature on smart cities has been expanding in the last few years; these include: Lombardi et al. (2012), Caragliua and Del Bo (2012), Kourtit et al. (2012), Lazaroiu and Roscia (2012), Albino et al. (2015); Marsal-Llacuna et al. (2015), Hara et al. (2016), Marsal-Llacuna (2016), Manitiu and Pedrini (2016), Caird (2017), Kummithaa and Crutzen (2017), Grossi and Pianezzi (2017), Dall'O' et al. (2017), García-Fuentes et al. (2017), Girardi and Temporelli (2017), Klopp and Petretta (2017), Anthopoulos (2017), Fernandez-Anez et al. (2018), Ruhlandt (2018), Yigitcanlar et al. (2018).

Many recent studies have focused on the analysis of smart and sustainable cities (United for Smart and Sustainable Cities, 2017), benchmarking approaches and differences between them. Ahvenniemi et al. (2017) review smart city assessment frameworks and compare them with international standards such as ISO 37120, LEED and BREEAM, and assess the composition of the smart and sustainable indicator sets. Huovila et al. (2019) illustrate differences between indicator-based standards from the point of view of the presence of smart and sustainable indicators in a detailed manner. Our previous research proposed several novel elements including differentiation between weak and strong sustainability (Shmelev, 2017), varying policy priorities (Shmelev & Shmeleva, 2018), and using a strong sustainability assessment framework involving outranking ELECTRE MCDA methods (Shmelev & Shmeleva, 2019a).

Smart and sustainable connections

The most well-known international smart city rankings include European Smart Cities Ranking (Giffinger et al., 2007), Ranking of Smart Global Cities (Institute of Information Science, 2014), Cities in Motion Index (IESE, 2016), Smart Cities Index (EasyPark, 2017), and Global Smart City Index (Juniper Research, 2017) (see Table 11.1).

In Chapter 1, we reviewed the leading sustainable cities indicator frameworks including UN SDG Framework, UNECE ITU, UN Habitat Global Prosperity Initiative, Global Livability Index, Resilient City Index, Global Power City Index, etc. (Shmelev & Shmeleva, 2019b) and the Environment Europe Sustainable Cities Index is discussed in detail in Chapter 2 of this book (Shmelev & Shmeleva, 2018).

Most publications dealing with smart and sustainable indicator frameworks have a theoretical focus. As opposed to that, our contribution is in the empirical assessment of smart and sustainable connections based on the Environment Europe™ sustainable cities database. In the following section, we will explore a research question, which can be formulated in the following way: what is the extent to which smart and sustainable urban performance is connected empirically?

In this section, we will consider the key measures of smart urban performance: registered patents per 1,000 inhabitants, Internet speed and diversity of the underground system, illustrating the smartness of the public transport network in relation to key measures of social (unemployment and life expectancy) and environmental (CO_2, PM_{10}, water use, waste recycling) urban performance.

Table 11.1 Global smart city rankings

Year	*Key organization*	*Ranking*	*Reference*	*Number of cities*	*Number of indicators*
2007	University of Technology Vienna	European Smart Cities Ranking	Giffinger et al. (2007)	70	64
2014	Institute of Information Sciences	Ranking of Smart Global Cities	Institute of Information Science (2014)	20	13
2016	IESE	Cities in Motion Index 2017	IESE (2016)	181	66
2017	EasyPark	Smart Cities Index	EasyPark (2017)	100	19
2017	Juniper Research	Global Smart City Index	Juniper Research (2017)	100	58

It is clear from empirical data on 140+ global cities collected in the Environment Europe™ database that higher levels of education in cities are capable or stimulating innovation, which is also helped by high-speed Internet available in global cities.

We see cities that are extremely successful both in the level of innovation expressed in the form of registered patents and levels of tertiary education including San Francisco, Boston, Stockholm, Tokyo and Copenhagen. This corresponds to the results of our global smart and sustainable assessment presented in Chapter 2.

Internet speed seems to be important for the generation of innovations. High levels of Internet speed are observed in such cities as Stockholm, Tokyo, Copenhagen, Seoul, Singapore and Hong Kong, and this connection is most apparent in Stockholm, Tokyo and Copenhagen.

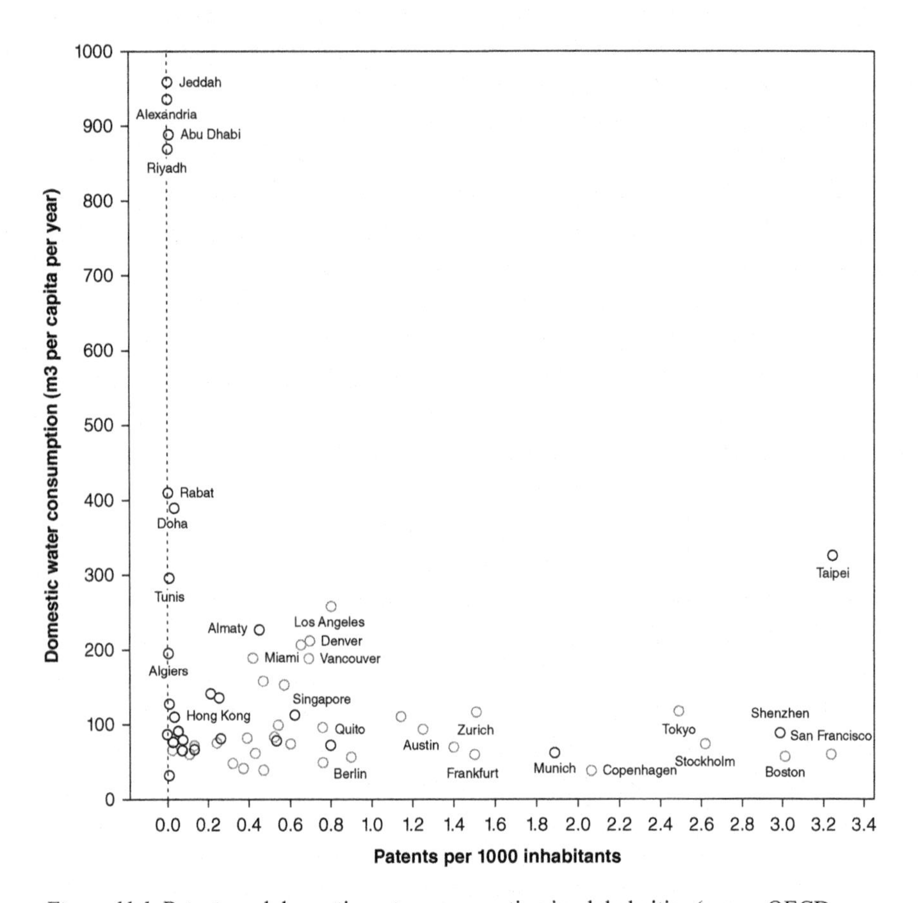

Figure 11.1 Patents and domestic water consumption in global cities (grey – OECD; black – non-OECD)

Source: Environment Europe™ Sustainable Cities Database

It is also clear from the data we have that such innovation is having a positive spill-over effect in the social dimensions, and is associated with lower unemployment.

Low unemployment is observed simultaneously with high levels of registered innovation expressed in patents in cities such as Taipei, Shenzhen, San Francisco, Boston, Stockholm, Tokyo, Copenhagen, Munich, Seoul, Portland, Frankfurt. On the other hand, in cities where the number of registered patents is much lower per capita – Johannesburg, Madrid, Barcelona, Tunis, Los Angeles, Alexandria, Istanbul, Algiers, Rabat, Cairo, Delhi and Warsaw – unemployment is much higher.

Interestingly, patents are connected with reducing water consumption (Figure 11.1), increasing recycling rates (Figure 11.2) and lowering urban PM_{10} concentrations (Figure 11.3), effectively increasing life expectancy.

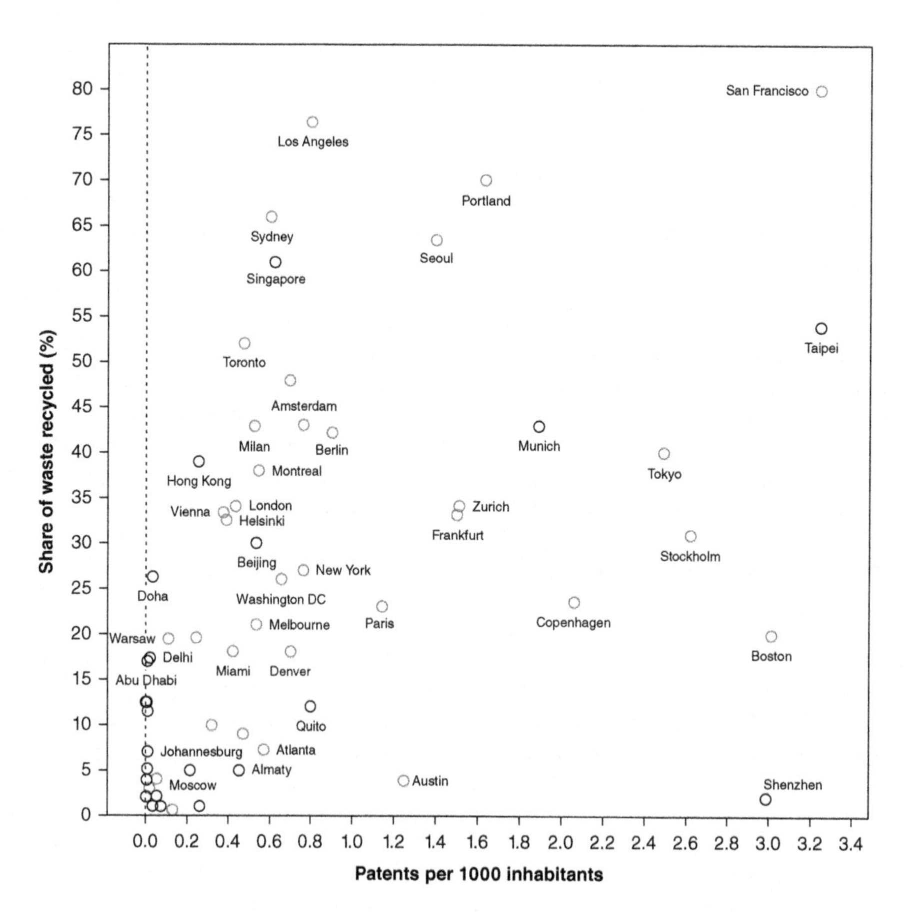

Figure 11.2 Patents and share of waste recycled in global cities (grey – OECD; black – non-OECD)

Source: Environment Europe™ Sustainable Cities Database

Figure 11.1 illustrates that cities with high levels of smart performance expressed in innovations – San Francisco, Boston, Taipei, Shenzhen, Stockholm, Tokyo and Copenhagen – have much lower water use than cities with lower levels of innovation – Jeddah, Alexandria, Abu Dhabi, Riyadh.

Figure 11.2 illustrates that cities with high levels of smart innovation – San Francisco, Taipei, Portland, Seoul, Munich and Tokyo – have significant levels of waste recycling. Contrary to this, cities with low levels of innovation – Johannesburg, Moscow, Almaty, Abu Dhabi, Delhi and Atlanta – exhibited low recycling rates. The absence of innovations despite high performance on some smart dimensions, such as Internet speed in Moscow, may result in a situation where waste management solutions turn into the search for the region, where waste could be landfilled. On the other hand, the situation in Shenzhen illustrates the fact that although the number of patents is very high, recycling levels are low, which clearly illustrates that innovation has not yet reached the urban sustainability sphere there.

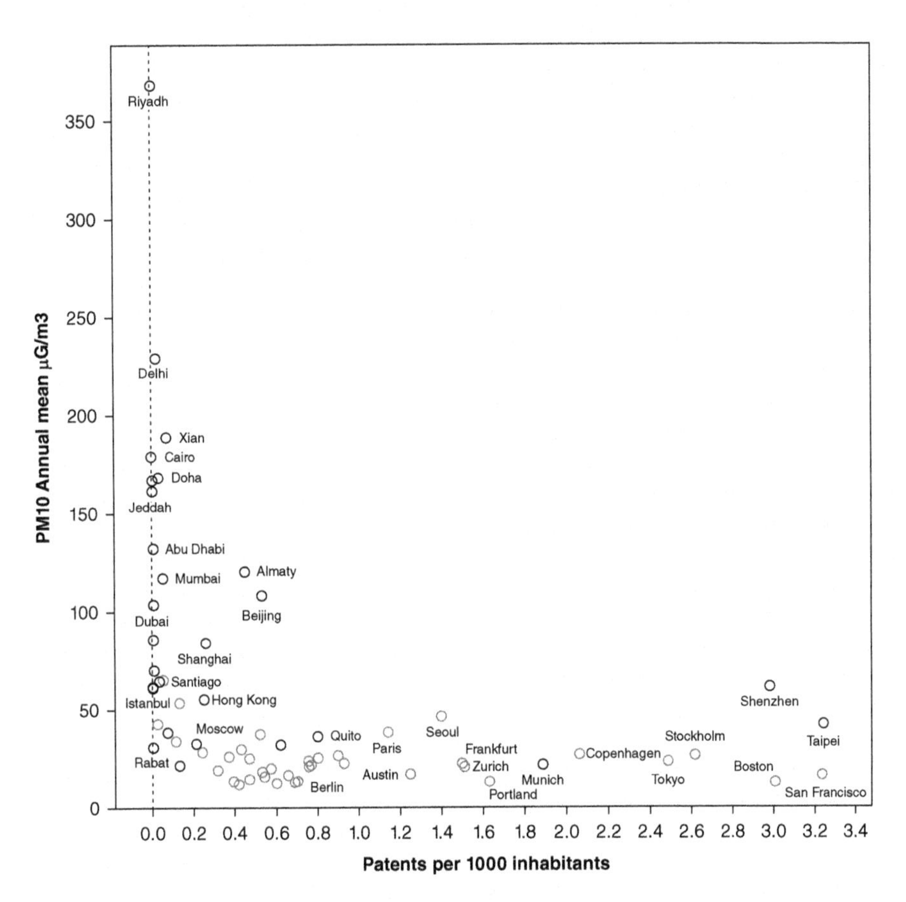

Figure 11.3 Strong negative correlation between patents and PM_{10} pollution in global cities (grey – OECD; black – non-OECD)

Source: Environment Europe™ Sustainable Cities Database

As regards PM_{10} air pollution, cities such as Taipei, San Francisco, Shenzhen, Boston, Stockholm, Tokyo and Copenhagen show very low levels of PM_{10} pollution and very high levels of innovation. At the same time, the opposite is true for Riyadh, Delhi, Xian, Cairo, Doha, Jeddah, Abu Dhabi, Mumbai, Dubai, Almaty and Beijing. A possible hypothesis explaining such situation could be the uptake of hybrid and electric cars. High levels of education connected with patents and innovation in advance cities could also influence pro-environmental behavior patterns, including using cycling and public transport.

It should be added that of course patents are not the only factor for low water consumption, low air pollution and high waste recycling.

Availability of smart infrastructure in the form of underground systems is capable of reducing PM_{10} pollution (Figure 11.4) and effectively improving citizens' health and life expectancy (Figure 11.5).

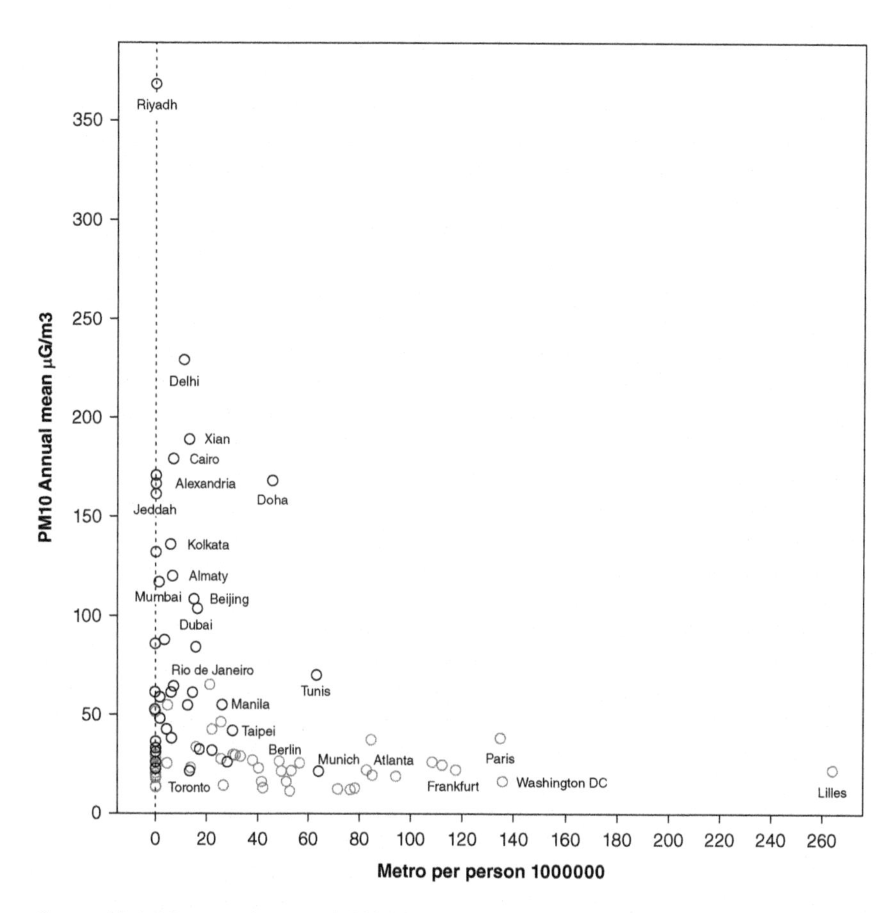

Figure 11.4 Metro stations per 1,000,000 population and PM_{10} pollution in global cities (grey – OECD; black – non-OECD)

Source: Environment Europe™ Sustainable Cities Database

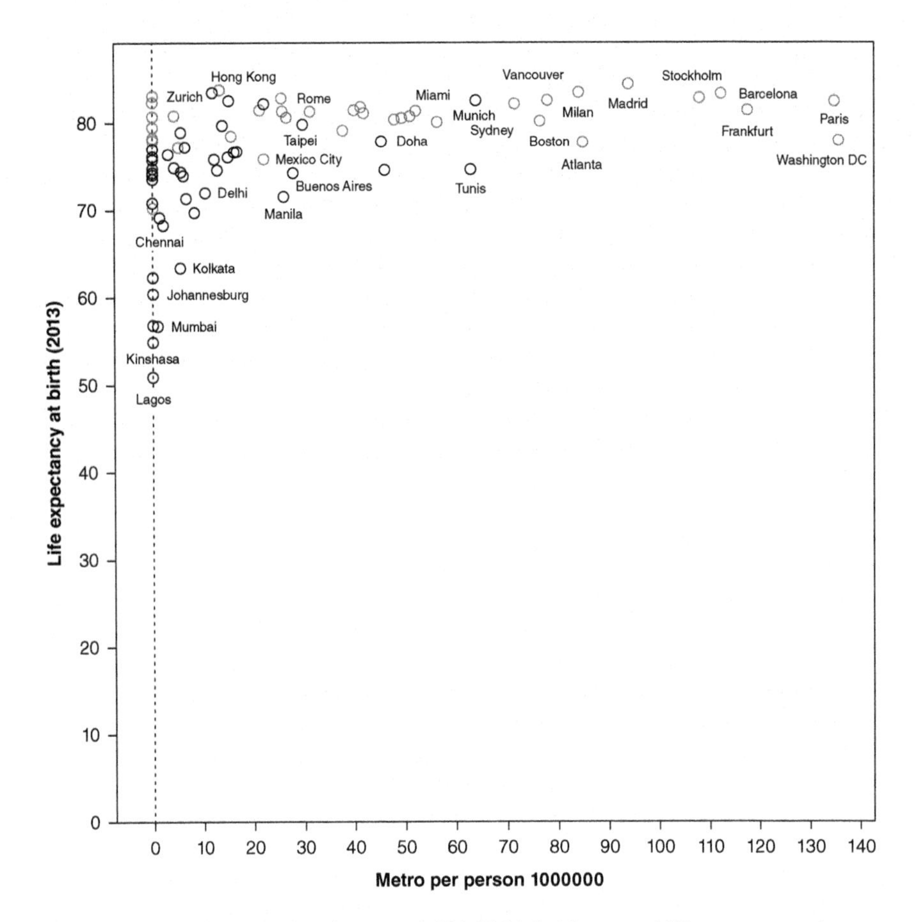

Figure 11.5 Underground stations per 1,000,000 inhabitants and life expectancy in global cities (grey – OECD; black – non-OECD)

Source: Environment Europe™ Sustainable Cities Database

As can be seen from Figure 11.5, availability of underground stations facilitates avoiding unnecessary emission from using private cars and is therefore capable of reducing PM_{10} pollution in cities. Availability of underground stations is also connected to higher life expectancy (Figure 11.5), which corresponds to our earlier and WHO results on the effects of PM_{10} pollution on health.

In this section we illustrated only some empirical connections between smart and sustainable urban dimensions. This subject undoubtedly requires further research.

This leads us to another related question. Is smart equally powerful in solving the global mitigating climate change impacts and addressing CO_2 emissions? The answer turns out to be not so straightforward. Although renewable energy is the crucial factor in reducing CO_2 emissions, neither Internet speed nor patents in cities are correlated with the reduction in CO_2 emissions, as the next section will illustrate.

Urban CO_2 function

Cities are contributing a significant share of global CO_2 emissions (75%) and will experience tangible effect from its consequences. Bai et al. (2018) asserts that by 2030, millions of people and US$4 trillion of assets will be at risk from climate-change-induced extreme events. Cities are making significant efforts in mitigating climate-related emissions. Thus, founded in London in 2005 at the summit of representatives of 18 leading megacities, the C40 partnership currently numbers 90 cities from over 50 different countries and is aimed at taking action against climate change. ICLEI, Local Governments for Sustainability is uniting over 1,500 cities towns and regions to build a sustainable future.

Based on the global data covering 71 cities, contained in the Environment Europe™ Sustainable Cities Database we were able to generate a regression that captured 80% of the variation in urban CO_2 emissions across the whole world using the factors presented in Figure 11.6. Urban CO_2 emissions tend to decrease with the increasing daily mean temperatures in the city (Table 11.2). On average, higher temperatures result in reduced need for heating and associated CO_2 emissions. Unfortunately, at the moment, we cannot take into the account increased electricity consumption due to air conditioning. Cities with an OECD capital status tend to exhibit significantly lower CO_2 emissions possibly as a result of higher technological development in public transport systems, electric cars, cycling and pedestrian mobility as a new trend in urban planning and design.

The large share of renewables in the energy mix tends to reduce urban CO_2 emissions according to our results. On the other hand, the share of coal in the energy mix tends to increase urban CO_2 emissions. An additional behavioral variable, representing the share of trips made by walking, cycling and using public transport is shown to reduce urban CO_2 emissions (Table 11.2), which compared to petrol-based combustion engine cars generate less harmful GHGs. Paradoxically, higher recycling rates under everything else being equal, tend to increase CO_2 emissions as additional amounts of energy are needed for complex recycling. The next variable, CO_2 tax, is reflecting the existing structure of incentives globally and shows an effect to reduce CO_2 emissions; however, statistical significance of this factor is lower.

Overall, this equation 'explains' 80% of the variance in the global CO_2 emissions. Such model can be used for out-of-sample forecasting.

One of the more interesting variables in the model is the OECD capital status (Shmelev & Shmeleva, 2019c). OECD members include 34 countries: Australia, Austria, Belgium, Canada, Chile, Czech Republic, Denmark, Estonia, Finland, France, Germany, Greece, Hungary, Iceland, Ireland, Israel, Italy, Japan, Korea, Luxembourg, Mexico, the Netherlands, New Zealand, Norway, Poland, Portugal, Slovak Republic, Slovenia, Spain, Sweden, Switzerland, Turkey, the United

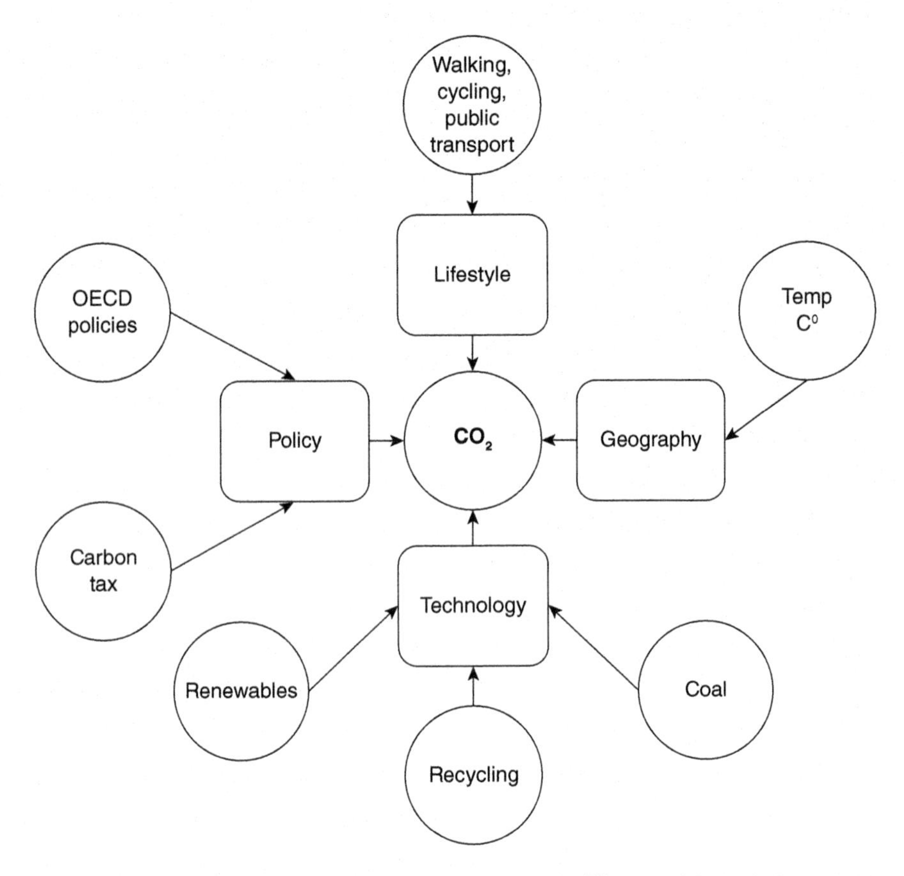

Figure 11.6 Conceptual framework for Environment Europe™ Urban CO_2 Emissions Model

Source: Environment Europe™ Cities Database, 2014 data

Table 11.2 CO_2 regression coefficients in the linear urban regression

Variable	*Coefficient*	*Std. error*	*t-value*	*t-prob*	*Part.R^2*
Constant	15.2640	1.023	14.9	0.0000	0.7794
Daily mean temperature	−0.234784	0.04427	−5.30	0.0000	0.3087
OECD capital status	−2.29855	0.6474	−3.55	0.0007	0.1667
Share of renewables in the energy mix	−0.0376761	0.01115	−3.38	0.0013	0.1534
Share of coal in the energy mix	0.0486420	0.009920	4.90	0.0000	0.2762
Share of trips made by walking, cycling and public transport	−0.113082	0.01036	−10.9	0.0000	0.6543
Recycling rate	0.0692216	0.01286	5.38	0.0000	0.3150
CO_2 tax	−0.0306765	0.01428	−2.15	0.0355	0.0683

Source: Environment Europe™ Cities Database, 71 observations, $R^2 = 0.805394$

Kingdom and the United States. Our research shows that in capital cities from OECD countries we observe an average reduction of over 2t of CO_2 per person per year.

One hypothesis for explaining such a phenomenon could be the OECD policy leadership in reducing the effects of urban transport on the environment and the creation of public spaces and pedestrian zones in cities. The cases of Copenhagen, Stockholm, Lund and Gothenburg where new pedestrian shopping centers were introduced in the early 1960s paved the way for the new OECD trend emerging a few years later.

The OECD has established a new Urban Environment group within its Environment Committee in 1969, focusing on the following research topics devoted to new policy instruments: 'Vehicle-free areas in cities' (1970–1972), 'Policy instruments for influencing the form and structure of urban development' (1972–1974), 'Urban noise abatement' (1973–1975), 'Low-cost improvements in the outdoor urban environment' (1973–1975), 'Management of publicly owned land' (1975–1977), 'Traffic policies for improvement of urban environment' (1975–1976). Such historical role of the OECD in promoting pedestrianization in urban areas gives us confidence to conclude that it might have something to do with the significant reduction in per capita CO_2 emissions in OECD capitals.

Another important development taking place more recently is the C40 partnership, the cooperative initiative of cities aimed at reducing greenhouse gas emissions (Shmelev & Shmeleva, 2019c). Launched by the Mayor of London, and OECD capital in 2005, by 2006 the number of member cities reached 40, hence the name, C40. Today, C40 includes 90 of the world's greatest cities, bringing together 650 million people and one quarter of the world's economy. Six out of thirteen members of the C40 steering committee: Amman, Boston, Copenhagen, Durban, Hong Kong, Jakarta, London, Los Angeles, Mexico City, Milan, Paris, Seoul and Tokyo are OECD capitals. This hypothesis requires further verification.

According to our results, none of the smart urban indicators in our database (Internet speed, patents, etc.) were statistically significantly connected with CO_2 emission reductions, which leaves open a question about the capacity of smart tech to mitigate CO_2 emissions. The brief review of literature reveals a relative lack of comprehensive studies looking at the full life-cycle of smart technologies constituting a new macrocosm. There is very limited research on full life-cycle impacts of smart technologies: impacts of smartphones (Suckling & Lee, 2015), Internet in general (Müller, 2013), e-books (Moberg et al., 2011), music delivery methods (Weber et al., 2010) and electric cars (EEA, 2018). More research is needed to determine how new smart technologies, the Internet of things and the world of gadgets are assisting us in mitigating climate change.

Conclusions

Although there are clear links between technological developments manifested in the form of high-speed Internet or innovation registered as patents and improvements in urban sustainability performance in waste recycling or water consumption and PM_{10} pollution, the effect of smart technologies on CO_2 emissions has been limited. CO_2 emissions are responding to the reduction of coal in the energy mix and introduction of renewables but life-style variables of proportion of trips made by walking, cycling and public transport turns out to be more powerful than reducing coal and expanding renewables combined. Carbon tax played a role, as did the average temperature or the OECD status; however, higher Internet speeds or more rapid innovation expressed in patents alone were not sufficient to affect urban CO_2 emissions.

As we see, further research is needed to find out how innovation and smart performance indicators influence urban sustainability performance. We see sustainability as the overall goal of urban development and smart tech as a potential means to reach this goal without placing unnecessary emphasis on smart solutions as a goal in itself.

References

Ahvenniemi, H., Huovila, A. Pinto-Seppä, I., and Airaksinen, M. (2017) What are the differences between sustainable and smart cities? *Cities*, 60, 234–245.

Albino, V., Berardi, U., and Dangelico, R.M. (2015) Smart cities: Definitions, dimensions, performance, and initiatives, *Journal of Urban Technology*, 22(1), 3–21.

Anthopoulos, L. (2017) Smart utopia vs smart reality: Learning by experience from 10 smart city cases, *Cities*, 63, 128–148.

Bai, X. et al. (2018) Six research priorities for cities and climate change, *Nature*, 555, 23–25.

Caird, S. (2017) City approaches to smart city evaluation and reporting: Case studies in the United Kingdom, *Urban Research & Practice*, 11(2), 159–179.

Caragliua, A. and Del Bo, C. (2012) Smartness and European urban performance: Assessing the local impacts of smart urban attributes, *Innovation – The European Journal of Social Science Research*, 25(2), 97–113.

Dall'O', G., Bruni, E., Panza, A., Sarto, L., and Khayatian, F. (2017) Evaluation of cities' smartness by means of indicators for small and medium cities and communities: A methodology for Northern Italy, *Sustainable Cities and Society*, 34, 193–202.

EasyPark (2017) *Smart Cities Index 2017*. Stockholm. https://easyparkgroup.com/smart-cities-index/

EEA (2018) Electric vehicles from life cycle and circular economy perspectives. TERM 2018: Transport and Environment Reporting Mechanism (TERM) report, No. 13/2018. doi:10.2800/77428

Fernandez-Anez, V., Fernández-Güell, J.M., and Giffinger, R. (2018) Smart City implementation and discourses: An integrated conceptual model. The case of Vienna, *Cities*, 78, 4–16.

García-Fuentes, M.Á., Quijano, A. de Torre, C., García, R., Compere, P., Degard, C., and Tomé, I. (2017) European cities characterization as basis towards the replication of a Smart and Sustainable Urban Regeneration Model, *Energy Procedia*, 111, 836–845.

Giffinger, R., Fertner, C., Kramar, Kalasek, R., Pichler-Milanović, N., and Meijers, E. (2007) Ranking of European medium-sized cities. *Vienna UT*, October: Centre of Regional Science. www.smart-cities.eu/download/smart_cities_final_report.pdf

Girardi, P., and Temporelli, A. (2017) Smartainability: A methodology for assessing the sustainability of the smart city, *Energy Procedia*, 111, 810–816.

Grossi, G., and Pianezzi D. (2017) Smart cities: Utopia or neoliberal ideology? *Cities*, 69, 79–85.

Hara, M., Nagao, T., Hannoe, S., and Nakamura, J. (2016) New key performance indicators for a smart sustainable city, *Sustainability*, 8, 206.

Huovila, A., Bosch, P., and Airaksinen, M. (2019) Comparative analysis of standardized indicators for Smart sustainable cities: What indicators and standards to use and when? *Cities*, 89, 141–153.

Institute of Information Sciences (2014) *Ranking of Smart Global Cities*. Shanghai Academy of Social Sciences, Shanghai.

IESE (2016) *IESE Cities in Motion Index*. IESE Business School, Barcelona.

Juniper Research (2017) *Smart Cities: What's in it for Citizens?* Juniper Research, Hampshire. https://newsroom.intel.com/wp-content/uploads/sites/11/2018/03/smart-cities-whats-in-it-for-citizens.pdf

Klopp, J.M., and Petretta, D.L. (2017) The urban sustainable development goal: Indicators, complexity and the politics of measuring cities, *Cities*, 63, 92–97.

Kourtit, K., Nijkamp, P., and Arribas, D. (2012) Smart cities in perspective: A comparative European study by means of self-organizing maps, *Innovation: The European Journal of Social Science Research*, 25(2), 229–246.

Kummithaa, R.K.R., and Crutzen, N. (2017) How do we understand smart cities? An evolutionary perspective, *Cities*, 67, 43–52.

Lazaroiu, G.C., and Roscia M. (2012) Definition methodology for the smart cities model, *Energy*, 47(1), November, 326–332.

Lombardi, P., Giordano, S., Farouh, H., and Yousef, W. (2012) Modelling the smart city performance, *Innovation: The European Journal of Social Science Research*, 25(2), 137–149.

Manitiu, D.N. and Pedrini, G. (2016) Urban smartness and sustainability in Europe: An ex ante assessment of environmental, social and cultural domains, *European Planning Studies*, 24(10), 1766–1787.

Marsal-Llacuna, M.-L. (2016) City Indicators on social sustainability as standardization technologies for smarter (citizen-centered) governance of cities, *Social Indicators Research*, 128, 1193–1216.

Marsal-Llacuna, M.-L., Colomer-Llinàs, J., and Meléndez-Frigola, J. (2015) Lessons in urban monitoring taken from sustainable and livable cities to better address the Smart Cities initiative, *Technological Forecasting and Social Change*, 90(B), 611–622.

Moberg, Å., Borggren, C., and Finnveden, G. (2011) Books from an environmental perspective. Part 2: e-books as an alternative to paper books, *International Journal of Life Cycle Assessment*, 16, 238–246.

Müller, E., Widmer, R., Coroama, V.C., and Orthlieb, A. (2013) Material and energy flows and environmental impacts of the internet, *Journal of Industrial Ecology*, 17(6), 814–826.

Ruhlandt, R.W.S. (2018) The governance of smart cities: A systematic literature review, *Cities*, 81, November, 1–23.

Shmelev, S.E., and Shmeleva, I.A. (2018) Global urban sustainability assessment: A multidimensional approach, *Sustainable Development*, 26(6), 904–920.

Shmelev, S.E., and Shmeleva, I.A. (2019a) Multidimensional sustainability benchmarking for smart megacities, *Cities*, 92, 134–163.

Shmelev, S.E., and Shmeleva, I.A. (2019b) Methods and indicators for urban sustainability assessment, in Shmelev, S.E. (ed.) *Sustainable Cities Reimagined: Multidimensional Assessment and Smart Solutions*, Routledge.

Shmelev, S.E., and Shmeleva, I.A (2019c) Indicators for measuring the performance of smart and sustainable capital cities, in Orttung, R. (ed.) *Capital Cities and Urban Sustainability*, Routledge, pp. 47–65.

Suckling, J., and Lee, J. (2015) Redefining scope: The true environmental impact of smartphones? *International Journal of Life Cycle Assessment*, 20, 1181–1196.

United for Smart and Sustainable Cities (2017) *Collection Methodology for Key Performance Indicators for Smart Sustainable Cities*. www.itu.int/en/publications/Documents/tsb/2017-U4SSC-Collection-Methodology/files/downloads/421318-CollectionMethodologyforKPIfoSSC-2017.pdf

Weber, C.L., Koomey, J.G., and Matthews, H.S. (2010) The energy and climate change implications of different music delivery methods, *Journal of Industrial Ecology*, 14(5), 754–769.

Yigitcanlar, T., Kamruzzaman, M., Buys, L., Ioppolo, G., Sabatini-Marques, J., Moreira da Costa, E., and Yun, J.H.J. (2018) Understanding 'smart cities': Intertwining development drivers with desired outcomes in a multidimensional framework, *Cities*, 81, 145–160.

Index